Nature's Management

Nature's Management

Writings on Landscape and Reform, 1822–1859

Edmund Ruffin

EDITED BY JACK TEMPLE KIRBY

The University of Georgia Press
Athens and London

Paperback edition, 2006
© 2000 by the University of Georgia Press
Athens, Georgia 30602
All rights reserved
Set in 10 on 13 Janson Text by G&S Typesetters

Printed digitally in the United States of America

The Library of Congress has cataloged the
hardcover edition of this book as follows:

Library of Congress Cataloging-in-Publication Data
Ruffin, Edmund, 1794–1865.
Nature's management : writings on landscape and reform,
1822–1859 / Edmund Ruffin ; edited by Jack Temple Kirby.
xxxi, 375 p. ; 24 cm.
Includes bibliographical references (p. 345–369) and index.
ISBN 0-8203-2162-1 (alk. paper)
1. Agriculture—Virginia—History. 2. Nature
conservation—Virginia—History. I. Title. II. Kirby, Jack
Temple.
S451.V8 R84 2000
630'.9755—dc21   99-039150

Paperback ISBN-13: 978-0-8203-2837-9
          ISBN-10: 0-8203-2837-5

British Library Cataloging-in-Publication Data available

Frontispiece: Edmund Ruffin—portrait from the governor's
mansion. The Library of Virginia.

in memory of
Ruth Kirby Daniel
(1904–1998)
daughter of the old country

Circles and right lines limit and close all bodies, and the mortall right-lined circle must conclude and shut up all. There is no antidote against the *Opium* of time . . . ; Our Fathers finde their graves in our short memories, and sadly tell us how we may be buried in our Survivors. Grave-stones tell truth scarce fourty years: Generations passe while some trees stand, and old Families last not three Oaks. . . .

[Yet] remembering the early civility [the dead] brought upon these Countreys, and forgetting long passed mischiefs; We mercifully preserve their bones, and pisse not upon their ashes.

—Thomas Browne, *Urne-Buriall*

Contents

Acknowledgments / *ix*

Introduction / *xi*

Part One. History

1. Production as Moral Imperative: The Morals of Agriculture (1822) / *3*

2. A Malthusian Crisis: A Letter from George Henry Walker and Ruffin's Response (1836) / *9*

3. The Old Dominion's Declension: Sketch of the Progress of Agriculture in Virginia, and the Causes of Its Decline (1836) / *14*

Part Two. Reform

4. On Fencing—Against the Open Range: Letters and Editorials (1833–1835) / *47*

5. Malaria—Against Mill Ponds: On the Sources of Malaria, or Autumnal Diseases, in Virginia, and the Means of Remedy and Prevention (1838) / *100*

6. Public Health and Recycling: Desultory Observations on the Police of Health, in Virginia—As It Is, and as It Ought to Be (1837) / *128*

7. Reclamation—Against Erosion: On Draining: Addressed to Young Farmers (1833–1834) / *170*

Part Three. Travels and Expositions

8. Farming the Great Dismal Swamp: Hasty Observations on the Agriculture of the County of Nansemond (1836) / *185*

9. The Great Dismal: "Jottings Down" in the Swamp (1839) / *193*

10. Carolina Swamps and Heroic Reclamation: Notes of a Steam Journey (1840) / *228*

11. Pines as Monuments and Commodities: Notes on the Pine Trees of Lower Virginia and North Carolina (1859) / *258*

12. The Geology of Low Country Progress: Agricultural Features of Virginia and North Carolina (1857) / *284*

Part Four. Valedictory

13. Toward the Managed Landscape: An Address on the Opposite Results of Exhausting and Fertilizing Systems of Agriculture (1852/1860) / *323*

Notes / *345*

Index / *371*

Acknowledgments

Edmund Ruffin probably craved and received affection from members of his family (though not from all of them) and from one friend. From the public and from posterity he wanted respect and obedience. Of these last, I—a son of Ruffin's beloved tidewater Virginia countryside who has spent several years as his obsessive reader—can offer only grudging respect and of course no obedience at all. For Ruffin's social and political principles were and remain abhorrent, and his imperial (or would-be imperial) politics of nature are not mine either. Yet for all my resistance to Ruffin, I confess to a measure of understanding that approaches affection and a larger measure of admiration for his curiosity and learning. Ruffin compels this much, and I thank him first.

Second, I thank my colleague Michael O'Brien, whom I enmeshed in my preoccupation with Ruffin and landscape until he urged me to undertake this collection. O'Brien also recovered from Cambridge University's libraries a number of fugitive British references that appear in this work. Third, there is Michael Thompson, who arrived at Miami University as a doctoral student just in time to become an invaluable research assistant on this project. He copied and transcribed thousands of pages of Ruffiniana from microfilm, always with critical as well as careful eyes. Jeri Schaner contributed proficient keyboarding and other, more complex, labors at critical moments, too.

I am also grateful for the constructive reflections of three anonymous readers who served as referees for the University of Georgia Press. David Allmendinger was a valued supporter early on, and I must acknowledge an inspiring conversation (and an exotic beer) with Bland Simpson as I composed the introduction. La Constancia talked with me all the time I selected and ordered texts, did notes, and wrote my own prose, always with inspiring effect. Last (and hardly least), I express admiration and appreciation to Barbara Ras, executive editor at the University of Georgia Press, for her poetic sensibility and editorial wisdom.

Introduction

Early in August 1862, as General George McClellan's enormous army fought toward Richmond on the north side of James River, Union gunboats reconnoitered the south bank and landed troops in Prince George County, birthplace in 1794 of Edmund Ruffin, fomenter of disunion and war. The old agitator had years earlier established a new farm for himself in Hanover County, above Richmond at the Pamunkey's head of navigation, and then retired to politics and writing. The war made him a farmer again. His sons' and overseers' Confederate army service obliged the elder Ruffin to manage not only his own Pamunkey farm but Edmund Jr.'s and Julian Calx's estates in Prince George. Edmund Sr.'s base of wartime labor became a cottage on the lawn of Beechwood, Edmund Jr.'s mansion, which stood on a bluff above the river. There Edmund Sr. kept his accounts, at least twenty years of private correspondence, and his great library of agricultural, "scientific and classical" works.[1] Warned of the Yankees' approach, the old man fled to Petersburg with his son's family and slaves. There Ruffin awaited word of the soldiers' departure.

When he returned late on Sunday, 17 August, Ruffin was relieved to find Beechwood and his cottage still standing. (McClellan's troops had fired a house near Edmund Sr.'s Pamunkey farm, mistaking the property for the notorious rebel's.) But the Yankees nonetheless wrought certain vengeance. In the mansion especially, every small, movable household possession not stolen was smashed; larger articles of furnishings were broken or used for firewood. Soldiers spat tobacco juice on plastered walls, and on these and other Beechwood surfaces they scrawled with charcoal their names and pointed graffiti: "Old Ruffin don't you wish you had left the Southern Confederacy to go to Hell" and, most famously, "You did fire the first gun on Sumter, you traitor son of a bitch." Ruffin calculated that "low-bred" soldiers had occupied the big house, and that officers—at least the commander—had claimed his cottage. Enlisted men lacked "transport" and so resorted to vandalism; officers, he figured, had means to commit larger robbery and carried off his personal possessions, including correspondence and the library, "probably for sale in New York."[2]

The plundering of Beechwood was but the beginning of a disaster of celebrity for Edmund Ruffin. The event confirmed friends' and family warnings that he was a famous traitor, a hunted man, in danger himself and a danger to his own family. The Yankees appeared at Marlbourne, his Pamunkey place, and returned to Prince George. First a refugee crowded with daughters-in-law and grandchildren in town, then at Redmoor, the safe piedmont farm Edmund Jr. had bought, Ruffin lost the privacy to read and write, his principal joys in life. During the Peninsula campaign of 1862 one of his grandsons, a mere teen, had been killed. Then amid the horrors of the Grant-Lee struggles of 1864, Julian, the second son and his father's favorite, was killed at Drewry's Bluff. According to his captain, Julian, a sergeant commanding an artillery piece, died instantly from a musket ball through the brain while facing the enemy. Ever the morbid lion, Ruffin took comfort in both details of his son's death. He eulogized his son privately and consoled his daughter-in-law, but Ruffin wondered in his diary "if age & decay have withered & dried up my affections & sensibilities, & hardened my heart.... I have not shed a tear." He did not mention that between 1853 and 1860 he had suffered no fewer than six deaths in his family. Then, as battlefield observer during 1861–62, he had witnessed macabre, heart-numbing suffering and death. So thirteen months after Julian's end, the war now lost, Ruffin brooded at Redmoor. Satisfied he need no longer fear captivity and a trial for treason, he faced instead humiliating poverty and dependency. Ruffin's physical self, so long taut, reliable, energetic, had rapidly degenerated. His hearing had been much diminished since Fort Sumter—near Charleston, where Ruffin first became a national celebrity, he had rested with his head against a Confederate mortar battery that fired a ten-inch round. Deafness progressed until by 1865 he was largely isolated. His night vision was gone, too, he reported, conceivably signaling ultimate blindness. Ruffin had also become incontinent, and he suffered dizziness, probably owing to elevated blood pressure. Left to himself one mid-June morning, the old man finished what must be one of the most remarkable suicide notes in American history. Most of his pages resolved religious doubts about self-murder; then Ruffin concluded with his much-quoted declaration of "unmitigated hatred to Yankee rule—to all political, social & business connection with Yankees—& to the Yankee race." At last, after an initial misfiring of his gun, he succeeded in blowing away much of his head. Edmund Jr. buried him at Marlbourne.[3]

Edmund Ruffin's legendary firing of the first gun against Fort Sumter, clothed as an honorary South Carolina militiaman, and his militant valedictory, leaving blood, bone, and brains on the walls and ceiling of Redmoor, capture much of his memory. He was a grim, driven, bitter, depressive, and unrepentant figure, brilliant and a bit mad. An admiring reader of ancient and modern antidemocratic texts, Ruffin condemned bourgeois liberalism and defended slavery in the abstract. He also asserted the innate inferiority of Africans and sought to prove his claim through social science. And deciding early to seek creation of a separate slaveholders' republic, Ruffin helped other so-called fire-eaters build an intellectually persuasive and, finally, a popular case for secession—all in an overwhelmingly rural region with very few decent universities, ephemeral journals, and merely local societies and other institutions on which public action is usually mobilized.

Ruffin was, in a word, a southerner. The descriptor evokes the fanatical critic of middle-class ideology and social relations, the white supremacist, the determined separatist, the bloodthirsty promoter of war glad to offer his own life—as well as the lives of his sons and grandsons—for the cause. Avery Odell Craven, Ruffin's first scholarly biographer, offered "Southerner" behind Ruffin's name in his 1932 title. Later, the Virginia Department of Highways employed the same legend on the historical marker beside Route 10 near Evergreen, Ruffin's birthplace. And at the end of the twentieth century, a neo-Confederate marketer based in Crawfordsville, Georgia, styles itself the Ruffin Flag Company. Among other useful totems, the company offers for sale T-shirts bearing Ruffin's likeness.[4]

There was another Ruffin, of course, who did not perish in helpless, embittered failure. This man is acknowledged by Craven and all other biographers and by the Virginia highway historical marker. This is Edmund Ruffin, innovative agronomist, discoverer of the problem of soil acidity and its cure, and the father (arguably) of soil science in the United States. Edmund Ruffin, southerner, remains a fascinating and appalling reactionary, essential to comprehending slavery's defense and disunion, but this Ruffin is also a curiosity of a time and cause both long lost, save to neo-Confederate cranks. Edmund Ruffin, agronomist, conversely, lives as genius ahead of his time, a very modern man who links the antebellum era to the twentieth century's land-grant universities, to the U.S. Department of Agriculture, and to what is called agribusiness.

Ruffin's first celebrity, indeed—proceeding from the local to the regional—was as agronomist. In 1818, when he was but twenty-four years old, he founded the Agricultural Society of Prince George County. Three years later John Skinner, editor of the most important farm journal of the time, *American Farmer*, published in Baltimore, printed Ruffin's first research report on soil acidity and its treatment with fossilized shells, or marl. And in 1822, Skinner published Ruffin's important theoretical essay, "The Morals of Agriculture." By the time the Virginian's full study of acidity and marl, *An Essay on Calcareous Manures*, appeared in 1832, Ruffin effectively left farming for a life as editor-writer-propagandist for scientific agriculture. He founded his own journal, *Farmers' Register*, in 1833 at Shellbanks, his second Prince George farm, but moved shortly thereafter to nearby Petersburg. *Farmers' Register*'s audience was not large—the subscription list rose to about 1,400 by 1835, then fell below a thousand by 1842, the year Ruffin closed the journal. But *Register* readers were influential and engaged Ruffin in a serious discourse he craved. The authority of his *Essay* and the *Register* brought Ruffin an invitation from the governor of South Carolina to conduct a geological survey of his state, which Ruffin partly accomplished in 1843. Public opportunities exhausted for the moment, Ruffin turned once more to farming, establishing his new base on the Pamunkey, called Marlbourne, and building sufficient family fortune to retire again to writing and celebrity. The latter was sealed in 1851, when J. B. D. De Bow of New Orleans published a long biographical sketch of the Virginian. Ruffin had become the South's premier "agriculturist."[5]

Thirty years after the famous suicide, W. P. Cutter, librarian of the U.S. Department of Agriculture, wrote in his agency's *Yearbook* that Ruffin's seminal 1832 treatise on soil acidity was "the most thorough piece of work on a special agricultural subject ever published in the English language." Nine years later, in 1904, the editor of Richmond's venerable *Southern Planter* paid homage to leading contemporary agronomists—experiment-station scientists and educators—but reminded readers "that the South produced the first man (Edmund Ruffin) who, in this country, endeavored to apply science to the advancement of agriculture." The editorial was illustrated with a portrait of Ruffin supplied by his grandson, Julian, of Marlbourne, and a reprinting of Cutter's 1895 essay followed.[6]

In 1914 Congress completed the institutionalization of agricultural science with the Smith-Lever Act, which created the Cooperative Extension Service. Administered from Washington through land-grant agricultural colleges in the states, the extension service dispatched trained agents to every rural county in the nation. Agents' mission was to teach farmers precisely to assay, test, and measure soil, drainage, moisture, seed, fertilizers, and pests. Applied science would yield higher production and profits. This phenomenon was little more than Ruffin's own message, decades before applied germ theory and genetics became the capital of agricultural educators.[7]

Perhaps no one understood this historical connection better than Avery Craven (1886–1980). An Iowa farm boy and country-school science teacher, Craven applied to Harvard's graduate program when his principal asked him to teach history as well. At Harvard, he met Professor Frederick Jackson Turner and wrote a master's thesis on soil exhaustion in New England. Instead of returning to Iowa with his new credentials, Craven entered the University of Chicago's doctoral program in history. Turner sent Craven to work with southernists on the agricultural history of the South. Craven would duplicate his soil history of New England in equally long settled Maryland and Virginia to test his Turnerian hypothesis that frontier nations tended to ruin landscapes as they advanced. The hypothesis was proved (of course) while Craven taught in a series of midwestern colleges, consulting with local agronomists in the extension service's most popular region, composing his dissertation. The University of Illinois Press published the dissertation (*Soil Exhaustion as a Factor in the Agricultural History of Virginia and Maryland, 1606–1860*) in 1926. Yes, Craven argued, by 1800 Maryland and Virginia pioneers had despoiled the earth (as had New Englanders). Soil exhaustion in the East hastened the pace of westward migration and extended exhausting agronomy throughout the antebellum era: frontiers defeated progressive agriculture. Craven, like his mentor, Turner, was an antifrontier historian. But enter Edmund Ruffin, his marl experiments, the *Essay* and *Farmers' Register*. Tidewater farmers were persuaded, and by 1860 Virginia and Maryland had more or less accomplished a recovery based on scientific practice similar to that of Craven's own day. "Book farming" had redeemed the antebellum mid-Atlantic (and, by inference, perhaps other eastern places); book farming's wisdom and legitimacy ought to be secure.[8]

Ruffin's other celebrity, as southerner, ultimately distracted Craven, however. A Ruffin biography became Craven's second book. Its subtitle—*A Study in Secession*—effectively announced the end of Craven's preoccupation with agricultural history and the beginning of his long and distinguished career as a political historian of the Civil War era. *Soil Exhaustion*, meanwhile, established the canonical interpretation of frontier process in southern left-behind regions and secured Ruffin's place in the pantheon of agronomic heroes. Exhaustion to redemption became a very long lived trope, repeated in textbooks and supposedly freshly researched monographs for no fewer than six decades.[9]

Soil Exhaustion has, since the 1970s, also been incorporated into the canon of the new environmental history.[10] And rightly so, for if environmental history examines the interactions of humans with the rest of nature, Craven did just that in his first book. It is a sort of agroecological work on a relatively homogeneous subregion in the long duration, bearing some resemblance to the landscape-and-society studies also undertaken during the 1920s and later in France by historians associated with the "Annales school." Except, of course, that unlike the French, Craven was uninterested in peasant proprietors and their folkish agronomy save as a great negative other to be reformed by Ruffin.[11]

Soil Exhaustion may indeed still be read with profit as a precocious experiment in what we now call environmental history. The work might be more specifically categorized as conservation history, however, or better still, as conservation advocacy. The distinction is important. Environmental history evolved from a 1960s political movement that included a systematic (and well-justified) attack on "conservation," which was an older political movement associated with Theodore Roosevelt; his chief forester, Gifford Pinchot; and many others. Conservationists valued nature as a commodity, feared depletion of woodlands and exhaustion of soils, and struggled for scientific wise use of natural resources. "Wise use" implied not only science but management and manipulation of nature by credentialed experts. Conservation achieved triumphalism during the Great Depression, when in the East the Tennessee Valley Authority reconfigured a vast landscape and in the West a thirty-year binge of high-dam building began. The New Deal's Civilian Conservation Corps also accomplished professional forestry's great ambition to suppress fire throughout the country. By the 1950s, then, when the green revolution was established in the United States and exported around the globe, the ultimate triumph of useful science seemed at hand. American geneticists, chemists, agronomists, and engineers had become masters of the Earth itself. Then came a reaction that continues to this day.

The most popular and enduring aspect of 1960s and early '70s counterculturalism was certainly Americans' disaffection with technology. The affluent, leisure-seeking majority discovered a public commodity beyond purchase at new malls: quality of life—clean air and water, unsullied wetlands, pristine mountains and forests. People discovered Aldo Leopold's *Sand County Almanac* (1949) and demanded paperbound reprints, which remain steady sellers. Leopold's beloved nature diary—neo-Thoreauvian rhapsodies—was probably no more compelling to readers than the famous "Land Ethic" section of *Sand County*, which eloquently demonstrated the interdependency of humans and the rest of the biological community. Big industry and its scientist spokespeople, meanwhile, were busily despoiling their own previously unquestioned authority, especially in their campaigns to discredit Rachel Carson (following the appearance of *Silent Spring* in 1962) and their promotion of nuclear power plants (at the time of the Three Mile Island accident in 1979). Lyndon Johnson and Richard Nixon signed environmental legislation; Americans began to celebrate Earth Day; the federal Environmental Protection Agency became an institution to reckon with; and the American Society for Environmental History was founded.[12] Late in this environmental era, Craven's paradigm and Ruffin's exalted place in agricultural history at last passed under scrutiny and revision.

First came the 1988 work of a British historian, William M. Mathew, *Edmund Ruffin and the Crisis of Slavery in the Old South: The Failure of Agricultural Reform*. Mathew rejoined the two Ruffins, southerner and agronomist, as a consistent and mutually supporting tandem. But first Mathew relentlessly demolished Craven's and Ruffin's "agricultural revolution." Except for a few hundred progressive planters blessed with convenient marl deposits and sufficient labor to dig and distribute it, tidewater farmers did not rejuvenate croplands à la Ruffin. Ruffin's own class failed him. Only when the old man reemerged from Marlbourne as champion of slavery and secession did he win wide celebrity among the same people who had rendered his agronomy a dead issue. Mathew's exhaustive research and militant tone resemble the somewhat earlier historiographical implosions of the myth of agricultural revolution in eighteenth-century England and France.[13] *Ruffin and the Crisis of Slavery* is not particularly an exercise in environmental history, although Mathew does occasionally employ environmental-era language—for example, Ruffin "challeng[ed] old ecological assumptions."[14]

Conversely, David F. Allmendinger Jr.'s remarkable 1990 biography, *Ruffin: Family and Reform in the Old South*, consciously addresses Ruffin's ecological significance. An unconventional, nonlinear arrangement, Allmendinger's portrait is preoccupied with death and demography writ small (the personal and familial) and large. Allmendinger benefits from colonialist scholars' research on early American families and communities as well as age of ecology sensibilities. He recognized relationships between sudden population takeoff and a landscape's carrying capacity. When Edmund Ruffin was born in 1794, his family had already long lived and farmed on Virginia's tidewater rivers. Yet infant and maternal mortality, as well as adult males' frequently early deaths, were still the rule, rendering family continuity as tenuous as life itself. Edmund was himself orphaned, and he married another orphan. Then came what Allmendinger appropriately terms a "Malthusian revolution"—the rapid births of no fewer than eleven Ruffin children, nine of whom lived to maturity. Father Ruffin's imperative to innovate on his long-worn crop fields is understood, then, as microcosm in extremis of a regional phenomenon. His success at farming secured his family, then his public, intellectual career as reformer and agitator. That Ruffin was a reader of T. R. Malthus who persistently engaged the economist's grim equations renders the environmental context historically appropriate—essential, I would say—as well as timely.

Allmendinger is less emphatic than is Mathew about Ruffin's "failure" as agronomic reformer. Yet Allmendinger's detailed confirmation of the tiny audiences for Ruffin's *Essay* in its various editions and his *Farmers' Register* as well as his other landscape writings buttresses Mathew and contradicts Craven and all his uncritical repeaters. More germane to Allmendinger's approach is Ruffin the successful re-creator of farms, the manager extraordinaire of nature, and the would-be manipulator of the entire South Atlantic. From a virtually forgotten 1852 Ruffin lecture in Charleston, Allmendinger revealed that Ruffin hoped not only to dredge Virginia rivers south of the James to make them navigable but also to drain every swamp from Richmond to (at least) Savannah and to crisscross the larger region with canals.[15] Ruffin becomes rather like a New Deal monumentalist or a water-obsessed Californian of almost any time—except, of course, that Ruffin would have accomplished this massive rearrangement of the South Atlantic lowlands with enslaved labor. Ruffin emerges from Allmendinger, in my reading, as the protoconservationist.

My own environmental history of the lowlands between the James and Albemarle Sound in North Carolina—*Poquosin: A Study of Rural Landscape and Society* (1995)—is indebted to Allmendinger. Although *Poquosin*'s chronological scope approaches four centuries, Ruffin fills a long chapter and lurks nearly everywhere else as a result of nineteenth-century progressive agronomists' critical role in fixing the landscape's fate in the long duration. Ruffin was, I think, the most important among articulate riverside planters and businessmen. I term them *cosmopolitans* not only because they were engaged in an international grain trade but also because they were consumers and practitioners of European science and technology. The majority of the low-country free population (*hinterlanders* in my typology) were also Euro-Americans, but they practiced a folkish agronomy that was as much Indian as European—they ranged hogs and cattle on the commons, burned the woods and shifted crop fields, and planted beans and peas in corn rows. Their animals looked scraggly, and their fields were crooked, jumbled, and sloppy. Lacking labor beyond family and perhaps a few slaves, the hinterlanders undertook few "improvements" such as ditching for drainage, since most crop fields were temporary. Ruffin took little notice of hinterlanders except to dismiss them or to attack the commons.

In *Poquosin* the "Ruffin reforms" become much more encompassing than previously represented. Ruffin was a desperate Malthusian facing a household Malthusian revolution during the 1810s and '20s, and the result was his experimentation and marling, as Allmendinger has shown. Failing to sell his property so he might immigrate westward, Ruffin also became a more ferocious antifrontierman than was Craven. Frontiers were poison not only to sound agronomy but also to the demographic stability essential to civilization itself. Like his contemporary, South Carolina writer and planter William Gilmore Simms, Ruffin understood that civil, prosperous communities were possible only when good people were committed to place.[16]

The first Ruffin "reform," then, was the promotion of Virginia's history, wellspring of place pride. Ruffin became a founder of the Historical and Philosophical Society of Virginia and delivered a critical address before the society on the colony and state's agrarian history. Then as editor of the *Farmers' Register* he promoted the society, reprinted colonial records, and published for the first time William Byrd II's "History of the Dividing Line betwixt Virginia and North Carolina" and other Byrd records. (Ruffin borrowed these texts from Byrd descendants who were his neighbors.) Later, about 1851, as he began an autobiography, Ruffin turned also to historical fiction, drafting a forty-three-page Revolutionary War saga that he entitled "The Blackwater Guerilla."

Ruffin's interpretation of Virginia's agricultural history was one of eastern declension, of course. So marling, ditching, and introduction of new commodities (such as silk)—the best-known Ruffin reform—were firmly set in historical context. So too was Ruffin's complementary campaign (under the name "fencing reform") against the commons, or open range. The open range was an institution appropriate to frontiers, not a civilized region. Two centuries after English settlement, tidewater Virginia's hardwood trees were much depleted, yet colonial-era legislation still required that every farmer protect every crop field from foraging animals with a high fence. The Virginia General Assembly was not responsive to this reform during the 1830s, but Ruffin succeeded in organizing voluntary ringfence associations in Prince George. The scheme permitted contiguous landowners to construct a single enclosure around their entire neighborhood, so they might abandon crop field fencing in favor of confining animals on their individual estates. These groupings were finally sanctioned by the legislature in 1857. Ruffin had anticipated by half a century the more general (and successful) movement against the commons in the South.

Ruffin wished also to improve and secure the public health of the low country. Having lost two children to disease early in his married life, Ruffin withdrew his family from the river to a new, inland farm—Shellbanks, where he founded *Farmers' Register*. His tribe now generally healthy, Ruffin remained nonetheless obsessed with the relationship between landscape, drainage, and health. He studied malaria in particular and became convinced that millponds harbored the causative "miasma." Ruffin also failed in his campaign to change the law of Virginia to permit the public easily to force owners of small ponds to drain them. But Ruffin's interest in public health was broader than malaria: he sought to clean up towns and cities as well as farms, particularly to dispose of—recycle, actually—human and animal wastes and dead animals. His proposals are included among the texts in this volume.

Finally, and most ambitiously, Ruffin proposed great public works in 1852. This reform might well be construed not only as securing civilization (and "conservation") through massive manipulation of landscape but also as Ruffin's ultimate (proposed) war on hinterlanders. For hinterlanders' folkish agronomy thrived in and on the fringes of great swamps and pocosins (upland swamps) and along unnavigable rivers. Ruffin would have brought all the South Atlantic into the cosmopolitan sector, where everyone would engage fully in markets and cash exchanges. Of course he failed at this scheme, too, even though the work was more or less accomplished early in the twentieth century, albeit with free labor.[17]

Ruffin—An Imperial Oeuvre

Edmund Ruffin was born to privilege and the assumption of mastery. He was male, white, and heir to Coggin's Point, a plantation (or "farm," as

Ruffin chose to call it) of more than 1,500 acres. He would begin farming with about fifty slaves, too.

In another important sense, however, Ruffin lacked agency. Those sections of Coggin's Point not tidal marshes, ravines, or pine/sorrel scrub—his crop fields—were tan to whitish lands, worn out by many years of stationary European-style culture, especially of corn, that American staple and notorious user of nitrogen. Tidewater farmers of Ruffin's great-grandfather's and grandfather's generations had adopted the European model of permanent fields, the plow, and carefully managed herds of cattle (principally for their fertilizing manure). By the beginning of the nineteenth century, the model was clearly failing. Young Ruffin's neighbors and contemporaries were decamping for the West and fresh cropland. The frustrated heir resolved that he would go forth, too, but there was no buyer for Coggin's Point, and Susan Ruffin gave birth to another white mouth to feed on average every eighteen months between 1814 and 1832. Then there were black mouths but more importantly black hands and backs constructed for commercial agriculture.

Ruffin's large and growing family and all those enslaved laborers obviated a different sort of farming. Even 1,500 acres could not support all these people if they had ranged hogs, burned sections of the woods every twenty or thirty years, abandoned new fields after perhaps six or seven crops—if Ruffin had tried, in other words, the semisubsistence mode of his hinterlander neighbors, producing only small surpluses of grain for the market plus surplus pork, if any.

Hinterlander agronomy was what environmentalist dreamers today might call sustainable. For as long as human populations remained sparse on a heavily wooded and well-watered landscape, shifting fire culture and free-ranging of cattle and hogs seemed a system of permanence with but temporary harmful disturbance.[18] Ruffin seems never to have considered adopting this version of sustainability. The assumptions of his class and the circumstances of his young manhood effectively forbade resort to what he later publicly condemned as frontierish and primitive. Ruffin's situation thus amounts to presentation by exogenous powers of no alternative to the European model. He was compelled first to restore and then to sustain permanent commercial fields.

Yet beyond dominative material forces there was an ever-laboring mind dedicated to learning, grasping, and bending circumstances to will. As a boy Ruffin was alone and bookish, preferring Evergreen's library to helping his grandfather and father farm. He acquired "a fondness for agricultural studies," Ruffin wrote much later, reading Thomas Hale's four-volume *A Com-*

pleat Body of Husbandry (London, 1758–59), and the American John Beale Bordley's *Essays and Notes on Husbandry and Rural Affairs* (Philadelphia, 1799). At fifteen young Edmund crossed the river for a year of halfhearted studies at the College of William and Mary: he ignored most classes, favoring only geometry, and read what he wished. Called home at sixteen when his father died in 1810, Ruffin's formal education was finished. For two more years, in effect a guest of his stepmother, he lounged and read again, now mostly fiction, as Ruffin later told an acquaintance, but perhaps also Malthus's *An Essay on the Principle of Population* (1798), in Evergreen's library.[19] Soon after the United States declared war on the British, Ruffin volunteered as a private in Prince George's militia and went off to Norfolk for a season of fortifying and waiting, returning home in 1813 to claim Coggin's Point and a bride—and to confront the dilemma of European-style sustainability.[20]

The young proprietor soon discovered the irrelevance of his previous agronomic reading to conditions at Coggin's Point. Both Hale and Bordley had written of British soils, crops, and methods. But then—perhaps in 1814—Ruffin acquired the work of a contemporary Virginia planter, John Taylor of Caroline County. This was *Arator* (first published in 1813), which offered (among other things) a European-style prescription for soil exhaustion in the Chesapeake region. Taylor reported recovery of his own crop fields via "green manuring"—that is, returning directly to the soil abundant crops of both grasses and corn and deep plowing to achieve appropriate mixture of soil and vegetative offal. Taylor also prescribed excluding animals to prevent them from consuming valuable vegetation in crop fields. Ruffin followed *Arator* for several years and failed. His land would hold neither animal nor vegetable manure.

Ironically, salvation came from an English book, originally published in 1813 with an American edition published in Fredericksburg, Virginia, in 1815: *Elements of Agricultural Chemistry, in a Course of Lectures for the Board of Agriculture*, by Sir Humphry Davy of the Royal Institution in London. Davy's book, which contained lectures delivered between 1803 and 1812, is—to reapply W. P. Cutter's extravagant encomium for Ruffin—arguably "the most thorough piece of work on a special agricultural subject ever published in the English language" to that time. Davy subscribed (like all his contemporaries) to the humus theory of soil building—that is, productive land is constructed of decomposed plant and animal manures. He introduced to a receptive British audience both the concept and means of precise measurement of soils' content.

Ruffin discovered a revelation in Davy's fourth lecture, on soil acidity in Lincolnshire. The American contrived an instrument to duplicate Davy's test, evaluated his own soils for "salt of iron," deduced the necessity of improvisation for Prince George conditions, and confirmed, finally, the necessity of calcium (Davy recommended quicklime) as corrective. An old slave identified an abandoned marl pit on Coggin's Point. Ruffin tested the fossilized shells and ordered his hands to begin excavating and carting marl to his fields. The redemption of his farm and family had begun through books and a transatlantic scientific discourse.[21]

Europeans produced science and agronomic applications; Americans such as Ruffin were attentive, receptive consumers, especially of British innovation, because the language was familiar. Davy's lectures, meanwhile, were well received in his own country because progressive agriculturists sought desperately to improve production during the Napoleonic Wars. In 1805 the Bath and West of England Society, having already established the first formal field experiments devoted to careful recording and management, founded its Committee for Chemical Research. Organic (as opposed to inorganic) chemistry remained undeveloped, however, for many years.

Then in 1837 a German baron, Justus von Liebig, pioneer of organic chemistry and professor at the University of Giessen (1824–52), arrived in Liverpool to attend the meeting of the British Association for the Advancement of Science. Several British scientists, Lyon Playfair among them, had already read in German Liebig's *Organic Chemistry in Its Application to Agriculture and Physiology*. They and the association received Liebig eagerly; and within four years Playfair had translated and published an English edition.[22]

Liebig boldly attacked the humus theory and offered instead what the English began to call a mineral theory. To Liebig, organic fertilizers (that is, manures) were irrelevant or superfluous because plants received nitrogen from the air via rainfall and phosphorus and other minerals (such as carbon) from the unimproved soil. When soils weathered, or wore down, Liebig advised, they required addition of inorganic elements rather than traditional forms of manure. Awed by Liebig's reputation and powerful argument, the English incorporated his theory into academic curricula, including that of the new Royal Agricultural College at Cirencester, founded in 1845. A young scientist and agriculturist, John Bennett Lawes (1814–1900), however, was an early doubter. Lawes had created a model agricultural experiment station on his estate at Rothamsted. There, in tandem with another scientist, Joseph Henry Gilbert (1817–1901), Lawes began to test the mineral theory, and by 1849 Lawes and Gilbert undertook an attack on Liebig. Atmospheric nitrogen was insufficient, they found, for nonleguminous plants to thrive. Nitrogenous manures were essential for persistent growth and abundant production. Well after Ruffin's death, much of Liebig's theory ultimately was indeed proved false. Soil chemistry was more complex than either Davy's or Liebig's theories comprehended.[23]

By the time Liebig went to Liverpool, Ruffin was already well established as a Davy-inspired author and as editor of *Farmers' Register* in Petersburg. In 1836 he had summarized in the *Register* "British opinions on the 'Essay on Calcareous Manures'...." They were reasonably respectful, albeit rather unsurprised, since marling had long been a familiar practice to many Europeans. And Ruffin, like other U.S. agricultural editors, took immediate notice of Liebig's *Chemistry* when it appeared on this side of the Atlantic in 1841. Ruffin's competitor, C. T. Botts, editor of the *Southern Planter*, which had been founded in 1840 in Richmond, was dismissive: "Mr. Justus Liebig," he opined, "is no doubt a very clever gentleman and a most profound chemist, but in our opinion he knows about as much of agriculture as the horse that ploughs the ground." Ruffin thought differently. He devoted an admiring ten large pages of *Farmers' Register* to Liebig and declared that the German chemist had deservedly received in America more "respectful attention and applause" than any other scholar.[24]

In that Ruffin was already long the champion of an inorganic "manure," marl, his endorsement of Liebig's mineral theory might be dismissed as self-aggrandizement through European authority. I think differently, however. Liebig was obviously a man of larger purpose. He was, like Ruffin, a Malthusian—or better, an anti-Malthusian in the sense that Liebig devoted himself to the defeat of Malthus's demographic arithmetic, even though the odds and logic seemed impossible. Lyon Playfair's translation of Liebig's prefatory declaration in *Organic Chemistry* must have touched Ruffin profoundly: "The most urgent problem which the present day has to solve, is the discovery of the means of producing more bread and meat on a given surface to supply the wants of a consistently increasing population." In addition, even more than Davy's lectures, Liebig's chemistry encouraged agriculturists everywhere to believe that precision of measurement, testing, observation, and record keeping would ultimately yield revolutionary results. So egoist that Ruffin surely was, he understood that Liebig's science was too important to be conflated with the politics of status among academics and editors. Science applied to production was the key to human survival and to civilization itself in eastern Virginia and the

South Atlantic. Still—and curiously, perhaps—Ruffin continued not only the practice but the advocacy of supplementing marl (or lime) treatments for acidity with vegetable and animal (organic) manures, for soil building and storage of the nitrogen that his inorganic talisman effectively fixed. Ruffin preached and practiced, in other words, a balanced organic and inorganic program, rather the sort Lawes and Gilbert ultimately proved out at Rothamsted.[25]

Ruffin's consumption of European theoretical and applied science hardly began or ended with Davy and Liebig. Unfortunately, there is no inventory of the library that the Yankees carried away from Beechwood. But Ruffin's writings, especially a decade of issues of the *Farmers' Register*, provide many clues to his adoption of European attempts to understand and manage the earth. Except for Davy and Liebig, Ruffin seldom directly addressed scholarly works but instead mentioned them casually, usually offering only the author's last name, perhaps with a partial title as well. Ruffin probably read early in his life Erasmus Darwin (1731–1802, grandfather of Charles), especially *Phytologia: or, The Philosophy of Agriculture and Gardening* (Dublin, 1800), a work enormously influential in Britain and subsequently America during Ruffin's youth. (He cited Darwin in one of the texts in this volume.) In middle age, circa 1839–40, Ruffin publicly regretted his ignorance of botany, especially when he encountered the Venus flytrap and other exotica in southeastern North Carolina. (See "Notes on a Steam Journey," below.) Yet in addition to new books by Americans that appeared during the 1830s and '40s, Ruffin seems to have carried about with him during the '50s the weighty, multivolume masterpiece of François-André Michaux, *The North American Sylva; or, A Description of the Forest Trees of the United States, Canada and Nova Scotia*, first published in 1810. Michaux's father, André, was famous as a globe-trotting botanist who had been trained in Paris by a critic of the Linnean system of classification. Michaux's Latin names for some North American trees competed, then, with the prevailing nomenclature. Ruffin sorted and judged brilliantly, I think, in his valuable "Notes on the Pine Trees of Lower Virginia and North Carolina" (1859, reproduced below).[26]

From his youth a learned reader of landscapes, Ruffin proceeded to systematic study of geologic morphology and stratigraphy, that which lies beneath landscape. He read Sir Charles Lyell's seminal historical geology, amassed a fossil collection that he shared widely, and as South Carolina's geological surveyor, created a fossil cabinet in the State House in Columbia.[27] Yet except for ancient fishes, contemporary feral hogs, and occasionally tame silkworms, the Virginian's writings almost utterly neglect animals' contributions to shaping the earth's surface. The lacuna seems stunning, considering Ruffin's expansive curiosity.

Unsatisfied with English subjects and rare translations of foreign-language scientific works, for example, Ruffin taught himself to read French. French was the preferred second language of American high society during Ruffin's time and long afterward, but Ruffin—always impatient with social pretension—used written French primarily as an instrument of learning and problem solving. He was familiar with works by Antoine-Laurent Lavoisier (1743–94), generally regarded as the founder of modern chemistry; Marc-Antoine Puvis (1776–1851), a famous agronomist; and Antoine-Augustin Parmentier (1737–1813), a pharmacist, agronomist, and beloved investigator of the potato. Ruffin must have owned a set of the *Cours Complet d'Agriculture Theorique, Practique*, a rural encyclopedia by François Rozier (1734–93) published in nine volumes between 1781 and 1793 and in thirteen volumes in 1809. Rozier's encyclopedia included a tour and description of rural privies (*fosses d'aissance*) that employed Lavoisier's friendly consultation. Ruffin's own interest in public health, rural and urban alike, brought forth excerpts of Rozier translated by Ruffin himself. Public-health problems and solutions (not the potato) brought Ruffin also to Parmentier, who late in life was inspector general of sanitation for the city of Paris. At about the same time, toward the end of the Napoleonic Wars, Puvis served as public-health officer in an occupied region. Both French scientists published observations and principles on this subject, too.

Ruffin read, reported, and finally merged French science and practice with his own advocacy of marl. His calcareous manure would govern efficient decomposition and productive use of wastes to generate nurturing life anew. Ruffin's preoccupation with the cycle of life, death, and life figured not only in his rationalization of his own suicide and the directions he left Edmund Jr. for disposal of his body—no coffin, the clothes he wore at death—but with what we call recycling on a grand scale. (In this sense Ruffin becomes environmentalist as well as conservationist.) At one point in his essay on public health, Ruffin seems too pleased to report the exhumation and rendering into fertilizer of the bones of British troops killed and hastily buried at Waterloo a dozen and a half years earlier. Here was humanity's genius added to nature's own economy, Ruffin suggested. While translating and publishing French works, Ruffin's English absorbed the word *alimentary*, cognate to the English but employed much less often, either in the vernacular or in scientific usage. He meant not foodstuffs or food stores, as in French today, but the process of ingestion, digestion, and elimination. The end of alimentation should not be lost, he fervently believed, but should complete an endless circle. Ruffin came to describe and recommend all these things from his own observations and necessities and from the ideas of Europeans, especially the great rendering nations of Britain and France.

The circle seems an appropriate metaphor also for Ruffin's anti-Malthusianism. There would not be catastrophe and discontinuity in human history—no chaos either in theory or fact—but a use of resources so wise as to ensure plenty and continuity, around and around forever. Otherwise, the circle is less evocative than the straight lines and right angles of the geometry that compelled Ruffin as student and, I suspect, all his life. Straight lines convey the instrumental directness of scientific and economic purpose and the shapes of chemically maintained monocultural crop fields in much of the United States and Europe today—crop fields shaped less to suit geological morphology than the movements of labor-saving machinery and the designs of maximum productivity.

Finally, the Ruffinian view of nature seems more imperial than that of his European mentors simply because Ruffin was an American and a southern tidewater man at that. While European agronomists had little more luck "revolutionizing" agriculture than Ruffin and his contemporaries— breakthroughs of applied science came toward the end of the century[28]— Europeans occupied relatively crowded, long-settled countries without frontiers or substantial populations of farm owners who resisted markets and farmed folkishly. Ruffin not only was left behind by advancing American frontiers but was surrounded by a majority who ignored or resisted his modern agronomy and who instructed the legislature not to collaborate in his reformism. The contrast with Europeans accentuates and at least partly explains Ruffin's zealotry, his occasional intemperance of expression, and his often-expressed despair.

Editing Ruffin

Yankee revenge at Beechwood notwithstanding, a great corpus of Edmund Ruffin's private and public writings survive. There are correspondence, diaries, some business records, and drafts of essays and speeches in Ruffin collections in Virginia and North Carolina. Ruffin's letters and copies of his pamphlets populate dozens of his contemporaries' preserved papers. Original copies of his opus, *An Essay on Calcareous Manures*, in all five editions (the last appeared in 1852) survive in libraries' special collections as well, along with various booklets and tracts he published between 1840 and 1861—for example, *Report of the Commencement and Progress of the Agricultural Survey of South Carolina, for 1843* (1843), *The Political Economy of Slavery* (1858), and *Agricultural, Geological, and Descriptive Sketches of Lower North Carolina, and the Similar Adjacent Lands* (1861). And thanks to the splendid American Periodical Series, there are legible and complete microfilm copies of Ruffin's *Farmers' Register* plus other journals to which Ruffin contributed significant essays—such as *American Farmer*, *Southern Planter*, and *De Bow's Review*—in virtually every decent research library. Perhaps Ruffin should not have despaired; he is immortal, although perhaps not as accessible as he might have preferred.

Accessibility of Ruffiniana has nonetheless grown. During the past four decades important segments of his writings have accumulated on acid-free stock between good covers and are widely distributed among libraries and individuals. First came J. Carlyle Sitterson's excellent 1961 presentation (from Harvard University Press) of the first edition of *An Essay on Calcareous Manures*. Hardly a third the size of Ruffin's last edition, the 1832 work remains the edition to read, I think. Ruffin's next four versions added minor corrections of the first edition but primarily incorporated elaborations on testing soils and assaying marl, more field trials, and rather undisciplined, discursive ramblings. The most dedicated agrihistorical specialists presumably will pursue these developments (and distractions) in special collections libraries: most readers will be best served by the original classic. An important exception, however, is Ruffin's accumulating "Desultory Observations on the Police of Health, in Virginia" (begun in the second edition of the *Essay*). This engrossing exposition, including Ruffin's own translations of French authorities, is presented below.

The weightiest published Ruffiniana is William Kauffman Scarborough's edition of *The Diary of Edmund Ruffin* in three volumes (1972, 1976, and 1989), together totaling 2,363 pages (including Scarborough's invaluable indexes). Ruffin began the diary at age sixty-two, he explained, because he feared he was losing his memory. So for the momentous last nine years of his life, Ruffin—frequently on the road and a participant in and eyewitness to the American union's dismemberment—saw much and recorded everything. The diary has become one of the most significant printed sources on the triumph of secession and the war.

Almost on the heels of Scarborough's last volume came two more published collections from Ruffin's middle years. These feature not the secessionist agitator but the practical scientist and private family man. First (in Ruffin's own chronology), the year after Ruffin closed *Farmers' Register* he undertook a geological survey of South Carolina, whose government published his famous report. During the survey Ruffin also kept a copious diary. William M. Mathew, earlier the relentless debunker of Ruffin's marl reform, published in 1992 a somewhat abbreviated version of the diary that still exceeds three-hundred rather closely printed pages. Entries more than elaborate Ruffin's formal report: they extend magnificently the demonstration of exuberant curiosity so familiar to readers of *Farmers' Register*.

Following the Carolina survey, Ruffin returned to Virginia and developed his new estate at Marlbourne. Then in 1851 Ruffin supplied details and documents of his life to William Boulware, who wrote the brief biography of the Virginia "agriculturist" published that year in *De Bow's Review*. Having begun the business of organizing his past at Marlbourne, Ruffin started to write privately an autobiography addressed to his children but probably intended for ultimate public distribution. "Incidents of My Life" was contained in three manuscript volumes written in 1851 and 1853. The first was lost (apparently during the Civil War); the third was never completed but included Ruffin's appended "remembrances" of two of his daughters who died in 1855. The two surviving volumes came finally to the Ruffin collection at the Virginia Historical Society in Richmond in 1967.[29] A number of scholars examined these invaluable additions, but biographer Allmendinger not only used them best but undertook (with society sponsorship) the difficult transcription that resulted in their publication in 1990. To the manuscripts that Ruffin composed during the 1850s, Allmendinger added invaluable appendixes: Boulware's sketch; Ruffin's moving recapitulation of the 1840 suicide of his best friend, Thomas Cocke; and three essays on marling that are also autobiographical.

This new collection of Ruffiniana includes essays from late in Ruffin's middle age. Most of the writings offered in this volume, however, represent labors of his youth and early-middle years, when he edited *Farmers' Register*. The aspect of presenting a somewhat younger Ruffin is nonetheless subordinate to the thematic. This volume seeks to offer Ruffin's own exposition of his sense of historical destiny, his protoconservationist reformism broadly construed, and his mature observations and summary program of landscape manipulation. The youngest Ruffin reveals the imperative to manage nature and produce food. The editor elaborates the context of the imperative and emerges as determined anti-Malthusian and comprehensive historian-reformer. The protoconservationist having clearly emerged before 1840, Ruffin's ambition to improve landscape by manipulation mounted with his wider learning and, perhaps illogically, with the defeats of his reforms. At last came the monumentalism of the 1852 Charleston address.

The principle of selecting and ordering these editorials, letters, notes, and essays is demonstration of the ecological interpretation of Ruffin's life and career suggested earlier in this introduction. My reading is late-twentieth-century green, induced of course by Ruffin's own texts. Other readers may decide for themselves.

Most of the Ruffin texts included here were first published in his *Farmers' Register* between 1833 and 1842. Good fortune alone permits confirmation of Ruffin's authorship, for he seldom signed editorials or essays. Soon after Ruffin closed down the journal, his friend N. F. Cabell asked Ruffin to identify authors of all unsigned or pseudonymous contributions. Ruffin complied, Cabell preserved, and early in the twentieth century Earl G. Swem, the greatest of Virginia bibliophiles, discovered and published. Swem confirmed that Ruffin indeed composed many of the articles in the journal, though far from a majority. There were relatively few reprintings from other farm magazines.[30] Letter writers are nonetheless important below, especially in the segment on Ruffin's fencing reform. I have included several non-Ruffin texts there to illustrate elite support and contextural parameters for this particular campaign. The Ruffin-Cabell-Swem list enables the bracketed and noted attributions of authorship in these instances.

These and Ruffin's own writings are reproduced entirely. Authors of non-Ruffin texts are identified when known, while all other texts can be assumed to be by Ruffin. Square brackets in the original texts have been changed to parentheses; any material printed here in square brackets represents my editorial insinuations. The order of presentation of texts here is sometimes not chronological, because my own sensibilities led me to arrange essays on similar subjects from the more particular to the larger in scope; thus, Ruffin's treatise on malaria (1838) precedes his opus on public health (1837). Readers of different sensibility are invited to skip and backtrack. Readers not accustomed to nineteenth-century exposition—long sentences broken by now nonstandard comma placement, occasional odd word order, overweening (perhaps) seriousness, and certitude—are advised to persist. There is a mostly comfortable rhythm as well as clarity in Ruffin's prose. Both man and prose are ultimately engrossing. And his seriousness—this was the business of civilization itself—was not unleavened. There occasionally appears a memorable snippet of pique, an aesthetic opinion, a quirky but revealing usage. Ruffin loved Charleston, for example, from his first visit in 1840, yet just before his first departure from the city he noticed two young men with mustaches, nearly ruining the port's repute as far as the ever-shaven Ruffin was concerned. The sightings occurred well before the onset of the age of mustaches and beards, partly explaining, conceivably, Ruffin's growing crankiness later on (see "Notes on a Steam Journey"). In his first essay "On Draining," Ruffin made an outrageous pun, his material being flooding creek waters, Virginia's property law, and lawyers. And his insistence on calling northeastern North Carolina's Lake Phelps (in "'Jottings Down' in the Swamp") "Lake Scuppernong" seems to me a notion good enough to adopt ourselves. Then there are the French—Ruffin's translations of their writings on privies and urban sewage systems, gases, and explosions and his incorporation of wastes into not only his own language but metaphorically into an attitude that approaches a philosophical system.

Part One

History

1
Production as Moral Imperative
The Morals of Agriculture (1822)

Mr. Editor,

I wish that some of your correspondents, who have more leisure and more ability than myself, would take into consideration the subject on which I shall submit a few desultory remarks. If the morals of agriculture deserve no such attention on account of their importance, the subject is at least worth the notice, and is properly within the province, of all authors of *addresses* to agricultural societies. Most of these gentlemen appear to be so much at a loss for subjects, that their addresses would not be badly designated by the title of *"Essays on things in general."* I, therefore, recommend this subject to any person intending to prepare an annual address, unless he really should have something else to lay before his society and the public.

The Hindoos believe that whoever plants a tree, digs a well, and begets a child, is sure of admission into heaven. As ridiculous as this part of their religious creed my appear, it shows the wisdom of their priests and rulers by whom it was instilled—who thus brought the strongest motives to induce every individual to increase the productiveness, population, and wealth of his country. When our ancestors emigrated from Europe, they wisely left behind them all their elfs, fairies, goblins, &c., and as it is impossible that we can long remain as we now are, free from popular superstitions, it would be a blessing to our posterity if we were to adopt, as one, the Hindoo tenet, so modified as to suit our different situation. We have no want of growing trees, nor of fresh water; and all experience proves that children will always be furnished fully as fast as the food necessary for their support. Population is always precisely proportioned to, or limited by, the means of subsistence, and in an agricultural country, must increase with the improvement of the soil, and decrease with its exhaustion. The farmer who makes his land capable of producing annually 500 bushels of grain more than before his improvements commenced, increases permanently the population of his country, by as many persons as his increased product will support. Another, who spends his life in reducing the fertility of his soil by the same amount, diminishes population as much; and that diminution is more effectual and permanent, than if he had confined his exertions to cutting twenty throats of every successive generation.

"To increase and multiply" is a divine command—and perhaps is the only command which all persons strive to their utmost ability to obey. But though, the usual means may be the most agreeable, I beseech your readers to believe that they are far from the most effectual. It is true, that no harvest can be reaped unless seeds are first sown; but every child knows that it is not the greatest number of grains planted which ensures the heaviest crop of corn, but the means afforded for the support of the plants, by the degree of fertility in the soil. Just so with population. Only let bread, or the means of obtaining bread, be increased in any country, and its population will soon be equal to the increased supply of food. On the contrary, if bad farming, or bad policy in the government, lessen the production of food, the inevitable consequence must be a diminished population. These positions (which every sound political economist will sustain) show what vast effects the labors of a single individual may have on the welfare of his country; and what beneficial effects might be produced, if it was believed (more especially by all law-makers,) that he who directly or indirectly lessens the productiveness of the earth, is guilty of sin, which, if more pardonable than murder, is far more injurious to the country, and more destructive of its population, than would be many murders.

But seriously—this subject deserves to be reflected on by all; it will give additional gratification and encouragement to the improving farmer, and furnish an impressive lesson to him who is pursuing a contrary course. It would be visionary to expect that the public good, alone, would induce improvement of the land at the sacrifice of private interest. Nor would it be desirable. A farmer can in no way do as much good for his country, as by pursuing precisely that course which is most profitable to himself. But though many attempts to increase the fertility of the soil are ill-judged, yet there are means enough which are profitable; and there is no case in which the owner of a farm can be most benefited by its exhaustion. The many, then, who waver between the two opposite cases, could scarcely remain uninfluenced by the *moral* consideration, that on the course of farming which shall be pursued by each individual, the comfort, nay, even the existence of thousands of human beings will depend.

For the purpose of illustration, I will compare the course of two cultivators of my acquaintance. N——, inherited a farm and stock, capable of well supporting an industrious and economical man, but which, if left to the sole management of an overseer, and treated according to the then usual practice, would not have paid the expense of cultivation for many years. Fortunately he knew what course would most promote his interest. For thirty years, he has not ceased striving to make two blades of grass, where one only grew before, and he has met with the success which his exertions deserved. He rejected all improvements (improperly so called) which promised not to return some clear profit on the capital invested, but considered no improvement too laborious or expensive, from which he could, *with certainty*, derive the principal and interest of the first cost. He bought no land which he was not fully able to stock, or that would not yield more clear profit, on the purchase money, than he could have obtained from investing the sum in making additional improvements on the land already in his possession. At this time, by means of improvement of the soil, and extended tillage, he makes crops six times greater than when he commenced. Though N——, has thus eminently promoted the public weal, it was without caring for it: his views were exclusively directed to the advancement of his own private interest. He is obedient to the laws of his country, and just and honest in all his dealings, because he knows that such is his best policy; but in no case does he allow his interest to yield to that of others, and perhaps never performed an act of real generosity in his life.

F——, is directly the reverse of N——, in disposition, character, and habits. Indolent, and having no fondness for farming, his business has been entirely conducted by his overseers; and according to the usual maxims which very naturally govern such gentry, they have exhausted his land as fast as they could clear it. Nothing but the immense fortune which their employer possessed, prevented him from living as most landholders in lower Virginia have done, on all of his annual increase, and part of his capital. But F——, is moderate in his desires, and therefore not of expensive habits; and notwithstanding his bad management, his income has allowed him to continue purchasing land, until he owns almost as much as a German principality. By these means, his annual crops are not materially lessened, though every field is in its turn destroyed, and deserted for a new one. Though he does not obtain two *per cent.* from his capital, yet as still less suffices for his support, he considers his wealth increasing as rapidly as the number of his acres. According to the usual calculation of profit, injury to the land is not taken into consideration. It is evident however, that the mode of cultivation pursued by F——, is merely abstracting the whole fertility of one field, in the form of tobacco, wheat and corn, and applying it to another in the form of purchase money. What was said of the famous conqueror and destroyer, Attila, "that the grass ceased to grow where his horse placed his foot," applies with more truth to my friend F——. Notwithstanding his many virtues, he has to the fullest extent which his means permitted, been the destroyer of grass, of grain, and consequently, of men. Famine marches after him, and will not commit the less havoc because he himself is able to keep beyond her reach.

F——, is remarkable for his kindness and liberality to the poor. Besides frequent occasional acts of charity to others, he has long supported several families, who would perish without such aid. I know how to estimate the motives, and according to them, to respect these two individuals. But their private virtues and vices, have nothing to do with my subject, except so far as the consequences of them affect the public good. F——, has destroyed the means of subsistence for 500, which in effect, is equal to starving, or preventing the existence of as many. N——, has given nothing in charity, but has given in the *wages of labor* more than F——'s wages and alms together: he has increased the production of the earth enough for food for 500 persons, and therefore he has increased population to that amount, though not at all by the Hindoo mode, as he has no children. It is very true that these people must work to obtain N——'s increased product; and so much the better. His improvements will not die with him, nor will the corporeal powers of this laboring population, and their descendants or successors which will continue to earn and consume it. The country is not benefited only by having its population increased by 500 persons; if they were all drones, they would rather be an evil. But the people who eat N——'s corn are field laborers, mechanics, manufacturers, sailors, and merchants, all of whom are continually increasing the national wealth by their industry, as well as its strength, by their numbers. F——'s charity has served not only to support several families, but has doubled their number, by the births which have taken place since they partook of his bounty. After his death, they must still be supported by others, or starve. They are not able to add any thing by their labor to the public stock, and though the children will hereafter be able, their present situation is the worst of all schools to acquire habits of industry. Were all our land holders like N——, the wealth and population of the state would quickly be doubled. Were all like F——, with all his virtues, wealth and population would rapidly diminish, until the country became a desert. Thousands are pursuing the ruinous course of the latter; very few cultivate so as alike to increase the natural resources of their own.

My opinion on this subject, taught me to expect but little increase in the population of Virginia, and not to be disappointed in the report, of the last census, which shows a gain of but ten per cent. in the last ten years. But for the recently awakened spirit of agricultural improvement, (the impulse to which, we owe principally to the author of Arator,)[1] I think that the tidewater district would have suffered a considerable diminution. As much vacant land as this district contains, there is but little uncultivated, (which until enriched) will yield any clear profit. Therefore, eastern Virginia, in its present state, is *fully populated*, and no increase can be expected *except from the improvements of the soil, and the consequent increase means of subsistence.* We export provisions, it is true; this may at first seem to indicate a surplus of the means for subsistence, and a fund for additional population. But such a conclusion would be incorrect. Our surplus food is exchanged for clothing and other commodities, which in fact, or from custom, are as necessary as sufficient food. Our only consolation is, that our excess of population emigrated to the west; instead of starving, as in most fully populated countries.

If private individuals can exert so much influence on the population and strength of their country, how much more extensive must be that of the government! A member of the legislature, by a single vote, may retard population more than by destroying the productiveness of all the land in his possession. A single bad law, which cramps ingenuity and industry, or destroys their honest gains, or what is worse, puts them into others' pockets, causes more poverty and depopulation than a thousand exhausting cultivators. Many are the sins of this description, which have been committed by our legislatures, both state and federal; it is enough to name as examples, the protecting duty policy, banking, and laws for the compulsory support of the poor. The last, though not the least of such evils, will hereafter become the heaviest. Poor laws impose taxes and penalties on honest industry, and offer rewards for idleness, extravagance, drunkenness, and debauchery—and their inevitable consequence will be to increase those vices, until their support shall have absorbed the whole income of the industry of the nation. England has already drawn near to that dreadful situation, and with her example before us, we are pursuing the same course to the same end.

APPOMATTOX.

2 A Malthusian Crisis
A Letter from George Henry Walker and Ruffin's Response (1836)

You still appear (by your letter of Dec. 11th,) to be the most desirous about *practical* operations. Now, farmers are not in much repute amongst other "learned" and "polite" professions, and consequently, not so with each other: when this is the case, is not a little mystery and delay the best?—better not to tell all at once—excite the curiosity—endeavor to state the *true* principles of agriculture *first*—then people will say, "I did not think of this before—this is a sensible fellow—I should like to know what his practice is, and *more* of the results." We are thought too meanly of, and we think too meanly of ourselves; that is, of our profession—its value and importance. The remarks of Col. Bondurant upon this subject, in your December No. are excellent.[1] I want to say something about this in my notes upon Brooks' letter from London, and about the true principles of agriculture—and we can poke in a little *practice* as we go along. Practice, by itself, is like recipes, prescriptions, good advice, &c., in the newspapers—soon forgotten, and little attended to.

Don't be afraid of some of my heterodox opinions: they will come in, and tell by and by; for I have a strong hold against, at present, high established authorities. *Between ourselves*, I have made the greatest and most important discovery ever made by man; this is in agriculture, and in political economy, as regards food and population. It is of no use to myself and society to tell this *publicly*, all at once. I am perfectly confident in the truth of my positions. I call my practice the anti-Malthusian system of agriculture. A word about it. Do you not see that man has never yet lived according to the *natural* law of his subsistence—which is, according to his organization; that Malthus and his followers never thought of ascertaining what this law is, and the corresponding law of agriculture!!!—and hence all their errors. Do you not see that man has hitherto deemed it of no importance whether nations subsist upon bread, potatoes, rice, or Indian corn, &c., with little or no animal food; and hence the utter exhaustion of the soil; because vegetable matter, (grass, roots, &c.) and animal manure (with calcareous matter as the governing power, in duly proportioning stalk, leaf, fruit, grain and seed—) have, under these circumstances, been almost wholly wanting; and hence the "decline and fall" of Rome, (and all other empires,) with accomplishing which, Caesar had no more to do than Gen. [Andrew] Jackson had. Individuals have no such power as *that*, either for good or evil—particularly the latter. The fall and permanence of nations depend upon the non-observance and observance of the laws of nature, by the *mass* of the people—not upon the acts of individuals.

I consider that man was formed to subsist upon animal and vegetable food, in certain proportions; the proportion of animal food increasing with decrease of temperature—obviously for the wisest and best purposes; and that the true principles of agriculture are conformable to this law of subsistence, and that this law of subsistence is conformable to the true principles of agriculture. England, and the United States still more, have observed those laws the best, and hence their prosperity and advancement above all other nations. How strange the political economists have never studied this part of the question! They have never yet even thought of it!!

I have endeavored to form my system of farming upon those principles. I told you of my superb crop of turnips without manure—after grass, top-dressed the year before. The barley after them last year (without manure,) will be 50 bushels per acre, if not more, (not quite all thrashed yet,) and a fine sample. I intended to top-dress the grass, in the barley stubble, last fall, but it appeared too rich, and I put the manure upon a poorer stubble. A few years ago this land would not grow more than five bushels of barley per acre, of the worst quality. The field was three years in grass previous to the turnips (ruta baga). With this system, deficiency of food and excess of population is an idle fear. The first year I mowed six wagon loads of hay per acre; the second, four loads; the third, top-dressed, three loads. It ought to have been top-dressed the first year. Manuring old grass is of little service to the coming crop; it is like trying to renovate the youth of an old man. But it is of great benefit to the succeeding arable crop.

As grass and arable cattle crops are duly proportioned to, and alternated with grain and other crops for man, according to the natural law of his subsistence, so will the fertility of the soil be the greatest, and the labor the least, leaving the *mass* of the population for manufactures and commerce, and all other professions. The practice of mankind has hitherto been perpetual tillage, with separate or little or no grass; or grass very partially attended with arable crops; or arable cattle crops and some grass, occasionally alternated with arable crops for man; with separate permanent pasture and meadow, as in England. Hence utter exhaustion of the soil—partial improvement, and stationary medium fertility; and hence under these circumstances, more or less deficiency of food, and excess of population; and hence again, under these circumstances, the full truth of the Malthusian theory. But Malthus evidently knew not why or wherefore, nor how to remedy the evils and errors, but by *breaking a third law of nature!*—for of the *natural* law of subsistence and of the corresponding true principles of agriculture, he knew nothing—nor even thought of them. Had he investigated them, and studied how far they had been observed, and how much they have been broken, neglected, and unduly observed, he might have been led to infer far greater perfection, wisdom, goodness, beauty and harmony in the laws of the supreme and all-wise Creator, than his sad and disheartening theory implies. There are the means of subsistence for *all*, if the laws of nature are duly observed by *all*. I should be sorry to believe it otherwise. To regulate the numbers of mankind is no part of the prerogatives of man. This can only be in the hands of that power which created him. But Malthus has opened the door to the truth, and that is a great deal. His theory, with other circumstances, and a knowledge of the condition of the people of England and the United States, led me to the discovery of what I am quite confident is the real truth in this greatest and most important of all questions, affecting the condition, well-being, happiness, and civilization of man.

GEORGE HENRY WALKER.

Ruffin's reply to Walker follows. It was originally presented as a footnote to Walker's sentence above, "I call my practice the anti-Malthusian system of agriculture."

The theory of Malthus of the laws of population and subsistence, may not be familiar to all who will read the remarks of our correspondent—and therefore we may be pardoned for presenting the following concise statement in explanation.

The doctrine of Malthus,[2] (which has obtained the assent of almost all political economists,) is, that population naturally increases in a geometrical ratio, while food can only be increased in an arithmetical ratio. Suppose that marriages on an average produce 4 children, who may themselves live to marry: then each generation will serve to double the population, according to the ratio of 1, 2, 4, 8, 16, 32, &c. Let the rate of increase be ever so much slower, it is still in a geometrical ratio, and, in a longer time, will produce similar results.

In a country thus increasing in population, the quantity of food, or means of subsistence, will of course be increased also—but however rapidly, it will not be in geometrical, but in arithmetical proportion—not by doubling—but by regular additions of equal (or more often of decreasing) quantities.

Thus food may increase at first, (and generally will, in new and fertile countries,) even faster than population—as in the ratio of 1 to 4, (or by additions of 3)—but then it will be only at the rate of 1, 4, 7, 10, 13, 16—and of course the increase of population will be rapidly overtaking, and then outstripping the means for its support.

The rule, however, can only work freely and fully so long as there are no checks obstructing propagation—and there is no country, no condition of man, in which such checks do not operate with more or less force. *Perhaps they have less influence on the slave population of most of the southern states, than on any other class in the world.*[3] The checks are either *prudential, moral, or physical*—the latter being the state of starvation and other extreme suffering following want of food. The first two checks to population serve to restrain the natural tendency to propagation—and if *they* are not exerted, the *last one* serves effectually to destroy the redundant increase and to reduce population within the limits fixed by the amount of subsistence then furnished by the industry and products of the country.

The inferences are—that the most prosperous and fertile country, with the best regulated society, contains the seeds which will surely produce an abundant harvest of misery. There is no effectual remedy—and the only hope for mitigating the weight of these approaching inflictions, is the rigid enforcement of the operation of the three checks to population. This stern, repulsive, and stubborn doctrine, wars with the strongest as well as the most general natural impulses of the human race, and even forbids the exercise of acts of charity, by continually warning us that in giving present aid and comfort, we are producing a ten-fold amount of future misery. We have not read Malthus for nearly twenty years—and now state his views, and our own inferences from recollection—but the lapse of time has not lessened our submission to his reasoning. Though we doubt the ability of our correspondent, and of all other persons, to show the theory of population to be false, we heartily wish him success, and should be rejoiced to yield to his arguments our present unwilling conviction of the truth of the heart-benumbing, hope-stifling doctrine of Malthus. Ed. Farm. Reg.

3 The Old Dominion's Declension
Sketch of the Progress of Agriculture in Virginia, and the Causes of Its Decline (1836)

Though fully sensible of the honor conferred on me by the request of the Executive Committee of this Society, to prepare an address on agriculture, it is with reluctance that the effort to comply is made—and nothing would have induced the attempt, but the fear that my declining the performance, would be attributed to any but the true motive. Heretofore, I have been accustomed to treat of such subjects as needed only to be rendered intelligible—of which, the manner and style were of far less importance than the matter conveyed—and for which a plain and even rustic mode of expression was not altogether inappropriate and inexcusable. The task now imposed on me requires abilities of a different and higher order; but still, for its execution, no aid will be sought in the grace and beauty of language. Being sensible of possessing no powers to draw from these sources, the fruitless attempt will at least be avoided.

But if no difficulty of this kind existed, others would be presented in the vast field embraced by the subject, and the impossibility of compressing it, or of doing justice to even any one distinct part, in an address as concise as the occasion requires. It is necessary to choose some limited portion only, for consideration: and therefore, my remarks will be confined to a hasty view of the progress of agriculture in Virginia, and the most efficient causes of its depression, and continued decline. Even for this detached portion of the subject, far more time and space would be required than this occasion will afford: and scarcely more can be presented than a copious table of contents—or the mere outline of a picture, which will be left for others to fill up and complete.

In a sketch so rapid, it is impossible even to name the many incidents and changes which government caused, during the early times of the colony of Virginia. These however would furnish many matters curious and interesting to the reader of history, or useful to the political economist, as illustrations of the effects of legal mis-directions of industry. (A.)[1] Afterwards, when separate individual rights in lands were fully established, and the direction of industry was no longer materially influenced by law, nearly the whole agricultural labor of the country, and for a long series of years, was given to the culture of tobacco. In Europe, tobacco has never been raised in perfection; for however luxuriant the vegetation, and abundant the crops, the peculiar flavor which gives value to the product, has been always found deficient. This inferiority is certainly not owing to want of sufficient heat, or to any other defect of climate, in every region of Europe—and therefore, I infer that is caused by a general difference in the constitution of soil. Perhaps the deficiency of calcareous matter in the soils of Virginia, and generally also of her sister states, and its more usual diffusion through the best lands of Europe, have caused this one difference in our favor—a superiority however, of which the effect, as well as the cause, is subject for regret, rather than rejoicing.

Until times comparatively recent, there was but little skill used, (or required by purchasers,) in the tillage and after-management of tobacco. Steady and regular work, and nearly all of it of one kind at each separate period of the crop, occupied all the laborers on a plantation—so that but little intellect was required, either to execute, or to superintend, the several processes. The clearing of woodland engaged the labor of the whole force, every winter, and for months together, all at a time under the eye of the supervisor. The tillage was mostly performed by hand labor, because (under the existing circumstances,) it was cheaper than the extended use of the teams and expensive implements, and the accompanying varying and dispersion of labor, which belong to the most improved and perfect husbandry. During the time of general tobacco culture in Virginia, there was no thought of increasing products by enriching the soil. New clearings of forest land were the only means used, both for extended tillage and increased products—and each new piece of land was planted in tobacco as long as it was able to bear that crop, and then in corn, until reduced to such a state of sterility as to induce the owner to throw it out of cultivation altogether, to be recruited, by nature's care alone, under a new growth of trees. The amount of cultivation for corn was barely enough to feed the laboring force—and that also was carried on by nearly or quite all the hands being employed together, and in the same operations. This is nearly the same kind of tillage that is necessarily adapted in every new "planting" country, whether the crops be of corn, tobacco, cotton or sugar—and however rude and imperfect, and even impoverishing to the soil, may be such a mode of tillage, it was wisely adopted under such circumstances as formerly existed in Virginia. It was the existence of similar circumstances in still greater strength, which forbade the general use of the plough, and even the cart, on the sugar estates of the West Indies, and which still makes the hand-hoe almost the sole utensil on some of the cotton estates of South Carolina. The error consists in continuing such rude tillage after the circumstances which required it have ceased—and such has been the error probably in all the cases cited. In a new country, where land is cheap, and labor dear, it is good economy so to direct tillage that each *laborer* may be able to yield the greatest product, (without much regard to the land—) and also, in such manner that as little intellect as possible may be necessary in the operative class. This rude kind of culture, and the employment of slave labor, are well suited to each other: and there can be no doubt, but that by the general use of slave labor, Virginia, while a *planting* country, was greatly benefited—and that her wealth and prosperity were greatly more promoted, than they would have been without slaves. Nothing will make the lowest class of laborers industrious, in any country, but *compulsion*—whether it be presented in the form of hunger and cold, or the power of a master. Where land is fertile and cheap, and the climate mild, and the necessaries of life attainable with little effort, more than half the laboring power of the country would be wasted in idleness, without the institution of personal slavery. It is this institution (though not so named,) that has made the rapid growth and prosperity of the *convict colony* of New South Wales a theme of admiration and astonishment—though the bondsmen were of the most worthless kind: and the recent destruction of slavery in the British West Indies will surely and speedily reduce these heretofore productive and flourishing islands to a worse condition than New South Wales would have exhibited, with a *free* population, composed (as her's was) of vagrants and felons. In every country of modern Europe, personal slavery has existed—and it has no where ceased, by operation of law or otherwise, until manifestly opposed to the interest of the masters. Long after its end in Scotland, and in comparatively a modern time, such was the state of poverty, idleness, and misery of a large portion of what ought to have been the *laboring* poor of that country, that Fletcher of Salton actually recommended the institution of *personal slavery*, as the best remedy for the evil. Yet the general poverty of soil of Scotland, and its severe climate, rendered it one of the most unfit regions for slavery—and Fletcher, however mistaken in this respect, was an enlightened patriot, and a true lover of liberty. These facts are some of the many proofs that slavery is not always a political evil—and that its cessation, even by universal consent, has not always been considered a blessing.[2]

But nothing in society can be considered as permanent; and the circumstances which make slavery beneficial are not exceptions to this general rule. These circumstances begin to change as the agriculture of a country becomes highly improved. In a "farming country," the dearness of land requires that the owner should especially economize that portion of his capital. Improved processes, and extended and varied culture, make it necessary to disperse the laborers on every farm, and greatly to vary their employments. Then mental power is no less wanting than physical force. The ignorant slave who could well and profitably wield the hoe or the axe, under continual supervision, and the ignorant director of the work, who had only to see that it was "kept moving," become equally unfitted for the new and more p[er]fect operations—and if they continue ignorant, they will always remain unfit agents for operations which, if properly conducted, would be greatly superior in effect and in profit. It is for this simple cause, that loss, instead of profit, so often attends the introduction of improved systems of farming—and the blame is generally laid to the defects of the system itself, when it is altogether attributable to the unfit agents and means employed.

But though I dissent from the very general condemnation of the earliest and rudest system of husbandry in Virginia, no one more reprobates its being continued beyond the proper time for a change to better farming. With the usual obstinacy of adherence to old practices, tobacco continued to be the principal market crop in lower Virginia, until the interruptions of commerce, which preceded and accompanied the last war with England, reduced the price to less than what hay sold for soon after, for the support of cavalry horses.[3] In this, as in other cases, political compulsion, more than reasoning and proper estimates of profit, drove us from the old general system of tobacco culture. It is now continued only in the middle and upper parts of Virginia; and the processes for its preparation for market have been there so greatly improved, that much skill and judgment are required, and are exercised, for the profitable management of the crop.

Wheat was long in taking its place as a large crop—and would have been longer, but for the very high prices, (sometimes $2, and even $2 50 the bushel,) caused by the general war in Europe. The crops were at first so foul with weeds, and the grain so badly cleaned, that such would not now be marketable. Then came the ravages of the Hessian fly, which diminished the quality of the crops, as much as slovenly preparation injured the quality of the grain. But improved tillage, and greater care, have served to compensate all such causes of loss, and have made the wheat crop of Virginia, and of the James River lands especially, a most important item of the total agricultural products of the state. This culture, however, which was so difficult to be established on a proper footing on our lands the best adapted to it, was next, by blind imitation, extended to almost every farm, even though the least suited to wheat, by the composition of the soil. Some proportion of lime is essential to constitute a good wheat soil—and very little land in Lower Virginia, naturally possesses that proportion. Still, so much and so peculiarly are the tillers of the soil the creatures of habit and example, that the culture of wheat was long persisted in, in spite of frequent failures, and general ill success, on soils altogether unsuited. At last the season of 1825, so fatal throughout the eastern half of Virginia, did not leave enough of good wheat on many farms, to again sow the fields—and that circumstance, more than reasoning and previous experience, induced wheat culture to be abandoned on the soils least suited to it, and its place to be occupied by the then novel field culture of cotton.

Indian corn has always constituted much the greater part of the food of the people—and with its offal forage, the food also of our live-stock. Hence it is the most important crop for its abundance and value, though its great home consumption makes its export but of small amount.

Cotton, as has been already stated, was introduced but recently in Virginia, as a field crop—and its long previous disuse, and its now successful and wide extended culture, present another striking example of the slowness with which new truths in agriculture make progress. During the last war, when our coast was blockaded by hostile fleets, all the cotton that supplied the northern factories was conveyed through Virginia, in wagons, over the worst of roads, and sold sometimes even here at 37 cents the pound. None of us then thought of making it as a crop, for sale, though neither our wheat, corn, or tobacco, would then bring remunerating prices. It has long been usual on almost every farm to cultivate a little "patch" of cotton for domestic supply—but such small culture is seldom otherwise than neglected, and therefore unprofitable—and thence grew the general opinion that it would be in vain to attempt the culture of cotton on a large scale. Since, it has become the principal market crop of several of our southern counties—and yielded fair comparative profits, even when as low as 10 cents the pound.

Still more slowly, after its first dawning, was a general system of good farming established any where in Virginia. Though there are now many farms that may vie with any in the world in good management, and profitable returns, (considering the difference of existing circumstances,) it must be confessed that over a large portion of our territory, the good farming which is founded on, and mainly consists in, economy of means and fertilization of the soil, has yet to be introduced. Much has been done for improvement, for profit, and as example, by many of our enlightened farmers—but a thousand times as much still remains to be done—and from the neglect of which, our country at large is losing, or wasting, continually and enormously.

Manuring, the basis of good farming, until a time comparatively recent, was almost entirely neglected—so much so, that the opinion generally prevailed that putrescent animal matters (the effects of which no one could avoid observing,) were alone worth being used for the purpose of enriching land. To the illustrious farmer and patriot John Taylor, Virginia is mainly indebted for the removing of this gross, but widely diffused error. From his writings, and from the practice of other still more successful, but less celebrated improvers of the soil, it became known that far greater resources for fertilization existed in the almost unlimited amount of vegetable matter. This was no more than was previously known to all well informed agriculturists: but a remarkable practice founded on that general truth was scarcely known in Europe, became inapplicable where land has been highly enriched, and the processes of agriculture have reached a high state of perfection, though most valuable in a region of poor and low priced lands, and where labor and floating capital are scarce. This is the "enclosing system," or the prohibition of grazing the fields when not under tillage, by which the land is enabled to manure itself by the decay of its own vegetable growth. This was a great step towards improvement, (on suitable soils,) even when the covering of the fields consisted merely of the natural growth of common weeds—and far greater was the benefit when clover was substituted, which furnishes the richest possible coat of vegetable manure. When this system is adopted on soils suited to clover by nature—or made suitable by the use of calcareous manures—and still more when the mysterious and wonderful aid of gypsum is added—the farmer has found resources for any amount of fertilization that his industry and intelligence will secure. As vegetables, especially of the clover tribe, feed on, and draw their growth and substance from air and water principally, if they are left to rot on the soil, they return to it much more enriching matter, than they had previously taken up for their support. This vegetable manuring is in itself a vast source of fertilization—but may be greatly increased by adding, in proper manner, the rearing of live-stock, and use of prepared manures. The attention which has but recently been directed to the use of calcareous manures, has already been productive of signal benefits—and every year's experience will serve the better to establish the necessity for their application, and the immense profit thus to be obtained, on our soils that are so generally deficient in that ingredient which is essential to a high grade of value and productiveness.

The observer of the progress of agricultural improvement in Virginia, may remark that the introduction of almost every new and profitable practice has been caused by means that amounted to compulsion—and rarely, if ever, has a general change been produced by the clearest reasoning, if not attended by the pressure of necessity. Thus, it required long continued and ruinous low prices to stop the general and injudicious culture of tobacco in lower Virginia—and the almost total destruction of the wheat crop throughout an extensive region, to prove that it was altogether unsuited to the soil. We may therefore hope that there still remain great blessings, in other new kinds of culture, to be *forced* on our acceptance by some supposed calamity, but real good. Thus, sheep husbandry, (which now the *dogs alone* can effectually prohibit,) and silk culture, may possibly be established with profit; and by such or other similar means, a prosperous population may be renewed in the central region, where emigration seems now to threaten a wide-spread waste. If it was possible that such a *calamity* could occur, as that we should be entirely shut out from the rich lands of the west, we would be *forced* to learn the value of our own position and resources, and the immense profits to be obtained by remaining contented at home, and making the best of the advantages offered by the land of our birth.

I shall now state more particularly the several causes which are most operative in repressing the improvement, and serving to prostrate the interests of agriculture.

I. The first cause in order, though not in grade, of injury to agriculture in Virginia, is the continuing too long the early and exhausting, as well as rude system of tillage, which was necessarily and properly adopted when our country was generally in the forest state. The error of too long a continuance of this course, was not so important (as is often asserted,) as to have blighted agricultural improvement over all the state. If no other and greater errors had existed, this one, the result merely of want of information, would have cured itself soon enough, in every case, to prevent any serious loss to individuals, or to the nation. This error has not operated alone, but has been aggravated, and endowed with its evil power, by other errors, which will be named hereafter.

II. A very mistaken opinion has prevailed generally as to the improvable nature of soils—and this mistake has served to misdirect effort, and to disappoint, and render profitless, the attempts of the most energetic farmers to enrich poor soils. It is but a late discovery that nearly all the soils of Virginia, (and it is believed also of all the Atlantic States,) are singularly wanting in lime as a natural ingredient—and it is yet but little known, that without a certain proportion of that ingredient, all efforts to fix new fertility in soil are vain. As a consequence of these positions, the adding [of] the ingredient which is wanting will be the more effective and profitable, in proportion to the previous deficiency: and if our tillage in general has yielded profit, under this great and remarkable defect, far more profitable must it be, after that defect shall have been removed. The ignorance of these important truths, is here and elsewhere one of the greatest obstacles to agricultural improvement. It is in vain to prescribe remedies for a disease, so long as its cause is totally unknown to the physician—and even if the only effectual remedy is found to be unattainable, it is better that the truth should be known, than to persist in a vain and profitless pursuit. But nature has not been niggardly in offering means for applying this remedy to the lands of Virginia. Large portions of the state are abundantly supplied either with fossil shells, or limestone—and recent discoveries give hopes that similar means may be found in, or cheaply conveyed to much of the most destitute region. Geological surveys and investigations may add incalculably to the one benefit—and the new and great facilities for transportation offered by canals and railways, will perhaps as greatly promote the other.

III. A third cause of injury to agriculture, and one which acts upon and aggravates all other causes, is the total absence of every thing like agricultural instruction, whether practical or theoretical. If in any other great and complicated business, it was the universal course for capitalists to commence as undertakers without any knowledge of the theory or principles—and to engage operatives and superintendents who were equally ignorant of the practical details—and that for all to acquire the knowledge wanted, reliance was placed solely on their subsequent untaught operations—in such cases, every voice would pronounce that the business must inevitably result in bankruptcy and ruin. Yet such is the case, nearly to the letter, in almost every agricultural business in Virginia, whether on a large, or small farm. Of the former, the wealthy young proprietor is educated far from his future property, and receives instruction almost exclusively in dead languages, and sciences designed never to have any practical application to his future pursuits—and perhaps afterwards he adds thereto more or less of study, (or the pretence of it,) of law or medicine—and having arrived at manhood, he throws all aside to undertake the new business of farming. Still, a rare and admirable aptitude for the pursuit may sometimes enable a proprietor to overcome all the disadvantages of his own ignorance, and that of his agents, and to succeed in spite of such enormous obstacles. But because some few such cases are before us, it is a great and dangerous mistake to suppose that good farmers can be thus formed, in all, or in many cases. It would be as rational to infer that all education in literary and scientific institutions was unnecessary, because, without such aid, a [Benjamin] Franklin could reach the highest station of celebrity and usefulness.

But however great may be these disadvantages of the wealthy farmer, they are as nothing compared to those of the far more numerous class of men of small possessions, and of more limited education. The former has leisure, and various facilities to gather information from others. Travel, books, frequent and varied association with his fellows who are better informed, (or who even if alike ignorant, are alike striving to gain knowledge,) all serve to aid his improvement. But the small farmer who is confined to his daily toil, has no such means, and can rarely acquire knowledge from others, if surrounded by such as are in circumstances similar to his own.

So much for the proprietors of farms, and general directors of farming. But of necessity, all extensive and complicated operations must be under the immediate direction of agents—and in our farming, these agents, the overseers, are without any previous training for their business, and of course deplorably ignorant, even if not in other things deficient. This alone would present a sufficient cause for failure, even if the proprietor is not wanting in ability as a general director—unless indeed, (as is usually the case,) such capable proprietors are in a great measure their own overseers, or have to undergo the drudgery of their business, and attend to those details which properly belong to the care of subordinates. But however low our overseers deserve to be ranked, as a class, there have been among them a few distinguished exceptions—men, who however humble was the commencement of their career, have earned, and well deserve to enjoy, the applause and high respect of their countrymen in general. The distinguished place in society to which some such men have honorably risen, by their knowledge and success as farmers, shows that the reward of reputation, as well as of wealth, may be surely attained by such meritorious exertions.

This cause of the degradation of agriculture—*ignorance*—is more subject to the control of government, than any other—and therefore it is more important to dwell on its baneful influence, and to suggest remedies for the evil.

If agricultural professorships were established in our principal institutions of learning, young landholders who are there acquiring liberal educations might easily obtain a competent knowledge of the general principles of agriculture, without sacrificing the other useful parts of scientific instruction. If this object would not be sufficient inducement to remain one more year at college, it would be an advantageous exchange in such cases, if the study of theoretical agriculture, and its connexion with chemistry and some of the branches of natural history, took up the time usually devoted to metaphysics and the higher branches of mathematics—the study of which will be of use to but few men, except as a good mental exercise—a kind of *gymnastics* for the mind. It will be easy to ridicule the agricultural instruction that could be acquired from the lectures of a professor—a mere man of books and theory. But though it is freely admitted that no such course of instruction, *alone*, could make a farmer, yet it would be the best preparation for the future acquisition of practical knowledge. It would be folly to look to the lectures of a professor for instruction in practical operations: but we might expect them to furnish the general and true principles of agriculture and its kindred sciences, (so far as they are connected,) without some knowledge of which no man can avoid committing continual blunders, and meeting with continual losses, as a practical tiller of the earth. For example: a farmer cannot know whether he is proceeding right or wrong in the very important operations of preparing, preserving, and applying manures, without some knowledge of the chemical ingredients of the materials used, of the changes produced by fermentation, and of the functions of the plants which are designed to be fed, and of the composition and properties of the soil intended to be enriched. Even a *little* knowledge on these points would serve to guard against serious waste and loss on every farm—the total amount of which makes an hundred fold greater national and annual loss, than all the expense of agricultural professorships, and of every other means for instruction that I shall advocate.

Experimental farms, under proper direction, would also serve a most valuable purpose for increasing general agricultural information. But it would be a mistake, fatal to the object in view, if such establishments were expected to present a system of pattern husbandry, or even to yield any clear pecuniary profit whatever. Such expectations would necessarily be disappointed—and thus would cast discredit on the whole plan. Experiments, if judiciously conducted, and accurately reported, would be more effectual than any other means for conveying valuable information to the agricultural community. They *cannot* be made extensively by private individuals—for the plain reason that they require too much expense of time, labor, and money, and in general, are attended with loss, even when the results are most valuable for the information they give. A farmer might lose $100 by making a series of experiments, of which the results might be worth $100,000 to the community at large. Hence, it is in vain to hope for such proceedings, unless induced and supported by the funds of the community—and it is foolish to count on deriving direct and immediate profit from experiments, whether conducted by public bodies or private individuals. Yet this foolish expectation is very general—or at least it is commonly deemed sufficient ground to condemn and ridicule any experiments as worthless, if the immediate result is loss to the inquiring and public-spirited individual who instituted them. Yet who ever counted on deriving direct pecuniary profit from any course of experiments in chemistry or natural philosophy? And without many such costly and losing experiments, the world would not have obtained the benefits of the steam engine, of the machinery for spinning and weaving cotton, the modern processes of bleaching and coloring, and hundreds of other improvements in the arts. If judged by the test of profit, as usually applied to agricultural experiments, Watt and Fulton, and Arkwright,[4] would have been pronounced mere fanciful schemers, if not fools, not only because of the expense of their experiments, but perhaps because neither of them could bring into operation the mechanical skill, and habits of business, necessary to the highest perfection and greatest profit of their splendid discoveries.

It has often been recommended, and by high authorities, that *pattern farms* should also be established, to teach agriculture by example. This I should strongly oppose—and for the reason that, from necessity, what might be *named* a pattern farm, would prove to be any thing but an exhibition of good and profitable husbandry. Instead of attempting such an establishment, it would be better to make use of, as an additional course of *practical* instruction, the now existing farms of many private individuals, which are truly patterns of good and profitable management. Many such farms might be named, deserving this character, in various parts of Virginia. It is evident that no one of them, even if managed in the most perfect manner, according to its location, and the peculiar circumstances of its proprietor, could serve as a general pattern farm. But one, for example, might exhibit a pattern for clover fallow and wheat culture, on the singular and valuable "red land" of the Southwest[ern Virginia] Mountain slopes—another for the same, on the rich, flat loams of James River—another, would show the profitable combination of tillage and grazing west of the Blue Ridge—another, the best mode of tilling and enriching the more sandy lands of Lower Virginia—and all might accord in some common points of resemblance, in addition to the merit of excellent general management. The proprietors of such farms might very properly be considered as adjunct and practical professors of agriculture; and would render important services as such, by each receiving in turn, as pupils, two or three of the young men who had previously passed through the course of theoretical instruction. On these farms, and under such instructions, a young man could learn more of practical and profitable agriculture in a few months, than from his own solitary and unassisted efforts continued throughout a long life. (B.)

IV. Another and very important source of injury to agriculture is presented in the frequent and extensive changes of the boundaries of farms, which are necessarily made by the conflicting "see-saw" operations of our law of descents and law of enclosures—the first, for the purpose of legal division among heirs, compelling single farms to be cut up into several distinct shares of far less *proportional* value, because each is totally unfit for a single property—and the second, compelling the impoverished owners to sell their shares as they successively come to possession, or can find rich and contiguous purchasers, and thus to form newly arranged large farms out of four or five small ones. I mean not here to dispute the alleged political benefits of this cutting-up and piecing system—nor to inquire whether national liberty and popular rights might not have every existing security, without paying this ruinous tax, this never ending tribute from individual and national wealth, for their preservation. It is enough here to state that this impediment to agricultural prosperity exists—and it will scarcely be required to maintain the assertion, by argument or illustration. It is not the old and long debated contest as to the comparative advantages of farms being either generally large, or generally small. Both kinds have their respective advantages, and also their disadvantages, and the agricultural interest of the country demands that there should be farms of every size, not too large for one man's superintendence, or too small for one man's profitable employment. It is not to the sizes and boundaries of farms, as existing at any one time, that the objection applies—but to the continual changes of both, not required by the interests and wishes of proprietors, but compelled by the operation of law. A well arranged farm, whether large or small, can rarely be divided, or be consolidated with others, without great loss of labor expended for previous arrangements, which are rendered useless by the change of limits. This loss is compensated by no gain whatever, to other individuals, or to the commonwealth. When one-tenth of the value of a farm is destroyed by its being divided, or of a cluster of distinct farms, by consolidation, (and how rarely is the loss of less amount?—) that loss is absolute and complete to all parties—to sellers, buyers, and to the nation—as much so as if one-tenth of the land had been swall[ow]ed by an earthquake. The labors of another generation may, by new improvements, create a new, and even greater value—but cannot recover any part of what was lost. It would be far better if every farm, which was not divided by the will of its deceased owner, was sold entire, with its stock of every kind, and the proceeds divided among the heirs.

V. The inducement and bounty offered in the cheapness of lands, for emigration, forms another evil of great magnitude to the agriculture and welfare of Virginia. Every dollar taken off of the price of the government land in the west, probabl[y] serves to take twice as much from the selling price of ours'. If the present *minimum* price, $1[.]25 for public lands, was reduced to nothing, the lands of inferior quality in Virginia would not sell at all. This long existing, and probably increasing evil, is not only beyond the control of individuals, but also of our state government. We have only to hope for the best, and fear the worst results, in this respect, from the future action of the federal government.

But it is not the cheapness of the new western lands, nor their rich products and profits, that alone would serve to prostrate the prices and estimation of ours', and to discourage the most profitable investments and improvements, if these causes were not given a ten-fold strength by a generally prevailing madness for emigration, compounded of discontent with the really attainable advantages of our native land, and a credulous and too confident faith in the remote and untried blessings that are promised in the west. It is in vain that rich blessings are offered to our acceptance at home, if they are undervalued and despised. The fact cannot be disguised—Virginia is losing by emigration, not only in the population and wealth that are pouring out as a flood, but because what goes out of the state serves to lessen the value of all that is left behind. The most signal improvements that have been made in our agriculture, and the greatest profits thence derived, do not cause any general enhancement in the market price of lands, or serve to give a new impulse to general agricultural improvement and prosperity. Such particular improvements have merely served in some measure to stay our downward career, and therefore, their effects, however great and profitable, are not properly estimated, and indeed are not extensively known. Virginia is likely to be reduced to a condition that will need all the patriotism, zeal, and talents, of her remaining sons, to prevent their country becoming waste, and almost abandoned. There is only one little spot in Virginia over which this mania for emigration has not spread its malignant influence, and among the residents of which there exists generally a true love for the land of their birth. This is the narrow peninsula east of the Chesapeake [the eastern shore counties of Accomack and Northampton]—and the remarkable difference there exhibited, shows the rich reward which follows mere contentment and industry. With a soil that is generally far from fertile, and where their valuable means for fertilization have been but little used, the lands sell high, rent high, and both landlords and tenants are satisfied, comfortable, and thriving. There is no peculiar cause for this state of things, except that for this people it may be said that *there is no "western country."*

But however heavily all these evils may press on the interests of agriculture, there also exist abundant inducements to exert our energies to throw off the burden. If even partial success should reward the effort, the agriculture of Virginia might reach a point of very high improvement and prosperity. Let us keep in view continually the causes which have depressed agriculture, and now prevent its rising from its fallen state—and let us also look to the glorious and rich reward offered for successful exertions for relief. Fallen as it is from its former high estate, Virginia is still a country to be proud of—a rich inheritance, for which we have reason to be most thankful for possessing. A large proportion of the soil of the state is naturally and highly productive—and there are many farms, and some districts, of which the general husbandry may be compared with the best of the United States. The ruggedness of a large part of the mountain region is well compensated by its fertile soil and mineral wealth—its beautiful scenery, and pure, invigorating air. The low country, though generally poor, is as susceptible of *profitable* improvement as any region whatever—and on that account, as well deserves as the rich west itself, to be the object of speculators, treasure hunters, and builders of castles in the air. The intermediate region possesses in some, but a less degree, the advantages of both its western and eastern neighbors. In addition—all parts of the state are now about to derive the advantages of a general system of inland transportation, which seems now, for the first time, placed on the only stable foundation—that is, of yielding early and sufficient profit to the constructors, or stockholders, of canals and railways, as well as of general benefit to the country at large. Above all—as a ground for hope—Virginia's best product has always been of *men*—and though her sons have deserted her soil in such numbers as to furnish half the population of the new western states, still the *stock* remains at home, in its original purity and value. There are abundant means still remaining for the high improvement and permanent prosperity of this commonwealth—and nothing will be wanting to produce that result, unless it be the *will*, and the *resolution*, to use the offered means.

The great evils which serve to prevent agriculture being prosperous in Virginia, may be summed up in the single word, *ignorance:* but however heavy may be this burden and curse on a nation, its importance cannot be appreciated, and therefore will never be complained of, by the great mass of any people. The more dark and widespread may be the cloud of ignorance which ages of misrule may have caused to overshadow a country, the more contented will be the mass of the people with their destitution of light. In benighted Spain and Turkey, rebellions for unnumbered wrongs, were matters of every day occurrence: but not a complaint has been uttered, or even the ground for it suspected, that the governments of these countries have permitted and caused the almost extinction of knowledge—which is the fruitful source of all the other evils felt. The few persons who may understand the cause, are utterly powerless for its removal, or mitigation.

In the legislature of Virginia only can be found combined the *intelligence* to know, and to properly estimate, the evils which are crushing our agricultural interests—and the *power* to guard against, or to remove them. To the sense of justice and of policy, no less than to the patriotism of that body, we have to appeal—and upon the success of that appeal, sooner or later, will depend whether Virginia is to rise—or to sink, without a remaining hope. The *will* of the legislature alone will be potential: the mere disposition to grant aid, if earnestly felt, and barely commenced to be put in action, will soon ripen into a judicious and effective general system for the resuscitation and support of agriculture, and of agricultural interests. And the cost of these measures, (the only, but great obstacle—) will merely be a sum utterly contemptible compared to the value of the least of the expected results—and much less than of many customary legislative proceedings of the least possible value. The legislative expense of selecting a person to perform the mere ordinary mechanical services of a printer, has, for two sessions, been as great as would well support a professorship for the two years. No richer endowments need be desired for agricultural professorships, and for all such other aids to agricultural knowledge as are now wanted, than to divert to these ends, the cost of barely three or four days of the *excess* of debate of every legislative session—or of the tenth part, only, of the average expense of discussing abstract political, or mere party questions, or resolutions, which have led to no *direct* results whatever; but of which the *indirect* effects, for twenty years past, have been becoming more and more destructive of the true interests of Virginia.

Appendix

Many of the circumstances which affected the early condition of agriculture in Virginia, and which were necessarily excluded by the proper limits for an address (and one of secondary importance) to a public meeting, are yet highly interesting. Extracts exhibiting such will be embraced in the first of the following notes—none of which (it is proper to say) were attached to the foregoing address, when sent in to the society by whose order it was prepared, and is now published.

Note A. page 748

From the first settlement of Virginia, in 1607, until 1613, there was no such thing as distinct landed, or indeed other private property which was the fruit of individual labor or enterprise. This was one of the many unwise provisions of the charter of the London Company, (the corporation that settled and owned Virginia until its dissolution in 1624,) or of the regulations made by that body—and this plan of having the settlers to labor in common, and without holding private property, would alone go far to account for the previous feeble and often suffering condition of the colony.

"Each individual put the fruits of his labors into the common stock, and industry and indolence were thus placed on the same footing. The disadvantages of such a procedure must be obvious. This was a property too vague and uncertain, to stimulate the enterprise and industry of the owner. The foresight of Dale put a stop to this evil; and though his remedy is partial and defective, it must still be considered as the dawn of civilization."—*Burk's History of Virginia, vol. 1. page 172.*[5]

"The most honest and illustrious would scarcely take so much pains in a week, as they would have done for themselves in a day; presuming, that however the harvest prospered, the general store must maintain them; by which means, they reaped not so much corn by the labor of thirty men, as three men could have produced on their own lands."—Purchas.[6]

"And now the English began to find the mistake of forbidding and preventing private property: For, whilst they all labored jointly together, and were fed out of the common store, happy was he, that could slip from his labor; or slobber over his work in any manner. Neither had they any concern about the increase; presuming, however, the crop prospered, that the public store must still maintain them.

"The five years also, prescribed in his majesty's instructions, under the privy seal, for trading altogether in common stocks, and bringing the whole fruit of their labors into common store houses, were now expired. Therefore, to prevent this inconveniency and bad consequence, Sir Thomas Dale allowed each man three acres of cleared ground, in the nature of farms. They were to work eleven months for the store, and had two bushels of corn from thence; and only had one month allowed them, to make the rest of their provisions. This was certainly very hard and pinching; but his new and favorite settlement at Burmudas Hundred had better conditions. For once month's labor, which must neither be in seed time nor in harvest, they were exempted from all other services; and for this exemption, they only paid two barrels and a half of corn, as a yearly tribute to the store. However, the prospect of these farmers' labors, gave the colony much content; and they were no longer in fear of wanting, either for themselves, or to entertain their new settlers."—*Stith, p. 131.*[7]

Having commenced at this great cause of previous mishaps, I now return to note some of the most remarkable or interesting events that relate to the early agriculture or general economy of the colony of Virginia.

The ships which brought the first settlers, entered the Chesapeake on April 26th, 1607; and the landing at Jamestown, to make a permanent settlement, was on May 13th. No corn or other crop was made until the following year—and then but a very scanty supply. Within one month after the return of the ships to Europe, June 22, 1607, fifty of the settlers died from sickness caused principally by want and unwholesome food. "The survivors were divided into three watches, and subsisted on crabs and sturgeon till September," when new supplies were received from England.

"About this time also, there sprang up a very troublesome sect of gold finders, which was headed by captain Martin, and warmly embraced by Newport. There was no thought, no discourse, no hope, and no work, but to dig gold, wash gold, refine gold, and load gold; and notwithstanding captain Smith's warm and judicious representations, how absurd it was, to neglect other things of immediate use and necessity, to load such a drunken ship with gilded dust; yet was he overruled, and her returns made in a parcel of glittering ore which is found in various parts of the country, and which they very sanguinely concluded to be gold dust: And in her they sent home Mr. Wingfield and captain Archer, to seek some better employment in England, for they had assumed many empty titles of offices here, as admirals, recorders, chronologers, justices of the peace, and of the courts of plea, with other such idle and insignificant pretensions."—*Stith, p. 60.*[8]

After the third crop season, and second actual harvest, the agricultural condition of the colony may be gathered from the following statement, which the historian evidently presents as highly prosperous: and indeed, so it was, compared to the state of things before, and after the end of the administration of Capt. Smith, but for whose conduct the very existence of the colony would have been lost.

"And thus, about Michaelmas, one thousand six hundred and nine, Captain Smith left the country, never again to see it. He left behind him, three ships and seven boats; commodities ready for trade; the corn newly gathered; ten weeks provision in the store; four hundred ninety and odd persons; twenty-four pieces of ordnance; three hundred musquets, with other arms and ammunition, more than sufficient for the men; the Indians, their language, and habitations, well known to an hundred trained and expert soldiers; nets for fishing; tools of all sorts, to work; apparel, to supply their wants; six mares and a horse; five or six hundred hogs; as many hens and chickens; with some goats, and some sheep. For whatever had been brought, or bred here, still remained. But this seditious and distracted rabble, regarding not any thing, but from hand to mouth, riotously consumed what there was; and took care for nothing, but to color and make out some complaints against Captain Smith."—*Stith, p. 155.*[9]

But the miserable crops made under the joint-labor and common-property system, and the larger supplies of food received from England, were insufficient without the corn obtained in trade, (or by fraud or force,) from the natives. War, soon after Smith's departure, cut off this supply.

"Famine now made its appearance, attended with circumstances at once melancholy and disgusting: Food, the bare idea of which, during better days, had created loathing and disgust; was now seized on, with greedy and bestial voracity."—*Hist. of Va.*, *p. 157, vol. 1.*[10]

"Those who had starch, made no little use of it, in this extremity; and the very skins of their horses were prepared, by stewing and hashing, into a dainty and welcome food. Nay, so great was the famine, that the poorer sort took up an Indian, that had been slain and buried, and eat him; and so did several others, one another, that died, boiled and stewed with roots and herbs: And one, among the rest, killed his wife, powdered her up, and had eaten part of her, before it was discovered; for which he was, afterwards, severely executed. In short, so extreme was the famine and distress of this time, that it was, for many years after, distinguished and remembered by the name of the starving year."—*Stith, p. 117.*[11]

"In the midst of the distresses of the Virginia colony, the remembrance of Smith obtruded itself, awakening in them bitter regrets for his loss, and calling out severe reproaches on themselves, for their baseness and ingratitude. From five hundred, their number was soon reduced to sixty men, women, and children; and this miserable remnant, could not reasonably calculate on the security of a single hour, from the assaults of the savages, even though by some miracle they should escape the agony of disease, and the torments of famine.

"In this forlorn condition, they were found by Sir T. Gates and Sir G. Somers, who, on the twenty-fourth of May, arrived in two barks, built with such materials, as they could find in Bermudas, assisted by the wreck of their own ship. It required them but little observation, to be convinced of the inadequacy of their means to remedy an evil so woeful and extensive; and after a short consultation, it was unanimously determined to abandon the enterprise. The colonists, with whatever was most valuable, being embarked, the ships dropt down the river to Mulberry Island: So near to entire extinction, was the germ of this mighty nation."—*Hist. of Va., p. 159, vol. 1.*[12]

"The colonists were importunate to burn the town and fortifications; but God, who did not intend that this excellent country should be abandoned, put it into the heart of Sir T. Gates to save it."—*Stith, p. 117.*[13]

A boat met them before they reached the mouth of the river, and announced that Lord de la War's fleet was near at hand, bringing reinforcements, and every needful supply. Under his command the wretched fugitives returned to the deserted walls of Jamestown.

"But while the Virginia establishment was thus almost miraculously preserved by the arrival of Lord de la War, a danger of no less magnitude awaited it in the impatience of the company in London, and their inordinate expectations of immediate profit. It appears, that the genuine commercial spirit, which works by bold enterprise and patient industry, was debauched at this day, by the bewitching reports of Spanish discovery; and the value of distant possessions, was estimated by the mines of rich metals they were supposed to contain. Disappointed in their expectations of discovering a Potosi in Virginia, the question was seriously discussed, whether the enterprise should not be abandoned. But the testimony of Sir T. Gates, solemnly given in at one of the quarter courts, backed by the representations of Lord de la War, who published a treatise on the occasion, removed the veil, which ignorance and misrepresentation had drawn before the eyes of the company; and it was determined once more, to prosecute the enterprize with spirit and activity.

"Sir T. Dale, with three ships abundantly supplied with all necessaries, arrived the 10th of May. He found the colony, as usual, indolent and improvident. To those vices their mode of living had added a disposition to mutiny, which being general and habitual, it was more difficult to repress."—*Hist. of Va. p. 163, 165.*[14]

These "representations" of Lord de la War, (made no doubt in sincerity, and what he thought truth,) will show that in that day "western country" descriptions were at least as highly colored as those which now are drawing away the people of *this* land of milk and honey.

"The substance of these representations was, that the country was rich in itself, but that time and industry were necessary to make its wealth profitable to the adventurers; that it yielded abundance of valuable woods, as oak, walnut, ash, sassafras, mulberry trees for silkworms, live-oak, cedar, and fir for shipping; and that on the banks of the Potowmac, there were trees large enough for masts, that produced a species of wild hemp, for cordage, pines which yielded tar, and a vast quantity of iron ore; besides lead, antimony, and other minerals, and several kinds of colored earths; that in the woods were found various balsams, with other medical drugs, with an immense quantity of myrtle berries for wax; that the forests and rivers harbored beavers, otters, foxes, and deer, whose skins were valuable articles of commerce; that sturgeon might be taken in the greatest plenty in five noble rivers; and that without the bay to the northward, was an excellent fishing bank for cod of the best quality; that the soil was favorable to the cultivation of vines, *sugar canes, oranges, lemons, almonds,* and rice; that the winters were so mild, that cattle could get their food abroad, and the swine could be fattened on wild fruits; that the Indian corn yielded a most luxuriant harvest; and in a word, that it was "one of the goodliest countries," promising as rich entrails as any kingdom of the earth, to which the sun is so near a neighbor."—*Purchas*.[15]

At the close of Sir Thomas Dale's administration, the colony had so much stability, that the grants of 100 acres, which before had been made to every new settler, were reduced to 50 acres—

—"and this alteration had its rise in the opinion, that the country being likely to flourish, and the difficulties of making settlements consequently having become proportionably less, it was no longer necessary or politic to hold out such strong inducements to emigration."—*Burk, p. 177*.[16]

The following gives some idea of the money value of lands at this time. Besides the general mode above stated of granting lands to settlers, there were two others. The first mode, probably, was the origin of the "aristocracy" of Virginia.

"When any person had conferred a benefit, or done service to the company or colony, they would bestow such appropriation of land upon him. However, to prevent excess in this particular, they are restrained by his majesty's letters patent, not to exceed twenty great shares, or two thousand acres in any one of these grants. The other was called the adventure of the purse. Every person, who paid twelve pounds ten shillings into the company's treasury, having thereby a title to an hundred acres of land, any where in Virginia, that had not been before granted to, or possessed by others."—*Hist. of Va., p. 178*.[17]

The habits of the colonists had now become industrious, and of course, there was no more scarcity.

"Nay, whereas they had formerly been constrained to buy from the Indians yearly, which exposed them to much scorn and difficulty. The case was so much altered under his management, that the Indians sometimes applied to the English, and would sell the very skins from their shoulders for corn. And to some of their petty kings, Sir Thomas lent four or five hundred bushels; for re-payment whereof next year, he took a mortgage of their whole countries."—*Stith, p. 140*.[18]

"The attention of (The London, or Proprietary) Company was directed with equal care to almost every subject of political economy; and as the country as yet held out no prospects of sudden wealth in the working of mines, agriculture was naturally resorted to as the means of trade and subsistence. Tobacco had in some degree grown into notice by the whim of the colonists, and the fashion of the times, unaided by the patronage, and indeed, in defiance of the repeated injunctions of the company. But a strange taste for this nauseous plant was rapidly gaining ground in Europe; and the king, notwithstanding his unaffected antipathy to it, tempted by the prospect of revenue, at length permitted it to be entered in 1614, as a regular article of trade. The colonists had learned the art of planting corn, together with the use of this valuable production, from the Indians. Vineyards were attempted; and experienced vine-dressers sent over for this purpose. The culture of silkworms was recommended with a like anxiety; whilst anniseed, flax, hemp, wheat, and barley, with various other productions, formed a large and judicious list for future essay and experiment. Colonies will, for a considerable time at least, reflect the manners and pursuits of the parent state. During the last years of the reign of James, a considerable taste for agricultural inquiry prevailed; and numerous treatises were published on the subject. The company sent over several of those tracts, for the use of the colony. It is not surprising, that at this time, a rage for speculative farming prevailed in the colony.

"The commerce of Virginia, from the nature of things, was for a long time of little value. Before the year 1614, she had no staple. But once, that she was legalized as a fair trader, and the industry of her citizens was excited by the prospect of wealth and the security of freedom, her advances were unparalleled and almost miraculous. In the year 1620, her tobacco was more than sufficient for the English market, and the continent was resorted to, as a vent for the superfluity."—*Hist. of Va., vol. 1, p. 305.*[19]

Tobacco was not only the staple crop, but it soon became, and long continued, the greater part of the legal and current money of Virginia. Payments and fines fixed by law were generally specified in tobacco—and that kind of money was not entirely displaced until the revolutionary war. Even after tobacco was no longer a legal tender in payment, habit and convenience continued to make tobacco notes, (or the certificates of hogsheads of tobacco, and their weights, being deposited in the public inspection houses,) a common currency. It is not 40 years ago since a tobacco note passed from hand to hand for the sum that the hogshead would sell for, and the possession of the note was the evidence of ownership. There was much convenience in this before other paper money had been authorized by law—and such a paper currency had the rare advantage of being the representative of *real value*. But this conversion of tobacco to current money, and at a fixed legal rate, doubtless aided in this respect the universal tendency of all legal inspections of the qualities of commodities—that is, of *deteriorating* the quality, and of course, ultimately, the market price. The legal inspection of tobacco remains to this day a prominent feature of the absurd and general inspection system: and the only redeeming merit of this is, that the certificates of the inspectors, as to the quality of the tobacco, are universally and totally disregarded as evidences of value.

I proceed to give some of the early rates of value as estimated in tobacco, and fixed by the governors' proclamations.

"During the administration of Captain Argall, tobacco was fixed at three shillings the pound. In 1623, Canary, Malaga, Alicant, Tent, Muskadel, and Bastard wines, were rated at six shillings in specie, and sack, nine shillings the gallon payable in tobacco. Sherry, and Aquavitae, at four shillings, or four and six-pence in tobacco. Cider and beer vinegar at two shillings, or three shillings in tobacco. Loaf sugar, one shilling and eight-pence per pound, or two shillings and six-pence in tobacco. Butter and cheese eight-pence per pound, or one shilling in tobacco. Newfoundland fish per cwt. fifteen shillings, or one pound four shillings in tobacco. Canada fish, two pounds, or three pounds ten shillings in tobacco. English meal sold at ten shillings the bushel, and Indian corn at eight."—*Hist. of Va., vol. 1, p. 307.*[20]

In 1607, an improvement was made in the mode of curing tobacco, which was deemed of enough importance to be stated in history. (*Stith, p. 147.*)[21] Previously, it had been cured in heaps. Mr. Lambert discovered that it cured better hung on lines, or on sticks, as is now customary.

"One hundred dissolute persons, at the express command of his majesty, delivered by his marshal, were sent over as servants, (in 1619,) much to the dissatisfaction and inconvenience of the company, who were obliged instantly, at the positive urgency of the king, to hire ships at an advanced premium.

"At the instance and advice of the treasurer, one hundred virgins were sent over as wives, for the purpose of fixing to the soil, the roving and inconstant spirits of the colonists."—*Hist. of Va., vol. 1, p. 206.*[22]

"Such of these maids as were married to the public farmers, were to be transported at the company's expense, but if they were married to others, that then those who took them to wife, should repay the company their charges of transportation."—*Stith, p. 166.*[23]

"The arbitrary conduct of the king, with regard to the persons ordered for transportation, was followed by one equally flagrant and unjust, respecting tobacco, contrary to the plain and express words of their charter, which exempted them from all custom and subsidy for twenty one years, excepting only five per cent. upon all such goods, &c., 'as should be imported into England,'

&c. The Spanish tobacco, which generally brought eighteen shillings the pound, and tobacco of Virginia, which was sold at three, were fixed by the financial logic of the farmers of the customs, at an average ratio of ten shillings the pound; while with a consequence perfectly consistent with the premises, a duty of six-pence the pound was demanded on the whole."—*vol. 1, p. 207.*[24]

"Tobacco had become the staple of the country; and with this article the colonists not only stocked the English market, but had opened a trade for it with Holland, and established ware-houses in Middleburg and Flushing.

"The king, notwithstanding he professed on all occasions the most marked dislike and aversion to this commodity, and had even labored to write it into disrepute,[25] did not see with indifference the diversion of a part of his revenue into foreign states, by the trade of the colonists. In vain the petition of the colonists, and the remonstrance of the company, attempted to soften or remove the obduracy of the monarch. Their deputies had to encounter the stern denial of justice from the privy council, in addition to the frowns and insolence of office. They were ordered to bring all their tobacco into England, in despite of their privileges as Englishmen, and the plain letter of their charter."—*vol. 1, p. 209.*[26]

"This year (1620) was remarkable for the introduction of negro slaves into the colony—an evil, than which, none can be conceived more portentous and afflicting. A Dutch ship bound homeward from the coast of Guinea, sold twenty of this wretched race to the colonists."—*vol. 1, p. 210.*[27]

In 1621, various things most needed by the colony were subscribed, or advanced on loan, by individuals of the company in England, to be sold in Virginia on certain terms, for re-payment. The most remarkable venture enrolled, consisted of—

"an hundred more maids, to make wives; and sixty were accordingly sent, young, handsome, and well recommended to the company, for their virtuous education and demeanor.

"With them was sent over, the several recommendations and testimonials of their behavior, that the purchasers might thence be enabled to judge how to choose. The price of these wives was stated at an hundred pounds of tobacco, and afterwards advanced to one hundred and fifty, and proportionably more, if any of them should happen to die; so that the adventurers might be refunded their original charge: And it was also ordered, that this debt of wives, should have the precedency of all others, and be first recoverable.

"And it was strictly enjoined, that they should be used well, and not married to servants, but to such free men and tenants, as could handsomely support them; that, by their good fortune, multitude, of others might be allured to come over, on the prospect of advantageous matches. And the company likewise declared their intention, that, for the encouragement of settled families, and securing a posterity, they would prefer and make consignments for married men, before single persons: and that as many boys should be sent as there were maids, to be apprentices to those who married them. They also granted adventurers, who subscribed to this roll, a ratable proportion of land, according to the number of maids sent, to be laid off together, and formed into a town, by the name of Maidstown."—*Stith, p.* 197.[28]

In the same year, the instructions to the authorities, brought out by Governor Yeardly—

"pressed upon them the raising [of] several useful commodities, as well as corn, wine, silk, and others heretofore frequently mentioned; as also making oil of walnuts, employing their apothecaries in distillation, and searching the country for minerals, dyes, gums, drugs, and the like: and they ordered them, particularly by the king's advice and desire, to draw the people off from their excessive planting of tobacco. To that end, they were commanded to permit them, to make only an hundred pounds of tobacco ahead; and to take all possible care to improve that proportion in goodness, as much as might be, which would bring their commodity into request, and cause a more certain benefit to the planter."—*Hist. of Va., vol.* 1, *p.* 226.[29]

The letters of the governor and council in 1621, to the company, state, "that tobacco was stinted to 100 plants per head, nine leaves to a plant," for each individual cultivator.

Among the laws made by the assembly in 1624, were the following:

"That whosoever should absent himself from divine service any Sunday, without an allowable excuse, should forfeit a pound of tobacco, and that he who absented himself a month, should forfeit fifty pounds of tobacco: That there should be an uniformity in the church, as near as might be, both in substance and circumstance, to the canons of the church of England; and that all persons should yield a ready obedience to them, upon pain of censure: That the twenty-second of March (the day of the massacre) should be solemnized and kept holy; and that all other holidays should be observed, except when two fell together in the summer season, (the time of their working and crop,) when the first only was to be observed, by reason of their necessities and employment: That no minister should be absent from his cure above two months in the whole year, upon penalty of forfeiting half his salary; and whosoever was absent above four months, should forfeit his whole salary and his cure: That whosoever should disparage a minister, without sufficient proof to justify his reports, whereby the minds of his parishioners might be alienated from him, and his ministry prove less effectual, should not only pay five hundred pounds of tobacco, but should also ask the minister forgiveness publicly in the congregation: That no man should dispose of any of his tobacco before the minister was satisfied, upon forfeiture of double his part towards the salary; and that one man of every plantation should be appointed to collect the minister's salary, out of the first and best tobacco and corn."—*Hist. of Va. vol. 1, p. 280.*[30]

"That, for the people's encouragement to plant a store of corn, the price should be left free, and every man might sell it as dear as he could: (for the governor and council did then, and long afterwards, set a rate yearly upon all commodities, with penalties upon those who exceeded it.) That there should be a public granary in each parish, to which every planter above 18 years of age, who had been in the country a year, and was alive at the crop, should contribute a barrel of corn, to be disposed of for the public uses of the parish, by the major part of the freemen; the remainder to be taken out by the owners yearly, on St. Thomas's day, and the new brought and put in its room: That three capable men of every parish, should be sworn to see that every man planted and tended corn sufficient for his family; and that those who neglected so to do, should be presented by the said three men, to the censure of the governor and council: That all trade with the Indians for corn, as well public as private, should be prohibited after the June following: That every freeman should fence in a quarter of an acre of ground, before the Whitsuntide next ensuing, for planting vines, herbs, roots, and the like, under the penalty of ten pounds of tobacco a man; but that no man, for his own family, should be obliged to fence more than an acre; and that whosoever had fenced a garden, and was ousted of the land, should be paid for it by the owner of the soil; and that they should also plant mulberry trees."—*vol. 1, p. 283.*[31]

It seems that the ladies of former days were at least as much given to coquetry as their fair descendants—but that they did not escape as easily as now from all ill consequences of their offence. The peculiar circumstances then existing—the number and variety of suitors, and the strong proofs of their love, offered by the readiness of each to exchange his whole market crop for a portionless bride—served to increase this natural propensity of the fair sex, and ought to have made the offence more pardonable. But the assembly took the matter in a more serious light, as by their enactment of 1620—

> "the governor was obliged, soon after, to issue a proclamation, forbidding women to contract themselves to two several men at one time. For women being yet scarce, and much in request, this offence was become very common; whereby great disquiet arose between parties, and no small trouble to the government. It was therefore ordered, that every minister should give notice in his church, that what man or woman soever should use any word or speech, tending to a contract of marriage, to two several persons at one time, although not precise and legal, yet so as might entangle or breed scruple in their consciences, should, for such their offence, *either undergo corporal correction*, or be punished by fine, or otherwise, according to the quality of the person so offending."—*Hist. of Va.*, vol. 1, p. [2]85.[32]

We have certainly improved greatly in this respect in Virginia. This amiable weakness of the best of nature's works, is no longer threatened by *law* with "corporal correction"—nor have we given the least countenance to the even baser modern fashion of the more northern states, of jilted lovers seeking alleviation of their woes, by bringing suits for pecuniary damages.

In 1629, it was required by the government that "every laborer should tend two acres of corn, or forfeit all his tobacco."

The quantity of tobacco which each person had been permitted to raise annually, was increased to "three thousand plants for men, and two thousand for women and children. This was afterwards restricted to two thousand, nine leaves on a plant, and no slips or seconds were permitted to be planted."

But no degree of restriction on the amount of cultivation, had served its intended object of improving the quality and the price of tobacco: and instead of such partial restrictions, it was now proposed that the extraordnary measure should be adopted of an entire cessation of the raising of tobacco, for a certain limited time. It increased the strangeness of the matter, that the people were more anxious to be thus restrained, than the government to impose the restriction. But now, to make the cessation effectual, it required the co-operation of the younger colonies of Maryland and Carolina.

> "Meanwhile the depreciation continued to such an extent, that the planters were scarcely able to clothe their families by the sale of their crops. An answer

arrived at length from the chancellor of Maryland, enclosing the lieutenant governor's proclamation, enjoining a total cessation for a given time, to all the subjects of that proprietary.

"These letters accompanied with his own correspondence, the governor laid before the house; and the question being taken whether this was a sufficient confirmation, it was decided in the affirmative. By this decision an act made during the former session restricting the planting of tobacco from the first of February, 1666, to the first of February, 1667, was declared to be in force; and the governor was directed to signify the same by proclamation to the several counties."—*Hist. of Virginia, p. [139], Vol. 2*.[33]

But however small the benefit found in this trial of the cessation of the culture of tobacco, its repetition continued to be urged as a favorite remedy for low prices. In 1681, the council wrote to Lord Culpeper, the governor then in England, respecting sundry grievances of the colony, and "entreated that his influence might be used to procure a cessation in planting tobacco."

"Meanwhile the rapid depreciation of tobacco, added to the operation of the commercial restraints imposed by parliament, had produced a general dissatisfaction among the people. They had vainly attempted to apply a remedy to the former evil, by procuring the co-operation of Carolina and Maryland. This plan had failed through the jealousy or avarice of those governments; and they were left to struggle with difficulties, which were daily accumulating, and of which they could see no prospect of termination. They could not even hope for the sanction of their own government, which formerly approved a cessation. Other maxims were now entertained by the executive, more suited to the views of the court; and the inquiry was not, what would be beneficial to the country, but how will it affect his majesty's revenue?"—*Hist. of Va. p. 228, vol. 2*.[34]

The popular desire for the *cessation*, added to the pressure of some other real (but less regarded) oppressive measures of government, produced much discontent, and finally broke out in the insurrection.

"The people of several counties, having lost all hope of a cessation by the dissolution of the assembly, ran together tumultuously, and proceeded to the entire destruction of the tobacco plants in the beds, before they were transplanted. Their proceedings were so timed, that the season was too far advanced to make good the loss by seed; and as the culture of sweet scented tobacco was almost exclusively confined to Virginia, they directed their efforts peculiarly to the destruction of this sort."—*Hist. of Va. p. 233, vol. 2.*[35]

"The governor now left to the exercise of functions purely executive, proceeded to a severe inquiry into the late insurrection. The king had instructed him that the plant cutters and their instigators, came properly within the purview of the statutes relating to treason, and had commanded, that the rioters should be proceeded against by the attorney-general, and punished with the utmost severity."—*Hist. of Va. p. 240, vol. 2.*[36]

The following is the earliest notice of cotton crops having been recommended.

"Sir Edmund Andross was a great encourager of manufactures. In his time fulling mills were set up by act of assembly. He also gave particular marks of his favor towards the propagating of cotton, which since his time has been much neglected."—*Beverley, § 142, p. 90.*[37]

The following extract will give some interesting information on the value of agricultural products, and other commodities.

"The price of corn and other articles of food, during this epoch, varied considerably, according to circumstances. Corn, at a medium, sold from ten to eighteen shillings the barrel. A bull was worth seven hundred pounds of tobacco, or eight pounds fifteen shillings. Poultry would naturally command a greater price, from the delays and difficulty of procuring a stock from Europe, and the inconvenience of their multiplication amongst cultivators, whose whole attention was almost wholly engrossed in clearing the forest for cultivation. A goose, during the administration of Hervey, cost twenty shillings, and we should conclude that other fowls were in proportion.

"The rates of ordinaries, established in 1666, by Sir William Berkeley, will throw more light on this head than is to be collected from the rates of separate commodities. As these rates were stated to have been fixed, in order to prevent the extortion of keepers of taverns and eating houses, we should conclude they are lower than the previous charges.

A meal for a master,	15lbs. tobacco.	
Ditto for a servant,	10	do
Lodging for either (per night)	5	do
Spanish wines, per gallon,	10s. or 100	do
French do	8s. or 80	do

Brandy, English spirits, or Virginia dram,	16s. or 160	do
Rum,	10s. or 100	do
Beer,	4s. or 40	do
Cyder or perry,	2s6 or 25	do

These two last are stated to be rated proportionally higher, in order to encourage the produce of the country.

"From the circumstances of the colony, an horse must have been an animal at once rare and valuable. In the year fifty-six the assembly ordered two thousand and five hundred pounds of tobacco to be paid to John Page, for a horse lost in the expedition against the Rechahecrians. The complaint of Page, and the wording of the order, show, that this sum was not thought equal to that which the horse might have commanded. If we estimate the tobacco at the market price only six years after, it will amount to an hundred pounds; a prodigious price, if we consider the rates in Europe during this period. In the same year, on the petition of Richard Nicholas, it was ordered, 'that sixteen hundred pounds of tobacco be paid him, for the charges and cost he had been at in recovering and finding a horse, which had been on the service in the same expedition.' At the same time Richard Walker was ordered five hundred pounds of tobacco 'for finding the horse of Henry Jupons, and four hundred more when he found that of Richard Eggleston.'"—*Appendix, Hist. of Va. p. xxxii. vol. 2.*[38]

The foregoing address was drawn up in the early part of January, to be delivered at the annual meeting of the Historical Society which was to have taken place on the 9th of February, but which was postponed from time to time, until March 2nd. When this task was undertaken, the attempt was about to be made through an Agricultural Convention, to obtain some legislative aid for the diffusion of agricultural knowledge; and the writer hoped, in this manner, to lend some feeble aid to that praise-worthy but fruitless effort. This intention, in some measure, directed the treatment of the subject, and induced the expression of hopes, which had not then been prostrated—as they were afterwards, and indeed before the meeting of the Society was finally held—by the total neglect of the whole subject, by the legislature of Virginia. When recommending objects for legislative encouragement, in aid of agricultural instruction, two principal, and very important subjects were designedly omitted. The one was the circulation of agricultural periodical publications; the other, the establishment and support of agricultural societies. The recommendation of the first was forbidden by the writer's private interest being connected with what he might advocate—and the second, because much space would have been required to explain his views of what agricultural societies *ought to be*, and the objections to their *usual* procedure, and because the subject had been treated fully upon several previous occasions. See Farmers' Register, Vol. I. pp. 147, 201, and Vol. II. p. 257.[39]

Part Two

Reform

4

On Fencing—Against the Open Range
Letters and Editorials (1833–1835)

Nicholas Herbemont, On Fencing and Other Enclosures (1833)

To the Editor of the Farmers' Register.

Columbia, S.C. Oct. 5th, 1833.

In order to add my mite to the great exertions now making by your correspondents, and particularly by yourself, for the promotion of agricultural interests in our country, I thought the great need to our country of permanent or at least durable enclosures would cause such hints as I might give, to be tolerated in your most valuable periodical, "The Farmers' Register." If you think so, I beg you will insert the following desultory observations.

It will not be disputed, I hope, that for the successful prosecution of any business whatever, due preparations in buildings, implements, &c. of the most durable and commodious kinds should be provided, and that the first expense ought to be incurred with judgment, but without stint, or fear of spending a few dollars more or less. We should always keep in mind that it is much more economical to lay out one hundred dollars for a useful thing than fifty dollars for one that is useless, or nearly so.—Make-shifts have always, I believe, been found the dearest way of proceeding to business, and should never be resorted to except in cases of the most absolute necessity.

One of the things in which we fail most generally in our country establishments, is in the due enclosing of our fields. The common worm-fence is certainly soon made, in a new country, where there is an abundance of choice timber for the purpose; but we all know, that except in very few favored situations, where the timber is of superior durability, such a fence answers but very imperfectly the object it is intended for; and that after a few years of continual trouble, in driving pigs and cattle out of the field and repairing the fence, it must be made anew. The post and rail fence is better, particularly where durable timber is easily obtained for the posts; but still, wood decays so rapidly, that I am certain it is much more economical to make at first a durable fence at double the cost, and that in a few years it will have proved the cheapest. The next best thing is live fences, and here also there is a considerable difficulty in the selection of the most suitable plant capable of answering the purpose most fully. There is scarcely any one plant that is not liable to some objection; we must, therefore, be careful to select that which has the least. The Nondescript, Cherokee Rose, the *Rosa Leonigesta*, of botanists, is used for hedges in South Carolina and in Georgia, where it has, in some places, succeeded admirably well, when in others it has failed. This failure may have arisen from the unsuitableness of the soil, the want of due care in the course of the planting and cultivating it when young; but more particularly, perhaps, from the aptitude of the rose tribe to die when least expected, without our being able to discover the cause. One of the objections to it is, that as a hedge, it covers a great deal of ground, and that cattle are fond of eating its young shoots. I have no means of judging whether it would thrive as far north as Virginia. It grows admirably well on the banks of low rich grounds, and wherever the soil is not very poor and dry. It is usually propagated by cuttings, which ought to be planted in the fall, and some say also in the middle of summer. The mode of raising it from seed would probably be preferable where the seed can be procured; but it is not easy to get it. I have not succeeded well with it, and I am inclined to give the preference to the *Mespilus Pyracanthus*, called in Europe, Burning Bush, (not the plant usually so called in this country, which is the Spindle Tree, *Enonimus Americanus*.) The Mespilus has very strong and sharp thorns, is an evergreen, or nearly so, and has the great advantage of never becoming naked at the bottom, and that its lower branches which grow horizontally, very readily take root. This last property gives it also a decided advantage over all the other hedge plants that I know of, except the Cherokee Rose, which is, that in planting the young plants for a hedge, they will be close enough at two feet apart, and even farther, if the ground is tolerably good; for the lower horizontal branches by being covered with a little earth will readily take root, and completely fill the spaces. Add to this that when the hedge is sufficiently full, two or three of these horizontal branches from each original plant may be layered and covered with a little earth, and taken up the following year, when they will be found to have good roots, and be sufficient to plant a hedge two or three times the length of that from which they are taken. It is usually raised from seed which it produces in the utmost abundance. It may also be propagated from cuttings planted in a nursery of good soil well prepared, when they may be removed to their final place of destination. It grows well in all sorts of soil, as far as I have tried it, and, like most other things, it grows best in good land well prepared to receive it. There are many other plants that are used, and still more that may be used for hedges; but I need not enumerate those best known and in use. A plant that is still too rare to be made into hedges with the probability of complete success, is the *Maclura Aurantia*, Osage Orange. It is very hardy and strong, with very sharp long spines. I am informed that, with due precautions, and planted in a good soil well prepared, cuttings of it take root very readily. If so, it may be easily propagated, and may prove very beneficial.

Hedges are certainly far preferable to wooden fences; but yet they are liable to many faults, and require very constant attention, and frequent work, to keep them safe, and a full protection against the intrusion of cattle, and particularly of hogs. When a farmer intends to do his duty to himself and to his fields by cultivating them properly, and gradually improve them in fertility, it seems to me that it is well worth his while to surround them in a more permanent way than can be done even by a very good hedge. A stone or brick wall may be thought too expensive, and so it is for a field that is yearly deteriorating; but where the materials are convenient, it would most probably be found in ten or twelve years to have been the cheapest. In some countries, enclosures are made of earth in a mode called in France "pisé," where I have seen handsome large houses, churches, &c. that had been built of this cheapest of materials upwards of two hundred years before. I am informed it is also the practice in many parts of South America to build walls and houses in pisé, and that their strength and durability are almost incredible. The best earth for this kind of work is that sandy or gravelly soil that contains just enough of clay to make it adhere together when pressed in the hand, so that after a slight pressure the marks of the fingers are imprinted in it, and the lump thus pressed does not fall to pieces without its being acted upon with some slight blow or lateral pressure. The full description of the manner of building pisé houses and walls is to be found in one of the volumes of the American Farmer of Baltimore, in Rees' Cyclopaedia, and also in the Cours Complet d'Agriculture of the Abbe Rozier.[1] It may probably be also found in several other books. This mode of building must not be confounded with what is called mud houses, which are inferior in every respect; for I am satisfied, that a pisé house well constructed of good materials, is at least as good, and as strong and durable as one made of brick. There is neither water, nor hay, or straw necessary to be mixed with the soil used; on the contrary, great precaution ought to be taken not to leave in the earth any vegetable substance at all, not a blade of grass, a straw, or a fibre of root. The reason of this is obvious; for vegetable substances will in time decay, and will thereby leave a weak place in the wall, or a want of connection of the earth. Water added to the earth prevents its being well pounded, and, in evaporating, may occasion cracks in the wall. The easiest and cheapest then is the best, and the earth such as is usually found about a foot or so below the surface, is in the best state possible. As soon as it is dug, it is thrown between two sides of planks forming the mould, which being properly secured from spreading open, so as to form irregularities in the thickness of the wall; this earth then being put into the mould, to the thickness of four to six inches, is, then pounded down with wooden pestles made for the purpose, until they no longer leave a mark on the pounded earth; more earth is then added, and again pounded, till the mould, which is about three feet broad and ten long, is filled to the top, and being loosened, is moved along the line of the wall, when the process of pounding the earth goes on as before, and so on till the wall is made. This is a very imperfect and insufficient account of the manner of describing the building with pisé; but I only meant to give an idea of the thing—for it requires plates to render it fully intelligible. It is, however, very easy, and such as can be done by any negroes under the superintendence of an intelligent man. A house or wall constructed in this manner, forms one solid mass without joints, and, if the earth did not contain too much clay, it will dry perfectly without the least crack, and be very nearly as hard as stone. A wall sufficient to enclose a field or garden, &c., may be made of the earth taken out of a ditch along side of it, taking care to throw away the surface that contains vegetable matter. So that a wall of this kind is made certainly at a less expense than would be required to dig clay, temper it, and mould it into bricks for the same wall; besides which, the expense of burning the bricks, hauling them, and laying them, are all saved. It is much more convenient to have the foundation of such a wall laid with brick or stone for four to six inches above the surface of the ground; but I have little doubt that a way may be contrived to build it altogether with pisé. Such a wall should have a coping of stone, brick, tiles, slates, shingles, or thatch, or boards; but experience will no doubt discover some cheap method of covering it, though by shaping the termination of it like a roof it would last many years without any covering. For building a house of this sort, if a cellar is wanted under it, dig the foundation the depth intended for the cellar, build it up with brick or stone to about one foot above the surface of the earth, and take the middle part of the earth for the purpose of raising your walls on the brick or stone foundation, and by a little experience it will be found that a very good house may be constructed at a very small cost, saving hauling, &c. A house built in this manner, if intended for a dwelling house, may be rough-cast outside, and plastered inside, and may be made as handsome and elegant as any house built of any material. For out houses, such as barns, stables, &c., as also for walls of enclosure, a thick white-wash of lime is sufficient. This is applied with a broom, beginning at the top. This looks rough, but it makes a very durable coating. I have read an account of such an enclosure of several miles in length, in South America, that had no coping, and had lasted so long that the oldest man in the country declared that when a boy the wall was then an old structure. Although it was not very sightly at the time it was seen by the inquiring traveller, it was yet an effectual fence, notwithstanding the want of a covering to protect it from the very great rains of that country. It appears to me from the truth of the above remarks, that where rail timber is not abundant, and has to be hauled any distance to the field to be enclosed, that a pisé wall would be the cheapest, and that it will be found so, particularly after the hands have acquired some expertness in this kind of work.

N. HERBEMONT.

Ruffin, On the Law of Enclosures (1833)

Justice and policy have concurred in fixing as a general principle in the laws of civilized nations, *that every individual should be compelled to refrain from trespassing injuriously on the property, or otherwise doing wrong, either directly, or through others under his control, to any other person under the protection of the laws.* In Virginia, however, there is one most important exception to the general adoption of this principle. The whole land of our country is, in effect, deprived of this protection against trespassers and wrong-doers.

Our law does not forbid A to suffer his cattle to eat and destroy the growing or ripe crops of B, but compels B to secure his fields effectually against the entrance of the cattle, or otherwise maintains A as innocent. In this case the rule just named is reversed, and so far as our fields and crops are in question, the principle adopted is this—*every individual shall guard and protect his property from depredators, and every one is permitted to consume or destroy all that may not be well guarded.*

The injustice of this reversed rule would be readily admitted by any one who considered the question abstracted from the circumstances from which our policy sprang, or who was not in some degree blinded by being accustomed and reconciled to the practical operation of this policy. But I am no advocate for aiming at theoretical perfection in government, or consulting arithmetic and geometry, rather than the actual condition of men, to make rules for their direction. I freely admit that at a former time, circumstances in Virginia required the general adoption of this policy—and that it is still expedient over a large portion of her territory. But in another large portion of the state, from change of circumstances, this policy has become as injurious in its practical operation, as it is unjust in theory.

When our country was newly settled, the extent of cultivated land was very small, compared to the forests and waste land, much of which, even in that state, was valuable for pasturage; and it was by far cheaper to fence a few and small spots of arable, than their extensive woodland pastures, or *range* for stock, and the range was much to[o] valuable to be abandoned. It was beneficial to every landholder that such should be the general plan of enclosures, although the open land thereby became one large common pasture, for the use of him who owned but little, or none of it, as much as for those who owned the largest shares. According to these views our legal policy with respect to enclosures was fixed—and from it all derived at first some benefit, though certainly the benefits were very unequally divided. The law, stated generally, requires that each owner of a field shall surround it with a good fence, of a certain height, or shall have no remedy against the intrusion of his neighbors' live stock.

But now the state of things in much the greater part of eastern Virginia is altogether changed. All the fertile land has long been cleared and made arable, and much more of such as is extremely poor, is also under tillage. In districts where the soil is, or has been, of good quality, very little timber for fencing remains, and, consequently, the expense of fencing is increased greatly more than in proportion to the increased extent of enclosures. The remaining woodland is poor, and so worthless for pasturage or *range* for stock of any kind, that many of its owners make no use of their unenclosed woodland for that purpose. But this abandoning the use of their woodland (which in fact is not worth using by any one,) does not relieve the owner of the expense of fencing his arable land—and this is required merely that other persons may not be deprived of the use of this woodland for range, which he does not consider worth using as such.

The great extent of fencing required bears especially hard on the tidewater region. There, on and near good lands, the original growth of durable pine and good oak has all been destroyed—and a choice has been already, or soon will be presented to every farmer, either to purchase suitable fencing timber at a distance, or to use the *old field* pines which form the second growth of his worn fields, and which are not more remarkable for their rapid growth, than for their rapidity in rotting, when used for fencing timber. There are no durable materials for dead fences, and the frequent divisions of lands, as well as the laws permitting stock of every kind to range at large, forbid or discourage the attempt to make live hedges. There are many farms which have five miles of perishable fences to keep in repair, that do not yield $500 of clear profit; and many other tracts are thrown out of cultivation, and yield nothing to their owners, (who in most cases are the especial *favorites* of our laws—widows or orphans)—because the expense of lawful fencing would exceed the whole rent.

Under this system and these circumstances, farmers who make suitable enclosed pastures for their own live stock, (as every one ought to do for his own interest,) are burdened with the expense of two kinds of fencing, for two different objects. His own interest requires that he should make an enclosure to keep his own cattle *in*, and the law compels him to fence his cultivated lands, to keep the cattle of other people *out*.

It is impossible to estimate with any approach to accuracy the enormous loss caused to the farmers of eastern Virginia, by this unjust policy. But if each person will calculate the amount of his own individual loss, it will be sufficiently evident that the whole annual amount is enormous.

It is admitted that the evil of this policy is not so felt by all who bear even its heaviest burdens. There are many, who because they have never known any different state of things, can scarcely conceive the advantage of making every man restrain his own cattle within his own bounds. Therefore, even if the legislature was impressed with the opinions I have advocated, it would be improper to make any general or sudden change in our system. Most of the benefits of a change might be secured, and the objections avoided, by partial changes of this law of enclosures, for such particular districts as would accept the offer by the voices of three-fourths of their proprietors, and they owning not less than three-fourths of the land within the boundaries of each district. If such a privilege was accepted by the owners of a district of five miles square, for example, they would still be bound to maintain at their joint expense a general enclosure in obedience to the general law, and to exclude the cattle of all persons residing without their limits. But the expense of such [a] general enclosing fence would be very inconsiderable, and would be diminished to each individual in proportion to its greater extent. Such a plan would secure to every considerable tract of country either kind of policy that its inhabitants were decidedly in favor of—and would be free from every reasonable objection. If the people of any one neighborhood were permitted to adopt the change proposed, under proper restrictions, all the risk and loss of the trial would be confined to them, and if it succeeded, the whole community would participate in the benefit.[1]

If our legal policy in this respect was altogether changed, there are many farmers who would still refuse to profit by the boon: and from their fondness for close grazing their fields by as many cattle as can be kept alive through the year, would keep up their present amount of fencing, though their fields would be no longer liable to the depredations of any cattle but their own. But most persons would soon learn the benefit of pursuing a different course. Each farmer having to maintain his own cattle, would keep a smaller number, and confine them generally to a permanent pasture well enclosed: and by being necessarily reduced to one-fourth of their present numbers, and treated as well as the change of system would permit, the live stock would yield more products of every kind (except *hides* perhaps,) than at present. The lands kept for tillage, thrice as extensive as the enclosed pastures, if too poor to be grazed, might be safely left without a fence, until their improvement in after time may make enclosures necessary for the owner's interest. There can be no doubt but if permitted to get rid of the burden of making and repairing three-fourths of our costly and perishable fences, that the change would be almost necessarily followed by greatly improved products from both live stock and tillage, as well as increased fertility to the whole country exempted from the usual impoverishing and unprofitable grazing of poor fields by poor cattle.

SUUM CUIQUE.[2]

W. J. Dupuy, On the Law of Enclosures in Virginia (1834)

To the Editor of the Farmers' Register.
You have but lately set out in the publication of a periodical for the promotion of the interests of agriculture; and truly the general circulation of

such a work is desirable. Most men are willing to receive instruction in the various farming operations, by which they cannot only labor with most ease and convenience to themselves, but make that labor most productive and profitable. Yet, how often are our labor and capital misapplied, or even made subservient to the use of others, because we have not been made sensible of the advantages of change in our plans and practices. Custom and the example of our fathers have rivetted upon us practices, which although they are injurious to our interests, are nevertheless unnoticed, because they are familiar. Such has been the blind devotion to old customs, that our lawgivers seem never to have thought of protecting by legal enactments, the rights, property and interests of agriculture; at least, they never seem to have thought it necessary to secure to the farmer the *full and entire* enjoyment of his rights and his property, which other classes of our citizens enjoy. The mechanic can lay aside his tools, and no one dare take them up. But the farmer, forsooth, can no sooner decide that a portion of his land wants recruiting and rest, than it is taken possession of, and ownership exercised by "Tom, Dick and Harry."

Our systems of moral philosophy, teach us that no man has a right to use *that*, by the use of which, he does an injury to his neighbor who has a better right to it than himself; but our code of laws have refined so much upon all moral obligations, and have extended their democratic principles so far, that no man's property can remain unoccupied by himself, without becoming the property and being subjected to the use of others. Is there any more reason that my land should be applied to the use of others, to my detriment, when I may choose for a season to dispense with its use, than that any article of my wardrobe, because not under lock and key, should be worn by another, when off my shoulders? Or that any of my teams should be harnessed by another, when not actually laboring for my benefit? Whoever leaves his land without an enclosure at once excludes himself (according to the provisions of our laws) from its use—at least to the full extent of the difference in number, between his own grazing stock, and all the stock in the neighborhood.

I am pleased that you have one correspondent at least, (in Fairfax,) who raises his voice against the present laws of our state, in relation to enclosures. I hope others will be found, who will dare to think and speak of this violation of private property, or perhaps, I should more properly say, the want of protection for private property. Why, as has been before asked, should my land which I may choose to turn out to improve by rest, be taken possession of, and impoverished by other people's stock. Or why should I be compelled for the protection of my own crop, to be at a greater expense to enclose it, than my neighbor's stock, which would otherwise prey upon it, would sell for? When, perhaps, my own stock, (as every farmer's should be,) are kept in an enclosed pasture, and are an expense to no one but myself. Why do not our laws protect our property, even though it may not be in our immediate use? In many parts of Virginia, and even in many neighborhoods in this county, it is notorious that those frequently have the largest stock, who have the least land to graze; and many are in the habit of buying up poor cattle at a reduced price, to sell out as beef, after being fattened on their neighbors' lands. I understand that in the northern states, where the rights of property have always been better understood, and better protected than with us, men are expected and compelled to support their stock at their own expense. If they are found on the commons, they are subject to a tax. I think it highly probable, moreover, that the disease called the Carolina distemper; which desolates many of our farms, may be kept up by the mixing of the herds, on the commons.

It is in vain that we improve our lands by marl or other means, without some better legal regulations, whereby the industry and enterprise of some, shall not be thrown away for the benefit of others, otherwise, Virginia must ever feel the inconve[ni]ences of improvement, however she may boast of her superiority.

It has always been a matter of surprise to me, that our legislature, composed as it is of so much intelligence, and of men too, taken from the farming interest, should while away so many long winters in proposing, and in gravely discussing the enactment of laws, many of which, with the well informed and practical farmer, only serve to excite his risibility. Examine our code, and the still more useless proposals for enactments, and you scarcely find one for the protection of the farming interests; and yet legislators and all are ever deploring the depressed state of agriculture.—When I look over the list of our representatives for the next legislature, I see men who could do much on this subject, and why may we not expect that they will? I know that the subject is a new one, and that it requires some independence to urge such innovations. But I hope that we have men who have other, and higher objects, than mere popularity. Such a scheme might at the first blush be thought ruinous and impracticable, but some reflection and calculation will convince the better informed of the contrary.

In the first place, on the score of right and equity, no man should be permitted to use his neighbor's property, because he does not choose to be at the expense to enclose it—for what is his neighbor's is not his, though it be nothing more than grass. It will be found, moreover, that every farmer spends *at least* one-twelfth part of his labor, (to say nothing of the waste of his timber) in keeping up fences, not for the exclusion of his own, but his neighbors' stock; for I will venture to say, if there were no other stock than his own, he would make some other arrangement, whereby he would save labor enough (and with a farmer time is money,) to buy double the pork he uses. I have no hesitation in believing, that the time which is now spent in fencing in Virginia, would be so employed as to buy double what the stock in the state would sell for. A farm, for instance, which would yield an income of $2,000 under the present necessities for expending labor, would, were those necessities removed, be capable of bringing in by fair estimate, one-twelfth more, or $2,166.66. This additional amount of $166.66, would buy 3,320 lbs. of pork, which under ordinary circumstances, would be sufficient to supply the family of such a farm.

I have gone upon the estimate that no hogs should be raised on our farms; but that is by no means necessary. It is believed, that every farmer will find it to his interest, if he will make the experiment, to enclose a certain portion of his lands for grazing and pasture, and leave unenclosed all that which is under cultivation. In that way, each man labors for himself, and reaps the profits of his own labor and of his own property. Although the thing may appear to be a novel experiment to many, yet it has been put into practice in other states, where the interests of agriculture are better understood and the profits more abundant than in Virginia.

The estimate which has been made of 1-12th of the profits of the farm which is now taken for fencing, any intelligent and practical farmer will allow to be moderate enough. Every farmer knows in what way he could spend that portion of his time most profitably. Were each one to devote 1-12th of the year to the making manure, in addition to the time now given, our farms would exhibit a very different appearance, and we could all very well afford to buy the whole of our pork, even if we should determine not to raise it. We should then not feel the absolute necessity of marl as a manure, however valuable it may be.

The value of the timber which is annually consumed by fencing; is no inconsiderable item in the calculation, particularly in the tobacco growing district of Virginia, and should the present destructive system be continued a few years longer, the timber alone will be considered a heavy expense in fencing.

No public enterprise has ever been undertaken, without first having it brought before the public mind for consideration and discussion. It is with this view, that these suggestions have been made, and I hope that some persons more competent than I am, will take up the subject and press it upon the attention of our legislature. It is time for justice to be done to that portion of our community, which is the very nerve and lifeblood of the commonwealth. By it all the burdens of government are sustained. By it the naked are clothed, the hungry are fed. And why should there be inducements held out by law for those who have not land, to prey upon him who has, or to compel him to labor at immense expense to secure to himself his crops, which the laws should secure to him? A farmer may not wish to raise stock, or even prefers enclosing them, and yet according to the existing state of things, he is compelled to labor to protect himself against those who have no right in equity to the privilege they now enjoy.

Nottoway, Nov. 25th, 1833.
W. J. D.[1]

[William Henry Foote?], *The Right of Grazing in Common (1834)*

To the Editor of the Farmers' Register.

If I were able, I would send your Register into every country [sic] in the state. Your correspondents are numerous, and they delight me; their spirit augurs well—for if wrong, pride and perseverance will soon put them right, the more especially as they put their proper names to their communications. Poor Jeremiah has no nerve for this: he must be permitted to speak as from his grave, and having said what he believes from experience, and poured his mite into the general treasury, let his remarks pass, with those concerned, for what they are worth. Men seeking profit, care not from whom it is derived, nor stop to ask for names;—rest assured of one thing—that what he states as fact, he will always be ready to prove before a court of record.

I tender my acknowledgements to many of your correspondents; but to Mr. Se[l]den[1] I am a great debtor. In 36 years of watching and toil, I have seen nothing, heard or imaged of nothing, like his *practice*. If I were an emperor, his practice, like the cut of a Chinese coat, should be fixed and unchanged for one thousand years; for sir, if in the sap and green tree it is thus productive, what will it not do in time? Like the queen of Sheba, I must pack up and regale my senses with the practical wisdom of Mr. Selden. My arrangements from necessity must materially differ from his. The *right of common* created by the General Assembly, and so long enjoyed by the good people of this state, puts it out of the power of any farmer in this country to enclose a standing pasture. His servants are not lawful witnesses, and the whole profits of an ordinary farm would not employ a sufficient number of competent witnesses to protect a fence enclosing three or four hundred acres of land. I may therefore, with strict propriety, charge the General Assembly with spoiling my estate full one thousand dollars per annum. As I am situated, I have to keep three fields, of one hundred and fifty acres each, in clover for pasture, and they are fallowed alternately. One more field of the same size is divided, upon which I make corn. Upon those two fields I put the greater part of my manure, and they are necessarily divided by a good fence, as I feed my cattle for the most part in the field. Now sir, look at the difference in expense and results between me and Mr. Selden. He cultivates under a single enclosure three fourths of his land, and I, by virtue of the act of General Assembly, two hundred and twenty five of wheat on fallow and corn land, and seventy five in corn, with the very heavy charge of four divisions fences;—out of four hundred acres, he drives [derives?] profit from three hundred, and out of six hundred, I can only have the same, for which I am indebted to the concentrated wisdom of the state. If it be true that the strength and virtue of an agricultural community depend upon its agricultural abundance, why legislate land into common? Why is one half of our land under civilized rule, and the other half entirely savage or Indian? To provide against unwise and evil legislation, I am constrained to use one hundred and fifty acres of land for pasture, or enter my own domains in common with those who would mark and appropriate every thing fat as their own.

JEREMIAH.

[Theodore A. Field], Enormous Losses Caused by the Fence Law of Virginia (1834)

To the Editor of the Farmers' Register.

Brunswick, Feb. 18th, 1834.

Sir,—If I mistake not, (for I have not the work to refer to,) John Taylor, in his Arator, estimates the time occupied on each plantation in repairing the fences, at two months; and I am convinced that much of this time might be profitably saved, to be devoted to other objects, by amending our law of enclosures. I must beg leave to submit the following inquiries to the public, through the medium of your invaluable columns. I should be happy to furnish a correct answer to them myself, but that cannot be done with any degree of accuracy, as we are totally destitute of agricultural statistics.

How many bottoms are turned out to be trampled [by ranging livestock] into mortar, producing neither corn nor grass, because they are too long and narrow to be fenced?

How often has the landlord to give up his lands rent free to his tenant, upon condition that he will "do up" the fences?

How often is a plantation objected to by one wishing to purchase, and what proportion of its value is lost to the owner, because though there may be a great deal of good land on the tract, yet he would be compelled to fence in too much poor land in order to bring the good into cultivation?

How many tracts of land are now valueless, for want of a sufficient quantity of timber to keep up the immense enclosures?

How often does he fail to get a tenant altogether because no one will take his land even upon the foregoing hard condition?

What proportion of his rents is lost every year on account of the difficulty of repairing the enclosures?

How much good land is now useless, in detached parcels, being separated from the body of the tract by a stream or road, on account of the immense labor necessary to enclose it by itself?

How much valuable tobacco land must now be continued in forest, to produce timber to repair our fences, which under a better law of enclosures, might be brought into cultivation?

How many rich, but small parcels of meadow land, lying on the borders of our crooked streams, are neglected, because the fence on the bank will rot in a few years, or be carried away by the freshets; or because the fence cannot well be made to follow the serpentine course of the stream, and if it be made at all straight, it would cut off so many nooks and corners, as to leave the balance unworthy of regard?

How great a loss is sustained in this, that we cannot clear 20, 50, or 100 acres of good land, here and there, where such may be found, and leave the poor ridges to produce fire wood and a small portion of timber, but must take it as it comes, high and low, rich and poor, in order to get our fields in a body for the convenience of enclosing them?

The rich man A, with 1000 acres of land and 40 hands, requires say one month to repair his fences: the poor man B, who is worth but one fourth as much, with 250 acres and 10 hands, must labor two months to repair his. That he must, is susceptible of mathematical demonstration—and the poorer he may be, the larger proportion of his labor has he to apply in this unproductive operation of fencing.

How many thousands of poor people who own small pieces of land, and who own, or could easily procure an ox and plough each, and who could with these means make a comfortable subsistence for themselves and their families, are deterred, improvident as they are, from making the glorious attempt, (I say *glorious*, as it might save them from the poor house or the penitentiary,) by finding their labors doubled by the unprofitable operation of fencing their respective little patches, or by contemplating the laborious task of cutting and mauling rails, which must be performed at the most inclement season of the year, or because they have not the team necessary to haul their rails to the ground, after they have prepared them.

The great number of rails to be hauled renders it necessary that the farmer should keep a cart or wagon, and twice as many oxen as would be necessary for all other purposes, and even with this double team, so painful is it to the feet to draw on hard frozen ground, and so laborious is this work over land recently thawed after a severe winter's frosts, or which has been lately ploughed, that it becomes a delicate question about the middle of February, to ask a farmer "how his steers hold out." Sometimes, with the blessing of God, (and I know no instance in which his goodness is more signally displayed,) they *do* "hold out," and the shift is fenced so as partially to protect the crop from the ravages of the hogs; but sometimes they do not "hold out," and consequently a large proportion of the ensuing crop is destroyed by the hogs, before it attains maturity. Yes sir, strange as it may appear to the inhabitants of all other countries, it is nevertheless true, that that portion of the year which elsewhere is devoted exclusively to taking care of the stock, and making manure, is consumed with us in hauling rails, which frequently renders our oxen totally unfit to carry out, in the spring, the little manure which unavoidably accumulates around us.

It may be worthy of inquiry, how much better, straighter, and shorter our public highways might be made, if our law of enclosures were altered. For, melancholy to relate, even our public highways are made to yield to the overwhelming consideration of fencing. I have been amused with the various curvatures described in some of our applications to the courts for orders to open new roads. For by way of rendering the measure acceptable to the court, and to avoid the multiplication of fences, we consent that the new highway shall be opened "around Mr. such a one's fence, thence back through Mr. ——'s lane, and thence around Mr. such another's fence," and so on. If I am asked, as I frequently am, what we should do with our stock if the law of enclosures were altered, I answer that we can do, what we certainly shall be compelled to do in the course of a few years, on account of the deficiency of timber, whatever that may be: that half a dozen neighbors can easily unite in purchasing and enclosing commons for their stock—and that by devoting the time and labor, which is now consumed in fencing, to raising and carrying out manure to be applied to our grass lands, we can grow more grass upon 10 acres of land, than is now produced by 200 of our trodden and worn out old fields. And lastly—that we can at least do what the inhabitants of all other countries in the world do—and surely the superiority of our stock over the stock of other states and countries is not so excessively great, as to attach us inveterately on that score to our old system.

In fine, our law of enclosures as it now exists, is a most intolerable nuisance to by far the largest, most wealthy, and populous portion of our state. I would suggest the propriety of altering it so far at least as to compel every one to keep his hogs in an enclosure. The effect of this alteration would be to enable us to adopt the use of live fences, which would save us a vast deal of labor. At present we cannot use them, as there is no sort of hedge that can resist the attacks of a thin, lean, tough-hided and hungry old sow.

PHILANDER.

Ruffin added the following commentary to Field's letter (Farmers' Register 1 [March 1834]: 634–35).

It is highly encouraging to our efforts to put down the monstrous evil which we all suffer from our legal policy respecting fences, to find so many correspondents either approve or sustain the views, already presented in the Farmers' Register, in denunciation of this system. The discussion, so supported, cannot but lead ultimately to a suppression of the evil: and if all those who think with us on this subject, will use their influence, a just and proper law of enclosures will be one of the first great gains from the exchange of opinion, and lending of mutual aid among the farmers of our country.

We lately had the pleasure of viewing the operations of an enterprizing and intelligent farmer from New Jersey, who has been induced by the low price of our land to purchase and settle a large estate on the lower part of James river. The most admirable care of his live stock was evident, and none enjoyed more of this care, or repaid it, with more profit, (as the owner believed,) than his hogs. Nevertheless, they were always confined to styes, and neither grazed the fields, nor derived the slightest benefit from the large body of "woods range," belonging to the farm. The fields (which were small, as is usual in the northern states,) were separated by neat, strong and durable post and rail fences, which though only three rails high, were as effectual in restraining all other cattle than hogs, as a "lawful fence"—and much more effectual against fox-hunters. But his outside fence was necessarily made (to suit the law and his neighbors' hogs,) "ten rails high, staked and top-logged." The owner was asked what estimate he placed on the difference of the fence law in New Jersey (where hogs are not permitted to go at large,) and in Virginia—and his answer was, that the whole purchase money paid for his estate was not a greater amount than the value of the difference. But if this estimate was too high, reduce it to any extent that can be asked for, and it will still be found that our fence tax is heavier than all others paid by the people of Lower Virginia. All that we pay for our 90 days' sessions of the Legislature, the support of public education, and for all the roads and canals in progress, does not equal the whole amount of what we individually pay in the making of fences, merely to enable the hogs of other people to live (or perhaps to starve) on our uncultivated lands. There would be some compensation for this system, if we gained in pork what we pay in fencing—even though the hogs of A gained what was lost by the land and labor of B: but it is notorious that there is no considerable gain even to those persons who gain all; and the immense annual importations of hogs for slaughter, sufficiently prove that the whole management of hogs in Virginia yields no nett profit, and but an inconsiderable gross income, to the community, notwithstanding all the enormous cost incurred.

James M. Garnett, Petition for a Change of the Law of Enclosures (1834)

To the Editor of the Farmers' Register.

Essex, Aug. 9th, 1834.

I now send you an article which I hope will be in time for your next paper. It is a petition about to be circulated in this county, and I hope in many others, for changing the law relative to enclosures. No "internal improvement,"—always excepting that in regard to the general education of the people, which is first and above all others in importance—is so much wanted; nor any, I believe, to the great utility of which, if once carried into effect, converts would more rapidly be made. The people in the counties bordering on, and comprehending our mountains, may not yet, perhaps, have so generally felt the evils of this law, as to wish for its repeal; and if that be the case it might be suffered, in regard to them, to remain as it is: although I greatly mistake the condition of that part of our State between the head of navigation and the north western limit of our highly improved valley country, if the people would not be nearly, or quite as much benefited as ourselves by the change. I would be among the last men in Virginia to advocate any alteration of dubious policy, in a general law; or that did not, in fact, promise most manifestly to benefit a very large majority of our fellow citizens. But in this case there seems to me not a shadow of doubt on any point involved in the proposal, it being one which requires only to be made, to gain numerous supporters. That it may meet not only your approbation, but that of all your subscribers, is the earnest wish of

Yours, with much esteem,

JAMES M. GARNETT.

TO THE LEGISLATURE OF VIRGINIA

The Petition of sundry citizens of Essex, respectfully showeth—

That having become thoroughly convinced of the daily increasing necessity for some change in our laws relative to enclosures, we have at last determined to apply to you, as our only hope, for some remedy of the existing evil. That this determination may not appear to have been made either hastily, or from considerations merely selfish, your Petitioners beg leave to state the chief reasons which have induced them to adopt their present course.

In the first place, the law of which we complain is a palpable violation of all other laws relative to property, which principle is, to compel every one "*so to use his own, as not to injure that which belongs to another.*" But the law of enclosures actually *permits* us to injure others by means of our stock, unless *they* make a fence of a certain height to guard against those very injuries which justice demands that the perpetrators should be prohibited from committing under any circumstances. Instead of compelling *stock-owners* themselves to restrain their stock from depredating upon other people's lands, which the above cited principle clearly enjoins, the *land owners* are forced to protect both it and its products, at an enormous expense of extra fencing—or, to suffer without the smallest compensation, all the losses they may incur from the want of this extra fencing.

In the second place, we complain of this law, as the very unnecessary cause of more quarrels—more ill-will—and more lasting animosities between individuals and often whole families, than any other law in our whole code: indeed, we believe it may be truly said, of more than all the others put together.

Lastly, it imposes upon every land owner and cultivator a far heavier tax than all his other taxes united; and what makes the matter much worse is, that not a cent of it goes into the public treasury; but is utterly wasted and lost, inasmuch as it consists in the extra labor which each has to bestow on his own enclosures to protect their products from other people's stock. This legal oppression is greatly aggravated by the fact, that the small land owner—the small cultivator—in other words, the poor farmer, suffers much more in proportion to his property than the rich one; for having to fence it in, and daily to watch it, so as to guard against every other person's stock, and being in general, not so well provided, as richer land owners, with a proportionate quantity of labor, more of his precious time must generally be spent in making such enclosures as this most unjust and oppressive law prescribes, to say nothing of the time lost in repairing and supervising the whole. We may venture to say that this time, upon an average, amounts to two months in every year, and that it certainly could be abridged one month or more, thereby saving at least one-twelfth, or more, of the whole labor bestowed throughout the State, *in the single business of fencing*, if the law on this subject were exactly the reverse of what it is. Compel each person so to take care of his own stock as to prevent their injuring his neighbors' property, and not a cultivator of land in our whole community, whether he be proprietor or mere tenant, but would be thoroughly convinced in less than a year, of the immense advantages of the change. Indeed, we know no reflecting persons any where, with whom we have conversed on this subject, who do not admit the truth of the foregoing arguments.

Your Petitioners beg leave further to represent, that the circumstances of the country when this law was passed, having entirely changed, the law itself cannot now be justified, if it ever could be even on the score of policy. We have no longer (at least in the tide-water part of our State) extensive tracts of uncultivated, unenclosed lands deemed of little or no immediate value to the proprietor: fencing materials are becoming comparatively very scarce, and thereby the actual expense of labor in collecting and applying them has been enormously enhanced, while the products of our impoverished fields can very illy sustain this most unnecessary deduction from their net proceeds: add to this, the opinion is daily gaining ground, that even uncultivated lands will much improve by excluding every kind of stock from them. All these circumstances combined, render legislative interference imperatively necessary; for the evils enumerated admit of no other remedy. They have prevailed so long indeed, as to give rise to notions in some parts of our country, relative to landed rights, which, if general in regard to other rights, would reduce us nearly to a state of nature. These notions make all land, not actually cultivated in some crop, a species of common property, free for all who choose to turn their stock on; and this too, whether they be under enclosure or not. Now, if our lands be really our property, it follows as a necessary consequence, that each land owner has an indisputable right to the exclusive possession as well as use of it, and a just claim to protection therein by the laws of his country. This is a universally admitted fact as to every other species of property—why not then, in regard to land also? When we rest our horses to avoid wearing them out; when we lay aside our clothes for the same purpose, he who takes and uses either without our knowledge or consent, may not only be prosecuted and punished by law for such act, but incurs the risk of being utterly disgraced by it. Yet, when we rest our lands to avoid wearing *them* out, and in hopes of improving them thereby, which we have an equally clear right to do, free from all interruption by others, we are compelled by the existing law of our State to keep an enclosure round them of a certain height, or any other person's stock may depredate upon them the whole year round, even if in cultivation, without our being able to obtain the slightest compensation whatever for such depredation: nay, if any injury be done to the trespassing animals, their owner may recover damages from the cultivator of the land, although the land itself, and every thing growing on it, in the language of the law, is styled, as if in mockery, "*his property!*"

These, may it please your honorable body, are crying evils,—of legislative creation too! and for whose benefit, we would respectfully inquire? For none whatever, but that very small portion of our community who may attempt to raise stock without sufficient means of their own to support them. This attempt the present law sanctions as fully as if it conferred the privilege in direct and express terms: sanctions too, without the least cost whatever to the stock-owners, while it compels all the planters and farmers throughout the State, annually to increase their labor—at the smallest computation, *one-twelfth*, if they would guard themselves against injuries which this law of the land empowers others to inflict on them with entire impunity. We venture to affirm that the code of no civilized nation on earth can show an instance of so very small and doubtful a benefit accorded by legislative enactment to the few, at so great and certain expense to the many. Indeed, we think it susceptible of the most satisfactory proof, that much more stock, and of far better quality would be raised than at present, if the law were changed, as we pray that it may be. Such would be the sure result of applying *that labor* which each cultivator would save from fencing out other people's stock, to making more food for his own, and taking better care of them. He would have at least one-twelfth more time to devote to the profitable employment of improving and cultivating his land, instead of spending that twelfth, as he now does, in the *unprofitable* occupation of extra fencing. Though last, not least, the peace, comfort, and harmony of every neighborhood would be incalculably increased by removing forever, this most copious, most pernicious source of contention and animosity. Make it, therefore, we intreat your honorable Body, no longer to depend upon the height of each man's fence whether his neighbors' stock shall be raised upon his lands or not; but let it rest, as it always should do, upon his own free choice, how far he himself will extend the privilege to others. It would often be granted, and would thus prove a bond of union, instead of being what it now is, a brand of continual discord.

All of which is respectfully submitted.

Ruffin, *A Commentary on the Law of Enclosures of Virginia* (1834)

For the Farmers' Register.

The agricultural interest of Virginia is already greatly indebted to the Farmers' Register and its correspondents, for commencing and continuing to discuss and expose the evils of the existing law of enclosures—and though it may require a long time to remove all the prejudices, and mistaken opinions, (of which enough will always be found ready to support every evil system of long standing,) still there can be no doubt of the ultimate removal of this burden, which agriculture has placed on its own shoulders. The decree has gone forth for the future destruction of this law—for it cannot resist the attacks of truth, reason and right, which have commenced, and which will not cease until the object in view is attained. Nevertheless, truth and reason have yet to fight a hard battle, and will find stubborn opponents. It will therefore be necessary for those who are convinced of the necessity for a change, to exert themselves, and to expose the deformity and iniquity of our existing policy as to enclosures, in every point of view. To this end, I shall offer some comments upon those provisions of the law which have either been passed over, or but slightly touched upon, by those who have preceded me in this discussion.

The author of the piece signed *Suum Cuique*,[1] (Farm. Reg. Vol. I, p. 396) and several others, have sufficiently exposed the great and main evil of our law of enclosures—the *principle* itself of the law, which instead of forbidding and punishing wrongful invasions of the farmer's property, requires of him to guard and protect it, or otherwise holds the aggression as justifiable and blameless. This principle, utterly adverse as it is to that of our laws in general, and of justice and the rights of property, is so important in its character and its operation, that it could not be counterbalanced, or justified, by any details or minor provisions of the law, however efficient they might be to secure the remnant of the farmer's right to the product of his land and labor, which the general principle of the law had not swept away. It is therefore not strange, that those who, in the pages of the Farmers' Register, have attacked the principle of the law, have scarcely noticed its particular features, and the utter worthlessness of even those provisions which were intended to operate as a means of restraining and punishing trespassers. This more humble task I will undertake. The law on the subject, which precedes these comments, will facilitate my references, and permit my remarks to be more concise.

To enable a farmer to recover damages, or obtain any legal redress whatever for injury, or destruction to his growing or standing crops, it is required by the law, that the field should have been enclosed at the time with a "strong and sound fence" of some one of the several kinds described, and of a specified height throughout: and that the field was so enclosed must be proved by the owner as a necessary preliminary step, before he can prosecute his claim for compensation for trespass. The trespasser (if I may dare to apply that term to one whom the law in effect *always* clears of the imputation,) has nothing to prove—the whole burden of proof lies on the sufferer. But suppose, that undaunted by difficulties, and determined to assert the few rights that the law leaves to him, he pursues the remedy offered, and calls for a survey of his fences, according to the provisions of the second section. Now in the first place, (and especially if the farmer is in Lower Virginia, and uses the perishable pine timber which is now the general resource for fencing,) no fence around a large field can be found every where "strong and sound," at twelve months old, even if it is permitted to remain so for twelve days after being built. It may be a fence abundantly good to restrain the owner's stock, or any other animals properly kept—but it is scarcely possible to avoid, from the progress of continual decay alone, some rails being found neither "strong or sound" enough to keep out a mischievous animal, who has been taught by hunger, and the habit of trespassing on neighboring fields for sustenance, to search keenly for all such defective places. But the regular progress of decay in fences, however rapid it may be, does not cause half of this particular evil. Wherever night-walking negroes, or other persons, choose to have a foot path through a field, they will certainly throw down, and keep down, some of the upper rails where they cross the fences—and all the owner's efforts we all know cannot prevent this being done. Any such defect in fences would suffice to prevent the owner's having any remedy or redress for the depredations of mischievous animals.

But suppose that the farmer has managed so well, and been so lucky, that no defect has been produced in his whole line of fence, either by decay, by storms, nor by design or carelessness of any neighbor, black or white—and that trespassing animals, notwithstanding, are found within the field, and that he is able to prove the fact: and that, on these (apparently) sure grounds, he makes complaint, according to law, before a justice of the peace, and obtains the order for the survey of the condition of his fences. These steps cannot be taken within a few hours, nor completed in secret: and even if the fences were perfectly good when the surveyors commenced their examination, the owner of the stock might secretly and safely cause gaps to be made before the survey was finished, and thus procure a condemnation of the fence. But without this, the mere fact that stock had entered the field would be enough to show that they had made the fence by their entry, otherwise than a "lawful enclosure" at that place: and unless the breach could be proved to be their act, its existence would serve to justify the trespass.

Even if by possibility neither this nor any of the other obstacles before named, stand in the way of the farmer's obtaining redress, there are still others that will prove sufficiently effectual. For the first offence, he is only permitted to recover the value of the actual damages sustained—and for this purpose the fact of every offence must be fully sustained by proof. Well—the farmer finds hogs in his corn field at earing time, or in his rank wheat just before it is fit to reap. That damage has been done, is unquestionable—and that unless the intruders are speedily removed, it will be enormous, is equally so—but how is it possible to establish by testimony the extent of the damage, and to fix it upon the particular animals that have produced it? The wheat field may exhibit evidence that hogs have concealed themselves under its heavy cover for at least a month, and have caused perhaps damage to the amount of $50: but how can it be proved that all this was caused by the one or more hogs found and caught at last? Besides, it is most likely that the hogs would not be secured so as to be identified, but would escape, to return and renew their attacks. If uncertainty existed as to the identity, or the ownership of the animals—or as to whether the damage had been actually committed by them, there could be no redress obtained under the law. If, however, the farmer has a legal witness present, and that witness actually sees a hog tear down and eat six ears of corn, and he can swear to the individual hog, either by his ear marks, or from former personal acquaintance—and the survey is made, and the fence found lawful, *then* all obstacles are at length overcome, and the farmer who has suffered the trespass may recover damages to the amount of the value of *six ears of green corn:* and if (by almost a miracle) he could a second time get through all the like difficulties of proof, with the same trespassing neighbor, and convict him of a similar offence, he would receive as compensation double damages, or the value of *twelve* ears of corn. I doubt, however, whether he would be entitled to double damages, unless it was the same animal that committed the first trespass.

It is difficult to be serious in discussing the pretensions of this law to the character of being remedial, and giving compensation to the sufferer, and adequate punishment to the trespasser. In truth, its provisions offer no redress whatever: they are a solemn mockery of the claims of right and justice, in the details, no less than in the general principle of the law—and it might more truly be entitled "*an act to authorise the trespasses of live stock on the fields of neighboring proprietors, and to prevent all remedy therefor.*" Its actual operation is according to this title and character, and not according to its title in the statute book—for in more than thirty years that I have resided in a part of the country where there has never been a lack of trespassing stock, I have never heard of any resort to the law for damages, nor of any sufferer who even thought of it as a remedy. This fact alone, (and I believe it is general,) is enough to prove that the law in every particular is utterly worthless for the purposes which it *professes* to serve.

It may be asked then, "if the law is so inoperative, how does it happen that fences actually serve in most cases to protect the crops, and that all fields are not frequently depredated on by the stock of men who are willing to take any privilege that the operation or omission of the law can give?" I answer, and the truth is notorious, that it is not fear of the law that restrains such acts of such individuals, (who, thanks to God, are yet but few in number,) but the fear of public indignation, and perhaps of such damages to their trespassing stock, as are not authorised by law, but which are far more effectual. It is to the general sound moral sense of the community—to the strong perception of right and wrong, and the different effects of this on those who act uprightly or unjustly towards their neighbors, which serve more to restrain trespassers than all the protection promised by the law.

From these views of the particular provisions of the law, and its general disuse, I reach the conclusion, that the law of enclosures is in every respect inoperative—in effect, null and void—except so far as it gives up the property of some individuals to be used or destroyed by others, without the possibility of compensation being obtained, or legal punishment inflicted.

R. N.

Ruffin, *On the Petition for a Change of the Law of Enclosures* (1834)

If the question of the good or bad policy of our present law of enclosures was considered fairly, either as a matter of right and justice—of public interest—or of private interest to nineteen-twentieths of all the individual cultivators of the soil of Eastern Virginia—there would be no doubt of the condemnation of the law, on either, or all of these scores, by any intelligent tribunal, having in consideration the circumstances and well-being of all. But, unfortunately, almost every question affecting public and general interests, is discussed upon false or improper grounds; and it is easy to foresee what will be the grounds assumed to oppose this beneficial change. It will be said by those who, for their own aggrandizement or profit, are ever attempting to delude the poor and ignorant by professing to be their exclusive friends and defenders, that this change is for the benefit of the rich, and cannot be made but by sacrificing the rights and interests of the poor. But the slightest consideration will show that the benefit to be derived by land occupiers (other circumstances being alike) will be the greater in proportion to the small extent of their possessions—as every one knows that the smaller the fields are, the greater extent of fencing is required to enclose them, in proportion to their size. Hundreds, perhaps thousands of small tracts of land in Lower Virginia, are left untilled, waste, and unproductive, because the whole rent or net product, would not pay for the expense of fencing—whether with our perishable "old field pine" timber, or with more durable and more costly materials. Of course, such land yielding nothing to the owners, is worth nothing for them to hold: but may be sold for a low price to some adjoining large landholder, who can afford to bring the land into part of a much larger enclosure. If he had not this advantage, he could not afford to give any thing for the land, for the same reason that the first owner could obtain nothing by its cultivation: but his owning a large body of land enables him to buy of his poor neighbor at $2 per acre, what will be to the purchaser worth perhaps $4 or more, and which to the seller was worth nothing. Such effects of our fence law are numerous and notorious. Indeed, if wealthy men and large landholders wished to buy up the small tracts of their neighbors' at the lowest price—and if they were so shortsighted as to expect to prosper by the general loss and distress of the community—they ought to sustain the present law of enclosures, as the sure means for reaching their object.

But perhaps it is still a poorer class for whose interests the "friends of the poor" will uphold the present law—those who own *no land,* and who by the present law enjoy the privilege of keeping stock on the unenclosed lands of others. Those who will venture to use this argument ought not to be ashamed to demand that no laws should protect our corn and meat houses, lest the poor should suffer for want of food. But should it be required to oppose this miserable petty larceny ground of argument, (it deserves no better name,) it would be better to buy up this existing privilege, at double its full value. The privilege of grazing the unenclosed lands of other persons in Eastern Virginia, is worth scarcely any thing to the few who use it, and costs at least an hundred times as much to those who suffer the grievance. All the stock raised by grazing at large, and the profits of every kind, just and unjust, derived from the existence of the law of enclosures, would not pay one-tenth part of its cost to the community. The change asked for, which is simply that each man shall secure his own stock, would serve to add one-fourth to the average net profits of cultivation throughout Lower Virginia. If the greater comparative extent of forest, and other land unfit for tillage, west of the Blue Ridge, makes a change there inexpedient, there is no reason why the existing law should not remain in operation there, as long as those circumstances may continue. It is unnecessary here to repeat the arguments for the change of this law, which have already been presented in several articles, to our readers. We have aimed to confine our remarks to those points, which, as being of minor importance, have been passed over, or but slightly touched by others.

The foregoing remarks were written for the 5th No. and were at first intended to accompany the petition for a change of the law of enclosures: but were kept back as not required, and because it was useless to anticipate objections which had not then been made. No considerable objections were to be expected, so long as the discussion was confined to men of intelligence, and to all those who *read,* and are thereby entitled to judge for themselves. The objections that *might* proceed from such persons (for none have yet appeared in print,) would be founded on reason and justice, and would be such as they who uttered them would have no cause to be ashamed of. There will always be some strong objections to a thorough change of any long established and widely operating policy—and the advocates for this, cannot claim the existence of the fence law to be a case of unmixed evil. Objections of this kind, whenever they may be presented, should be met with respect, and opposed as proceeding from friends who have in view the same general object, the preservation and promotion of the interests of agriculture. But since our publication of the petition to the legislature on this subject, the cause has been transferred to a very different tribunal—to court yards and places for warrant trials, where fifth rate demagogues, who read nothing that serves to increase their glimmering lights, can influence the opinions of those who do not read any thing. Judging from the verbal reports that have reached our secluded dwelling place, the pitiful ground of argument, which was anticipated above, is mainly relied on by these self-constituted and noisy guardians of the interests of the people. They take precisely the position that we expected before they had ever heard of the question. The ground of objection and the manner of its being urged, are similar to all other attempts to array the poor against the rich, which have at all times proceeded from those men, who were striving to build up their own dirty interests, by deluding and defrauding both poor and rich. There are few men, capable of exerting the smallest political influence, who are so foolish as to believe that even the poor can be benefited by measures which impair the rights of property. Those who urge such measures with any success, *know* that all classes, and the poor not less than the rich, are interested in preserving the rights of property: but the office seeking demagogue also knows that his own interest may be served by his country's loss, and his own interest is all he seeks. Whatever may be his rank in the despicable roll—whether his ambition aims at the highest or the lowest office in the people's gift, the general course of the demagogue is the same. When he talks about "honor and patriotism, pure disinterestedness, and devotion to the popular will," there is most reason for the people to expect an attack upon their political rights, and property.

Labor is the only source of all property. There is nothing that helps to constitute wealth which has not been created by, or received the whole of its value from labor alone. God has given to man in abundance the materials for wealth—but these materials require to be worked up, and brought into a useful and productive state by labor, and without it they are utterly worthless. The property of the richest farmer in Virginia, to the last cent, was acquired by his labor of body or mind—or by the labor of his forefathers, or of whomsoever he received his property by the course prescribed by law. Every man, from nature, derives the power to labor, and by its exercise, the ability to accumulate property—and in this happy country, there are few citizens having nothing but their labor, who might not pass from this class, to that which holds property accumulated by the previous exercise of labor. The preservation of the rights of property is as indispensable to aid the first conversion of labor to wealth, as to secure the possession afterwards—and every man is personally and deeply interested in preserving the sanctity which has generally been attached in Virginia to the rights of property. Without these rights, all would be as poor as the most destitute now are—and every violation permitted to exist, serves not only to lessen the security and value of accumulated property, but also to obstruct its accumulation.

[James M. Garnett], *Petition of Stock-Owners (1834)*

To the General Assembly of Virginia.
The petition of sundry stock-owners, respectfully showeth:

That one of our most precious privileges secured to us in the by-gone time, by some of your wise and just predecessors, is about to be seriously assailed by that all-grasping, pestiferous class of men denominated agriculturists. We mean—what may properly be called the right of common upon every man's land, who does not choose to enclose it with a fence fully five feet high, according to the requisition of our present most equitable law on that subject. This right you will soon be asked to take away; and our present purpose is, to beg of you as our guardians and friends, rather to extend than to abridge it, even in the slightest particular. We pride ourselves on belonging to the very ancient family called "the Goodenoughs"—famous from a period beyond which "the memory of man runneth not to the contrary," for being the staunch, inflexible friends of all old usages,—the irreconcileable enemies of all changes whatever, of a general nature; but especially of such as are comprehended under the new-fangled term—"*internal improvement.*" We therefore deem it entirely needless, in addressing your honorable body, to take the slightest notice of our adversaries' arguments about the prospective advantages to result from repealing the present law of enclosures, or its incompatibility with the rights of property, either real or personal. It is surely quite sufficient for gentlemen of your intelligence and patriotism to be reminded, that *such is the law of the land*—that *our rights* under it are, of course, *prescriptive*—of very long standing too—that their exercise, according to our free and easy fashion, is sanctioned by that well known law-maxim, "*Consuetudo loci est observanda*;"[1] and that "innovations are dangerous things." Why the law which confers the privilege we now enjoy did not go farther, we have always considered a most unaccountable thing, as it might easily have brought us several steps nearer to that admirable state of nature where "might gives right," towards which so many of our public actings and doings of late years have been most happily tending. For instance, as it gives us the right, in all cases, where a fence five feet high does not intervene, to feed and raise as many horses, cattle, sheep and hogs as we please, on other men's land—*a priori* it should also have given us the right to raise ourselves and our children in the same way, since human beings belong to a far nobler race of animals than any of the quadruped genera. Again, it gives us the right, at any time—nay, at all times, to bear off as much of the produce standing or growing on or in these lands, as we please, be the kinds what they may, provided they be not enclosed with a fence at least five feet high; and provided also, it be borne off in *the bellies* of our stock, instead of being carried on their backs, or in vehicles drawn by themselves, or their owners, which would often be much more desirable. Now, why this over-nice distinction—this very circuitous mode of appropriation? Since the law-makers evidently designed that A might, in the mode specified, transfer to himself, without any equivalent given in exchange, property which once belonged to B C or D E, would it not have been much better, directly to leave to A himself the choice of the *modus operandi*—in other words, of the particular method of making the transfer? He might sometimes prefer bearing off the produce in his *own belly*, (as in the case of all kinds of fruit,) rather than send his stock to bear it off *in theirs:* or he might choose to convey it away in bags, baskets, or wheel-carriages of some sort or other, either of which he surely should have been at liberty to do, if the privilege that he now enjoys was rightfully secured to him. Again, as we have under the law, this unquestionable right to take (by the agency of our stock as our forage-masters,) not only a part, but the whole of the produce of other men's lands, unless they will be at the expense and labor of protecting it by a fence fully five feet high, why should this right to take what does not belong to us, be thus most inconveniently limited? Why confine it to the immediate produce of land, and to the taking by the mouths and teeth of our stock? If land be really the most valuable species of property, as it is universally admitted to be, why should *that* which is of the greatest value be held by a tenure less secure and exclusive than *that* which is of the lesser value? In other words, why should we enjoy privileges in regard to other men's landed possessions, and not have similar privileges as to their other kinds of property? Why, for instance, should our law compel every land cultivator, who would enjoy the exclusive use of that land, which the law itself so generally calls *his own*, to make a fence around it fully five feet high to prevent other people from using it, and not also compel him to make all his horses, his oxen, his cows, nay, his slaves likewise, of some specified height, to prevent those who own neither from using them also? It would be nothing more than an extension of the same principle which gives us stock-owners the right of raising them on other people's land, and at their expense, whether they consent to it or not. If the latter privilege could justly and equitably be accorded to us, no imaginable reason can be suggested why the former should not also have been granted. It must have been, we presume, what the lawyers call a *"casus omissus."* We therefore pray, that your honorable body, instead of granting the petition of our adversaries, by repealing the present law relative to enclosures, will so alter it as to subject to our use, every land-owner's horse, ass, ox, cow, sheep, hog, or slave, until he will make each of them such a standard height as you may deem most advantageous to us to establish. And we further pray, that another provision be inserted in the new law, giving us the privilege of bearing off, in whatever mode we please, any part, or the whole of the produce growing on other men's lands, instead of restricting the method to deportation in the bellies of our stock: provided the said produce be not enclosed by a fence of the full height of *six feet*, carpenter's measure, the present standard of five feet having been ascertained by long experience, to be quite too low for some of our most valuable horses and cattle, which, from unavoidable starvation at home, have been driven to the necessity of learning to leap with inconceivable ease, over all such paltry enclosures as do not much exceed the height required by the present law. *This* was passed, as we hope your honorable body will remember, when lands were cheap, plenty, and productive; when our stock were not, as now, driven to their wits' end, to gain an honest livelihood; and when a leap of five feet high over another man's fence, was an achievement of far greater difficulty than one almost of seven feet would now be. Justice, sheer justice, consequently demands, that the law of enclosures, if changed at all, should be so altered as to accommodate itself to the increasing exigencies of stock-owners who have either no land at all, or not enough to support as many cattle as they wish.

All of which we respectfully submit to the wisdom and patriotism of your honorable body.

Ruffin, Petition for an Amendment of the Law Respecting Enclosures, on the Margins of the Navigable Tide Waters of James River (1834)

(Lest it should be supposed that the appearance of the following petition indicates some opposition to the former one asking for a more general change in the law of enclosures, it is proper to state that the two petitions originated with different persons, and in distant parts of the country—and though without concert, and asking for relief in different modes, the two petitions, so far from being in conflict, will serve to aid each other's object. Public notice was advertised of the intention of offering this petition, some months before the publication of the other.)

To the General Assembly of Virginia.
The petition of the subscribers respectfully showeth:
That your petitioners are owners or occupiers of land lying on, and partly bounded by, the navigable tide waters of James River, or some of its tributary streams.

That in addition to the hardship and injustice which, in common with most of our fellow citizens, we suffer from the general operation of the law respecting enclosures, (which general hardships we do not mean here to discuss,) your petitioners are subjected to other and peculiar injuries, from which they may be easily secured by legislative interposition, without any loss or disadvantage to the community in general, and without inflicting wrong on any individual citizen whatever.

That by the existing law, the navigable rivers and their creeks and tributaries, however wide and deep, are not *legally* a barrier against the trespasses of live stock from neighboring lands—and therefore that the water boundaries of farms must be kept covered by a "lawful fence," or the owner of the land be subject to suffer, without redress, any extent of depredations from mischievous animals.

The great width of the Lower James River makes it in most places, an actual and sufficient obstruction to the crossing of animals—and therefore the land owners may, and often do, dispense with fences along the shores, by making "water fences" jutting into the river, or creeks, to such distances as circumstances require and permit. But these jutting fences are not only very expensive and difficult to construct, and to secure from destruction by storms, but they are not legal substitutes for complete enclosures, and he who uses them does so at his peril—and he can have no redress for any depredations of animals trespassing on a field fenced partly in this manner, because it is not under a "lawful enclosure."

But greater evils are felt where the waters, though deep, are more narrow. There also, generally, all stock, except hogs, are effectually kept from passing, by the miry nature of the bottom and margins of the creeks. But hogs cannot be stopped by these obstacles, nor by any depth of water, nor even by a considerable width, as well as depth: and if a field of 100 acres was thus exposed to the encroachment of a single hog trained to mischief, there would be no legal remedy, or safeguard, except fencing the whole water line.

The fences thus required by the law along the margins of rivers and navigable creeks, are by far the most expensive in our part of Virginia, where fencing has become generally a most burdensome tax on agriculture. The adjacent lands in almost every case are divested of all timber suitable for fencing, which makes it necessary to bring it from distant places: most of the water lines are marshy, or otherwise not easily accessible to carts: and withal so crooked as often to need fences of double the length of straight courses; and after being constructed at such great expense, fences in these places are immediately and always in danger from storms, high tides, and the concealed attacks of mischievous hogs, or their owners who may desire to profit by such depredations.

If there was any compensating benefit (by whomsoever it might be received) for all the waste of fencing along the borders of deep tide water, we should hesitate to ask your honorable body for this boon, however great its value would be to us. But in fact, the granting of our prayer for relief will encroach on no man's rights, and scarcely in the slightest degree on any one's interest, whether that interest be in the use of his own, or of his neighbor's fields. The scanty supply of food obtained by live stock along the margins of rivers and creeks, and from which they might be debarred by the desired change in the law, is scarcely worth as much as the expense and risk incurred by stock being allowed to range at large in such places: and the whole profit thus derived from stock, is certainly not equal to the tenth part of the expense of the fencing which is now necessary for gaining that profit, and for no other purpose.

We therefor pray of your honorable body that the general provisions of the law passed at the last session, which makes the Upper Appomattox a lawful fence, may be extended to all parts of the tide waters of James River, its creeks and tributary streams, navigable by vessels drawing four feet depth of water—and that "stops," or "water fences," under proper regulations, may be made lawful fences between the adjoining lands on the same side of a water course.

All which is respectfully submitted.

Ruffin, *General Results of the Law of Enclosures in Virginia* (1835)

In every country, and in every age, there have been found existing some preferences given by government to the employment of particular kinds of capital, or of industry, at the expense of others: and however unwise, or unjust, may have been these legal preferences, they have, at least, rarely failed to produce their first and immediate object of an increased production of the commodities thus favored, or greatly increased profits to those persons engaged in the business—and more generally, both these ends have been more or less obtained. But if after a full trial of such legal preference to any particular branch of production, for many years, it should appear that both the amount of the production and its net profits had actually decreased, then (it might be supposed that) the condemnation of the legal policy might be safely left to the persons intended by it to be favored, or to the most thorough opposers of the principles of free trade and industry in general.

The law of enclosures and its effects, exhibit a striking example of the long continued operation of the preference given by our law to the business of cattle raising, at the expense of grain raising—of a preference of grazing, to tillage. If then the policy is of any value to even the favored class, whatever evils it may produce, there should certainly be seen this one good—increased products from live stock, and increased profits to the owners. Upon this issue, we are content to test the value of this part of the policy of Virginia, and to abide by the decision, even on this partial and limited view of the subject.

Virginia is not the only country in which the interests of tillage have been sacrificed to those (real or supposed) of pasturage: but generally, the effects have been beneficial to the business and the persons so favored, however injurious to other classes, and to the general interests of the country. Here, we are content to reap all the evils of this kind of policy, without deriving any of the benefits found elsewhere from such a departure from right and justice.

In the highlands of Scotland, owing to the peculiar circumstances of the country, the profits of sheep raising have been found to surpass all other returns of agriculture. The consequence has been, that tillage has been made to give place to pasturage. Thousands of small farms, formerly held by tenants tilling the soil, have been thrown together and converted to extensive sheep pastures, in which no human population remains save a few shepherds, and scarcely a single cottage, where there were formerly hamlets and villages, filled with numerous hardy laborers, with their wives and children. The sheep have eaten out the men. This state of things has served to drive from their homes, and compel the emigration of a people the most devotedly attached to their native land, and who would have preferred any fate short of starvation in their own beloved country, to banishment from its shores forever. But in thus exchanging a population of men for flocks of sheep, (however the change may be deplored by the philanthropist and patriot) there has at least been obtained the *gain* expected in wool and mutton.

In Ireland, where tythes are levied upon every product of tillage, and of the small land occupier who is necessarily a tiller of the soil, the products of grass land and live stock, have been kept free from this oppressive burden. The effect of this unjust exemption, (unjust if the nation is to bear the burden at all) has been, of course, to direct an undue proportion of land and capital to the raising of cattle. But though the effect of this state of things, is to keep out of tillage (the more profitable use, but for this exemption,) a vast extent of rich land—to increase the sufferings and the discontents of the oppressed and brutalized peasantry—still the expected gain is found to the graziers in large products of cattle, and large annual exports of beef, butter and cheese.

In Spain, the interests of the owners of pastures, as well as arable land, have for centuries been sacrificed to the established policy of the *mesta*, of which an account has been given in our last number. But though in this policy, as strange as it is unjust, the interests of tillage and even of resident stock-owners are sacrificed, to advance those of the owners of the migratory flocks—and though the general interests and public prosperity have been greatly impaired by this destruction of private rights—still the owners of the wandering merion sheep at least were fattened by the spoil, and their business prospered while all others suffered. If Spain remained poor, its soil untilled, the country but half populated—yet there was some little compensation found in the facts, that the policy which caused these evils also served to preserve the most numerous and valuable flocks, and to produce the finest wool in Europe.

Where, in Virginia, is the gain or advantage, private or public, to compensate in the least for the interests of tillage, and the rights of landed property, having been sacrificed to the interests of grazing, and the benefit of stock owners? The whole soil of the state is rendered by law a grazing common, for the use of every one, unless when secured by a "lawful enclosure," of enormous and useless expense, and which after all, can never be made a practical safeguard against trespassers. The cost of this policy to the owners and tillers of the soil, has been already and frequently discussed in this work, and is not now under consideration. The present question before us is, *what do the stock owners gain from the use of their legal privileges?* Let the general and notorious results speak in answer.

In the eastern half of Virginia, where the oppression of this system is most severely felt by land owners, there is not, and has not been for many years, any surplus product whatever derived from the grazing of live stock of any kind. So far from exporting any such commodities, we buy from other states very large quantities of salted meat, butter, cheese, leather, candles, and soap. The supplies of hogs and fat oxen brought annually from the west in droves are enormous, and have long been increasing—and most of our horses and mules are now obtained in like manner. It is a fact also worth notice, that the hogs thus driven to us 500 miles, are generally raised in clover fields, or within the farms and on the means of their respective owners, and not by the benefit of the "wood, range," or grazing in common, held so essential in Virginia. We also buy barrelled pork from New England, where the hogs are generally kept altogether in sties, and had no benefit whatever from grazing, or running at large.

It is unnecessary to carry these notorious facts more into detail. Every one knows the general truth, and has some idea of the magnitude of our import trade in live stock and their products, and that this trade has commenced, and has been regularly increasing to its present enormous amount, under the operation of our law of enclosures, and the preference it gives to stock raising. If then the system has brought so little benefit to the kind of property and its owners which it was specially intended to favor, and is still supposed to favor, we ask where are the benefits to be found, which are paid for so dearly by land holders, and by the general interests?

Fencemore, *The Policy of the Law of Enclosures Defended* (1835)

Those who advocate the repeal of this law have fallen into an evident mistake, in supposing that the expense of fencing would be diminished by the proposed measure. A simple reference to the most approved method of stock management in some sections of our own, as well as in foreign countries, will be sufficient to satisfy any one that the successful prosecution of this branch of husbandry requires infinitely more fencing than is at present found in the greater part of eastern Virginia. The northern farmers, for instance, have, with a view to this subject, found it to their interest to divide their arable lands into a great number of small fields, comprising from ten to twenty acres each, separated from each other by permanent division fences. This arrangement is adopted in order to introduce the artificial grasses, by which their arable lands are made to answer the double purpose of grazing and tillage. (See Far. Reg. vol. I. p. 542.)[1]

We find also that some of the most successful agriculturists in eastern Virginia have adopted a similar system. In answer to certain queries proposed by the committee of the Agricultural Society of Albemarle, Mr. John H. Craven states that he has five hundred acres under cultivation which are divided into ten fields of unequal sizes. This gentleman appears to be a successful raiser of stock of every description. His method of management as well as that of Messrs. Rogers and Meriwether may be seen by referring to the report of the committee. (See Far. Reg. vol. II. p. 226.)[2]

The same system prevails in western Virginia, where it has been evidently copied from the northern practice. Even in those regions where live stock is the staple commodity, a similar practice obtains. We understand that hogs are raised in the western country by turning them upon clover lots and grain fields alternately. And even a portion of their corn crop in its succulent state is used in the same way. It is apparent that the use of division fences are absolutely necessary under such a system.

As it is the avowed object of the enemies of the fence law to contract the present amount of fencing (Far. Reg. vol. I. p. 396)[3] it must of course be done by confining stock within as limited enclosures as possible. Take, for instance, an estate [of] 400 acres of arable land, and let 100 be enclosed as a standing pasture. Upon this field all the stock necessary for the support of the estate are turned, hogs, horses, cattle and sheep. It is impossible to conceive the complicated disasters of such a scheme. In short, these standing pastures have been tried over and again, and as often abandoned, at least, as far as concerns the indiscriminate introduction of every sort of stock upon the same field. Those who use them at all, are compelled occasionally to turn a portion of their stock upon other fields, or upon the *woods range*—a resource of which they would of course be deprived upon the principles of the proposed plan.

It is true that great advantages would be obtained by making more enclosures; but it must be recollected that every additional pannel of fence is obviating the objection to the present law, in a geometrical ratio. For the sake of illustration, we will suppose that these 400 acres are in the shape of a square, or a parallelogram. The enclosure of 100 acres allotted to stock can be divided into three equal fields by running two division fences. It is easy to be perceived that by removing these and joining them again to the outside fence they can be made to enclose 100 acres more. Two additional fields of the same size will afford a sufficient number of rails to enclose 300 acres. And with seven, there will be enough to comprehend the whole 400 under one fence. And with this the objection would end. We have no doubt that such an arrangement of lots would be of vast importance in stock management. It would afford the means of introducing the culture of the artificial grasses, by which this branch of husbandry has been rendered infinitely more profitable than under the old standing pasture system. And we would remark that this is the plan recommended by the author of Arator, to whom the agriculture of Virginia is so much indebted.

The method of management is simply this: some of these fields will be sowed in small grain to be consumed by turning the stock upon it after it has ripened; or they may be harvested, and subsequently gleaned, as the individual thinks best. A similar system would be adopted with regard to the grass lots. Some will be mowed for winter consumption, others will be appropriated to grazing. And as these artificial grasses require two or three years to arrive at perfection, prudence will dictate the propriety not only of enlarging these lots, but of increasing their number as far as practicable.

We admit that where lands are devoted almost exclusively to the purposes of tillage (as they are undoubtedly with us) that fencing becomes a very heavy and unprofitable tax. But if it be the object of the agriculturist to make stock management a subject of profit, he becomes amply remunerated for any amount of fencing which may be thought necessary to promote this object. And here in fact is the secret of those rock fences which we hear of in some parts of the state—one mile of which would absolutely ruin a tobacco planter to build.

The idea that a portion of our arable lands were to remain unenclosed has always appeared to me as a very preposterous one. In those countries where the law of enclosures is not known, and where fencing is infinitely more expensive than with us, their arable lands are all carefully enclosed. Conceive for a moment, a country intercepted in every direction by public roads, filled constantly with travellers and way-faring men. It would be impossible under such circumstances, to determine whether depredation, upon the unprotected crops of the country were the result of accident or design. A drove of hogs or bullocks would fill the country with consternation and dismay. The adoption of such a system would indeed be a happy contrivance for the drover. It would save him the expense of feeding his stock from the time he passes the Virginia border until he sells out. When the supplies on one route have been exhausted he will take another: and when the highways have failed, he will take to the byways. And after he has thus consumed the whole growing crop of the state, he will then sell his meat at his own price, to the very people at whose expense it had been fattened.

We insist that agricultural reform calls for no legislative enactment. The existing legal policy throws no obstruction whatever in the way of the individual who sincerely wishes to place his stock management upon a profitable footing. And we feel constrained to condemn all such attempts on the part of the legislature as gratuitous and uncalled for; and as oppressive in the extreme to the whole body of small farmers who constitute so large a portion of the agricultural community.

The foregoing remarks were suggested by an editorial essay on the law of enclosures, in the March number of the Register.[4] We concur fully with the author in the impolicy and injustice of giving legislative preference to any particular branch of industry. But when he asserts that the law of enclosures is of this description we beg leave to put in our humble dissent. This law says he "and its effects exhibit a striking example of the long continued operation of the preference given by our law to the business of cattle-raising at the expense of grain raising—of a preference of grazing to tillage." We think that by referring to the circumstances which have given rise to these laws they will be found to have their origin in motives directly opposite to those ascribed by the author.

It is a fact familiar to every tobacco planter at least, that the annual *clearings* which he finds leisure to make, are utterly inadequate to his immediate demands. We speak advisedly when we assert, that according to the practice which prevails upon tobacco estates no planter expects to release any part of his arable land until it has been cultivated at least eight or ten years in succession. We will explain the *"modus operandi."* And as the history of a single plantation will answer for that of every other, we will take, for instance, one of 500 acres, with a complement of ten hands. To each of these hands, the planter will assign 15 acres for annual cultivation; if he has a sufficient quantity of open land. The size of his whole crop will then be

150 acres; 20 of these are allotted to his tobacco, the remainder to his grain crop. All the newly cleared lands are invariably assigned to his tobacco, the first year they are cultivated: and where the lands are very good, they are tended in this crop at least three years in succession. This practice relieves the planter from the necessity of clearing land for his entire tobacco crop; which he cannot do, and tend a full crop of grain at the same time. Upon estates where a great portion of the lands have been opened, the annual "clearings" hardly ever exceed half an acre to the hand. But as our remarks have reference to rather an earlier period of operation, we will take it at a stage when they can accomplish at least a full acre each. At this rate then ten hands will be employed fifteen years in opening as much land as is allowed usually for annual cultivation. We have reduced the period to eight or ten years, because the three or four first years of settling a plantation are devoted almost entirely to the labor of clearing and fencing, as preparatory steps to the introduction of regular and systematic operations.

So then for the first ten years there is no open land to be spared for pasturage. By this time the fields that were first cleared are so much exhausted as to stand eminently in need of a little respite, if indeed they are not already irretrievably gone. In the mean time stock must be raised. During our earlier history the forest afforded an ample resource. Common interest dictated the propriety of enclosing all the arable land in order to secure the benefits of the *woods range*. And the legislature aimed at nothing more than to restrain the abuses to which the common worm fence of the country was liable.

But we will follow the history of the 500 acres farther, and see if there is any subsequent circumstance which can justify the inference of the author, that pasturage is preferred to tillage. We have seen that arable lands are consigned through necessity, to eight or ten years of constant cultivation. After this period, the three-field system is gradually introduced: viz. first year corn; second oats, or other small grain; third, rest, with pasturage: a system for a long time the boast of our agriculture, and the ultimate object to which the efforts of the planter were directed with the most anxious solicitude. It was indeed found eminently serviceable to stock raising, but it was ultimately discovered that lands deteriorated almost as fast under this system as they did from annual cultivation. Many planters, especially those upon thin lands, were therefore induced to relieve their lands altogether from the hoof, and to devote them exclusively to the purposes of tillage. And as a security against the temptation to pasturage whilst cross fences existed, they were removed, and the rails appropriated to strengthening the outward enclosure. A number of planters, however still persist in the three-field system. Others again have established standing pastures. But under every system yet adopted in the tobacco region, there is some radical defect which compels the planter to resort to the importation of live stock to supply deficiencies. This is a succinct but true history of agricultural operations, wherever at least the tobacco culture has been introduced. And we discover in every circumstance connected with the subject a palpable preference of tillage to pasturage.

But there is another circumstance which goes most forcibly to strengthen this conclusion. Wherever lands are cheap, and agricultural products high, the people are disposed from principles of obvious economy to apply their labor almost exclusively to tillage. The enormous sums which the planters once received for their tobacco crops did for a time justify the neglect of every other subject of husbandry. It was indeed ripping up the goose for the golden eggs, but it was with a success the reverse of that which happened to the man in the fable. We may mourn as much as we please over our impoverished inheritance, but had our ancestors been so prudent as to have saved even a tithe of what they made by this land destroying system, we should have little cause to shed tears over their steril fields. In connection with this subject, we will advert to a capital mistake into which they very generally fell. The majority of people in every country either live up to their incomes, or beyond them. But the Virginia planter who consumed annually the proceeds of his crop had the misfortune to mistake a part of his capital for his legitimate income. But to such however, there were innumerable exceptions. Wherever the planter was actuated by prudent foresight, or love of money, his descendants have had no occasion to complain of barren fields or empty coffers. Indeed the wealthiest planters have invariably been those who have sacrificed every thing for tobacco, and left ruin and desolation in their rear.

It is true that the author speaks of the preference of grazing to tillage as a legal one. But he leaves us altogether in the dark whether he considers it as the join[t] voice of the people expressed through their representatives—or whether like the "Mesta" of Spain it has been the result of a spirit of monopoly and arbitrary legislation. The examples however with which he has thought proper to illustrate the subject, leave us to infer that he looks upon the law of enclosures in the light of the arbitrary edicts of a Spanish monarch.

In Great Britain and Spain as well as in all other European countries, the people are found divided into a great many powerful interests. And according as one or the other of these interests acquire the ascendancy in the national councils, the opposite one must suffer the effects of a selfish policy. But in Virginia we know no such distinctions. We combine in one character that of farmer, grazier, planter and manufacturer. And how the legislature could separate the grazing interest from the planting, we are unable to discover. The members of the Virginia legislature are chosen almost entirely by the agricultural interest. It is not reasonable to suppose that a body so constituted would have persisted in a system of local policy which daily experience proved to be inadequate to the object proposed.

Nor can the law of enclosures be regarded as an attempt on the part of the legislature to direct a portion of the labor of the state into a channel which would obviate the necessity of importing live stock—because this law had its origin at a period many years anterior to the establishment of the western states, to whom we now look for supplies of this article of consumption.

FENCEMORE.

Ruffin appended the following commentary (Farmers' Register 3 [May 1835]: 49–50).

REMARKS

It is not the usage of this journal to bring forward in appended comments, editorial opinions, in opposition to any of different character maintained by correspondents—and though a more frequent resort to such comments has been more than once called for, it is still deemed most proper to leave every contributor free to speak, and every reader to judge for himself, of the various opinions and questionable points discussed. Whenever we have expressed objections to particular opinions, we hope that there appeared good cause for the exception, and that there was nothing in the remarks either disrespectful, or annoying to the particular correspondent. The adoption of a different course (although sanctioned by very general usage) gives an advantage to an editor, in controversy, to which he has no just claim: and still greater is this advantage made, when (as is also very common) a correspondent's argument is answered, or opposed, in *preliminary* editorial remarks.

Under the guidance of these general views, we should have submitted to the public the foregoing communication, without any accompaniment of our opposite opinions, and should have been the more inclined to pay that silent respect in this case, not only because the author well deserves respect, but also, because it is the first argument offered here in support of his side of the question, and therefore he is, especially entitled to a clear field and "fair plan." But as the writer has directed his strictures particularly to an editorial article, proper respect for him requires that we should depart from our usual course. Still it is not intended to discuss the general question.

There would be but little difference between our opinions as to the propriety of preserving the general principle of the law of enclosures in every such region as our correspondent describes, and which agrees very nearly with the general condition of Virginia in early times. No opponent of the law has objected to its early operation, nor to its continuance for any length of time that the same general state of things may continue in existence. The admission of the former good policy of the law was distinctly stated in one of the earliest arguments on this subject in Vol. I. p. 396, signed "*Suum Cuique,*"[5] and such we presume are the views of all. This removes all ground of controversy as to any region just emerging from the forest state, and therefore no farther observations on this head are needed.

But even for the cleared and impoverished portions of Virginia, (embracing, with some exceptions, most of the middle and low country,) our correspondent thinks the change of principle in the law would be improper, and that even if made, more fencing would be required by good stock husbandry than is now maintained. This part of the question we are in no manner required to discuss—and therefore shall pass it by, except to notice one or two mistakes, or omissions, which serve as part of the grounds from which our correspondent's conclusions are deduced.

It is readily admitted that a very high state of culture combined with grazing, and on very rich lands, requires, for its *perfection*, numerous and good enclosures. This is one of the first steps in agricultural improvement—agreeing, in results, (from our then peculiar circumstances,) with the earliest and rudest state of tillage in Virginia. But to arrive at this *perfect condition*, embracing enclosures on every field, it is necessary that the durable materials for fencing should be sufficiently cheap—that the landmarks on which to build walls, or plant live hedges, should not be changed in every generation, or oftener—and above all, that the profits derived from grazing should be sufficient to compensate amply the additional expense of enclosures. None of these circumstances exist in our naked and poor country—and the whole rent of the country, taking rich and poor land together, would not pay for keeping it enclosed in 20 acre lots, as may be good policy in England, and even in some parts of the northern states. But as to the latter, (at least so far as we are informed,) it should be remembered that hogs are not permitted to range at large—and the fences may therefore be made at half the cost. A fence of three rails at most, in mortised posts, is ample protection against cattle—and fewer will serve with a ditch and bank. If hogs were thus confined in Virginia, half the oppression of the present law would be at once removed—and besides gaining that benefit, there would be more hogs raised in consequence of this restraint. We even suspect that our correspondent and his neighbors, abundantly as they are supplied with woods' range, would make more pork, and have less need to rely on the supplies of meat from western drovers, which he seems to admit to be now required.

But between the forest state of a young country, and the highest state of improvement of an old one, there is an immense middle ground, in which enclosures of every field, and even of every distinct property, cannot be profitably kept up—and therefore, in most countries, are wisely dispensed with. That is, the owners are secured against trespasses, and therefore are free to enclose, or not, as they please. Most of the arable land on the continent of Europe is still in this unenclosed state—and almost all of that which is not more productive than the average of eastern Virginia. Even much of the richest and best cultivated land is unenclosed, because divided in such small shares that a fence around each would be too heavy an expense to be borne. And though we will not here argue the question as to the superior share of benefit which the small or the large landholders enjoy in the law of enclosures, we will observe, that so long as that law may exist, it will be impossible for land to be held in Virginia in very small shares. The law is perpetually operating to starve out, deprive of their little freeholds, and to banish from Virginia, the valuable class of small farmers whom it is averred the system protects. It is as much the operation of the fence law to accumulate many small tracts in few hands, as it is of the law of descents (however beneficial this may be in general,) to divide these accumulations: and from the frequent changes caused by the two opposite operations, dividing landmarks, (whether between large or small farms,) cannot be expected to remain long enough to mature a live fence on, or to permit a clear profit from its being raised to be derived.

As applicable to this branch of the subject, we will take the liberty of copying some sentences of a private letter written last winter by a very intelligent gentleman, than whom, no individual would be deemed better authority—and who, in addition, is a resident cultivator of the tobacco region of Virginia, and where, it is believed, that the mode of tillage, habits, and other circumstances, either are, or were but recently, precisely similar to those in the neighborhood of our opponent. "You are right," says this gentleman, "in saying but little can be done in the way of improvement, till there is a change in our law of enclosures. The present law is dissected in a masterly manner in the Register, [No. 6][6] and its absurdity completely exposed."—"You never said a truer thing, than that the poorer class was more interested in a change, than the more wealthy." There are hardly as many freeholds now as there were thirty years ago, and I verily believe that it is attributable to the present law. Small freeholds are amalgamated, brought under one fence and one owner."—"In my neighborhood there is thirty miles of fencing, and all to give to stock the benefit of promenading up and down a six-mile lane, where there is scarcely a blade of grass."

Having admitted fully the good design and policy of our law of enclosures, for the condition of the country when it was enacted, it will be unnecessary to disavow the intention of charging the present oppression to designedly partial, or worse than partial legislation. We spoke of the *operation* of the law, and not of the *intentions* of the legislators who enacted it. We never supposed that there was a deliberate design to sacrifice, or even to impair, the interests of tillage, for the benefit of the interests of grazing, as a matter of general policy. But this effect is not the less produced because of its not having been intended—though (as formerly stated) without even the poor gain of benefiting the favored interest.

But however much we may differ from our esteemed correspondent on this question, we are pleased that he has come forward to maintain his opinions. It is proper that every question affecting the interests of agriculture should be fairly discussed in this journal—and none of its readers ought to object to a full expression of any honest opinion, however opposed to their own conviction.

Fencemore, "Fence Less" and the Editor of the Register (1835)

Have each mistaken the import of our remarks with regard to the law of enclosures. We shall therefore with the permission of these gentlemen [Ruffin and Garnett], endeavor to make ourselves better understood. When we asserted (p. 47, Vol. III)[1] that agricultural reform called for no legislative enactment, we had some reason to feel assured that the tenor of our remarks would have exempted us from the charge of maintaining the good policy of the law. And the editor will recollect that the caption to the article we offered, was his own. The one proposed was simply, "remarks upon the existing law of enclosures," or something equivalent.

We distinctly admitted the necessity of agricultural reform, when we pointed out what we considered the most approved method of stock management. The connexion of this subject with profitable husbandry, is so well understood, that we deemed it unnecessary to offer any argument to prove it. And we were not a little surprised that the editor should deny he was called upon to discuss this particular question: particularly after the publication of an article from an English agricultural paper, from which we make the following extract: "Agriculture is divided into three great branches, *green cropping, white cropping*, and *stock management;* and they are mutually and severally dependent for success on each other. Without green cropping we cannot raise heavy crops of grain, and without great crops of grain, and consequently of straw, to be used as litter, and partly as fodder in conjunction with the green food for feeding stock through the winter months, we cannot make dung, and without plenty of dung we cannot raise green crops, and so on. And it is such a disposition of stock and crop as shall cause the one to be instrumental in promoting the prosperity of the other. A reciprocity of services, as it were, which ultimately converge to the general advancement of the whole, which in agriculture, constitutes a system, which system must be rigidly adhered to if any thing like *profit* is to be looked for in farming." (F.R. Vol. I. p. 674.)[2]

There are two ways of effecting agricultural reform—the one by legislation—and the other by individual enterprise. We are opposed to the legislature's interfering with this subject, for the following reasons. The present system of agriculture in Virginia, although as bad as it well can be, has become, by long usage, identified with the very constitution of society. This system is based upon the legal policy of enclosures—and whenever this law is repealed, the whole superstructure must tumble, and involve thousands in pecuniary distress. We know that this consequence will be denied. But we defy gentlemen to point out a single instance where force has been used to divert the labor of a nation from one channel into another without producing great pecuniary sacrifices. It is difficult in this particular case to point out the precise manner in which the loss would be sustained, because the nature and amount of this loss would depend very much upon the peculiar circumstances of the individual. But innumerable instances might be found in every part of the state in which injury would unquestionably be sustained by repealing the law. According to this view of the subject, we are limited to a choice of evils. On the one hand, the law of enclosures is unjust in principle, and oppressive in it operation. On the other, its repeal would be attended, necessarily, with great pecuniary sacrifices. In our opinion, this difficulty may be avoided by leaving the subject wholly to individual enterprise. We pointed out upon a former occasion, the manner of effecting the desired change. And we referred to instances where it was effected with signal success, both in this state and in different parts of the union.

It seems that the editor admits the policy of adopting the system of separate enclosures at certain periods in the history of agriculture, viz: when the country is just emerging from a *forest state*, and when it has arrived at the *highest state of improvement*. With regard to the intermediate space, he remarks: "But to arrive at this *perfect condition*, embracing enclosures of every field, it is necessary that the durable materials for fencing should be sufficiently cheap—that the land marks on which to build walls, or plant live hedges, should not be changed in every generation, or oftener—and above all, that the profits derived from grazing should be sufficient to compensate amply the additional expense of enclosures. None of these circumstances exist in our naked and poor country—and the whole rent of the country, taking rich and poor land together, would not pay for keeping it enclosed in 20 *acre lots*, as may be good policy in England, and even in some parts of the northern states."—(*Farm. Reg. Vol. III. p. 50,*)[.]³ From feelings of courtesy towards us, and an evident disposition to leave the discussion of the subject to others, the editor has declined going into the *general question*. But he has nevertheless thought proper to settle the whole subject in a manner more remarkable for its brevity than its accuracy. We never maintained that *rock walls, live hedges*, or 20 *acre lots*, were necessary to the proposed scheme of agricultural reform. They were merely referred to in order to illustrate our views of the subject. In the cases particularly cited as being worthy of imitation, the common perishable worm fence of the country was used. And the size of the separate enclosures was regulated according to the convenience of the individual. The editor certainly could not suppose that we recommended the construction of 20 acre lots upon an estate consisting of 500 acres of arable land? The profits of such an arrangement could under no possible circumstances, justify its adoption. Upon an estate of this size these small lots would not only be unnecessary for stock management, but would interfere very seriously with the other operations of the farm. The number of lots should be regulated principally by the rotation of crops. For instance, the four, five, or six-field system ought to have at least the same number of separate enclosures. With this arrangement, these separate fields can be gleaned or grazed, according to circumstances. As to the perishable nature of the fencing material, we could very properly remark, "sufficient for the day is the evil thereof." Why indeed should we trouble ourselves about the substitute until the period arrives when we shall be compelled to adopt it? Dead fences are certainly the best as long as they are the cheapest—and moreover, being easily moved from one place to another, are peculiarly suited to our law of descents, which is constantly changing our land marks. Perhaps the editor will object to the cases referred to, as coming under the specified exceptions. If so, he must exempt one-half of the estates in Virginia. But Mr. Craven[4] expressly states that when he took possession of the premises in question, the soil was reduced to a state of great exhaustion by the very system which has impoverished other portions of the state. Yet this gentleman has found the ways and means of adopting, with success, the system of separate enclosures upon the whole of this arable surface. We are but little acquainted with the general condition of the arable lands in other parts of the state, but with regard to those of Prince George, we have the following description. (we presume from the pen of the editor:) "Much land is planted in corn which does not produce more than one barrel of corn per acre; and about one-half of the arable land of the county falls short of two and a half barrels, which has been stated as the least product that will defray the expense of cultivation. *One-half of our land is not only cultivated without profit, but with certain and increasing loss*—and to this purpose our labor is devoted ninety days, the whole crop being supposed to require six months. Every consideration of profit demands that this portion of our soil should not be cultivated in its present condition." (*Farm. Reg. Vol. I. p. 234.*)[5] Yet the editor makes us embrace in our remarks, the whole of this unprofitable surface: and what is still worse, we are made to build rock walls, rear live hedges, and construct 20 acre lots, upon lands which do not pay the simple labor of cultivation. Our remarks were never intended to apply to soils so hopelessly impoverished, and we shall dismiss them with the remark, that the sooner they are abandoned the better. But to the owners of the small, but really valuable portion of the county, we point them to the neighboring forest, and entreat them to make separate enclosures, and rear the artificial grasses. This system has led to wealth under circumstances not at all more favorable, and will do it again with the use of capital and enterprise. In the words of the above report, "a farm which would yield a regular annual profit of $600 after paying all the expense of cultivation, would be thought cheap at $10,000.

But certainly it would be equally profitable to lay out 10,000 dollars on the improvement of land already in possession, if from that improvement, an additional clear profit of 600 dollars could be derived." Now we venture to affirm, that half of this sum applied in the manner recommended, would yield an additional income from stock alone, independent of their immense value as agents in fertilizing the soil.

We assert in the very face of the experiments with which we have been favored by "Fence less," that the proprietors of landed estates in Virginia, cannot afford to raise the stock necessary for domestic uses, in pens and small enclosures. We admit that where a few pigs only are sufficient for an entire family, as in New England, they may be cheaply raised upon the otherwise useless offal of the farm. But when the numerous mouths of a Virginia estate are to be filled, it is altogether a different affair. A different system of stock management must be adopted, or the proprietor will be ruined in his attempts to imitate practices unsuited to the condition of the country. The New England farmer derives no small portion of his profits from his stock, and he is therefore amply remunerated for the additional labor and expense he bestows upon the subject. On the other hand, we are tillage farmers principally in Virginia, and the most we ought to expect is a simple supply for domestic uses. Even with this difference in the circumstances of the two cases, this system we recommend as best for Virginia, is adopted, on account of its greater cheapness, by some of the very best northern farmers. This system must combine judicious grazing, in which stock become, in part, their own caterers. What becomes of the hundreds of bushels of small grain unavoidably left upon the harvest fields? Of the aftermath of the meadows? Of grazing lots at those periods when they cannot be mowed, and when they receive little or no injury from the hoof? Yet these are some few of the advantages to be wilfully relinquished for a system of management yet untried, and every way more expensive and laborious than the one proposed. "Fence less" tells us that he fed his ox in an old barn from the 1st of December until the March following—but does not inform us what becomes of him after this period. We conclude, however, in the absence of better proof, that he was introduced to the salting tub. The experiment of the cows is remarkable for the same omission—and we are left equally in the dark whether he made beef of them, or turned them to grass. The inference is fair, however, that he adopted the latter alternative. The third experiment of the ten shoats is changed to the summer months, for reasons sufficiently explained by the *clover hay* and the *morceau* of corn. These experiments would be very *stubborn* indeed, but for the omissions of a few very important months. "*Hiatus maxime deflendus.*" Now we ourselves once made an experiment of this sort upon several dozen shoats, commenced very honestly the last of October, and continued until we were threatened with famine. They were then turned out with the solemn injunction to root or die. We should have been obliged to "Fence less" if he had extended his experiments to his entire stock for one whole year, and had then presented us with the results—the state of the corn crib, &c. These he must admit are indispensable considerations in settling a matter of so much importance to the interests of agriculture. It is indeed to be lamented, that those who oppose the policy of the law of enclosures, have failed to point out exactly the method of management they propose to substitute in its stead. The injustice of the law is constantly deprecated: yet the planter or farmer is left at a loss to know what is to become of his stock. It is true we are favored with an occasional experiment or so—but even when fairly carried out for the whole year, they are unsatisfactory and inconclusive, because confined to too small a portion of the stock necessary for domestic uses.

As we have been particularly referred to a writer who signs himself "*Suum Cuique*" as proper authority on this subject, we will take the liberty of adverting, for a moment, to the plan he proposes to substitute in the place of the system imposed upon us by the law of enclosures.

"Each farmer having to maintain his own cattle would keep a smaller number, and confine them generally to a permanent pasture, well enclosed: and being necessarily reduced to *one-fourth of their present numbers*, and treated as well as the change of the system would permit, the live stock would yield more products of every kind (except hides perhaps) than at present. The lands kept for tillage, *thrice as extensive as the enclosed pastures*, if too poor to be grazed might be safely left without a fence, until their improvement in after time may make enclosures necessary for the owners' interest." (Farm. Reg. Vol. I. p. 398.)[6] Admitting that the fourth part of a farmers' stock would yield more products than the whole, under existing circumstances, (which by the by we don't believe) yet he is still left without the assurance that these diminished numbers will afford a competent supply. It is unquestionably the true policy of the people of Virginia to relieve themselves from the heavy tax they are compelled to pay annually for western meat. No plan of reform is worthy of a moment's consideration which does not profess to meet the whole difficulty. When the defect is radical it must be encountered with a remedy which goes to the very root of the disease. The proposed plan of reform, if not avowedly partial, leaves the subject in doubt and, uncertainty, even if we are sure of the anticipated results. But so far from gaining any thing by the plan of "*Suum Cuique*," we are firmly persuaded we should be losers by it. Let us see. One-third of the arable land of the country is enclosed for standing pasture—the remainder is appropriated to tillage. It would be difficult to state the exact amount of stock necessary for the use of any particular estate. But we can venture to affirm, that the live-stock necessary for the consumption of Virginia estates generally, would leave the portion devoted to standing pasture in a situation as destitute of vegetation, as the ordinary unenclosed commons of the country. If this be the case then, the farmer is left in a worse situation than before. He is deprived of one-third of his arable lands, for the advantage which the naked commons afford him, even if they are as destitute of vegetation as the summit of the Alps. His stock must be grain fed in both cases, but the proposed scheme deprives him of one-third of his means for feeding them. The farmers and planters of Virginia are sensible of this difficulty, and for this reason are but little disposed to favor the repeal of the law of enclosures. It is true they are saved the expense of fencing in two-thirds of their arable lands; but, on the other hand, they know that they must lose the profits of one-third of their estates, upon the standing pasture system. The choice of the two evils is too obvious to admit of a moment's deliberation. We will add, moreover, that so far from increasing the products of live-stock by diminishing their number, as "*Suum Cuique*" affirms, the proposed scheme, by diminishing the means of subsistence, would necessarily reduce their numbers, without improving their value. For instance, if an estate of 300 acres arable land will support 30 hogs in a certain condition, one of 200 acres will keep only 20 in the same condition. Here then we have a scheme if carried into practice, would deprive the people of Virginia of three-fourths of their stock and one-third of their arable lands, and leave them in a situation infinitely worse than before.

With regard to the effect of the law of enclosures upon the small farmers, the editor remarks, "The law is perpetually operating to starve out, deprive of their little freeholds, and to banish from Virginia, the valuable class of small farmers, whom it is averred, the system protects." That the law of enclosures imposes a great deal of unprofitable labor, under existing circumstances, we never denied, and it doubtless accelerates the tide of emigration, constantly carrying off so large a portion of this class of our citizens. But the true moving source of the evil, if an evil it be, exists in the peculiar circumstances of the country, and the same result would have happened though protracted perhaps to a later period, independent of any influence from the operation of the law of enclosures. This cause is to be found in the circumstance, that wherever land is cheap and labor dear, individual interest dictates the adoption of that hard and destructive system of cultivation which so very generally prevails in all new countries, particularly where the products of agriculture have borne enormous premiums. The first victims of this self-imposed, but land destroying system, are the proprietors of ordinary lands. As you advance in the grade of fertility, the tide of emigration is slower—but the cause is still operating until you arrive at soils whose recuperative energies defy the most pernicious agricultural practices. Another assisting cause is to be found in the law of descents. This law is constantly reducing estates to a size which forbids a fair remuneration for agricultural employment. This effect is most sensibly felt in tracts of inferior fertility; because soils of this description offer no inducement for re-uniting these scattered and valueless fragments. For this reason, the great proportion of small farmers are found upon worn and exhausted lands. But whenever lands present a fair prospect for remunerating labor, we find estates assuming a size in spite of these adverse circumstances, which fully justify schemes of profitable husbandry. We refer, for instance, to our alluvion lands and to those belts of extraordinary fertile highland to be found in every part of the state. These remarks are not made, to justify the policy of the law of enclosures, but to show that our agricultural evils are mainly attributable to far other causes than the one specified. The law itself is a mere accessary, the necessary result of our peculiar circumstances—and its repeal would have the effect of hastening the very catastrophe which is so much deprecated by its adversaries. Under the influence of these considerations, we cannot consent to the adoption of any legal measure which will add to the evils of a system which is already dragging its disciples to a point, when it must be finally abandoned.

It is farther objected to the law, that it has a tendency to "amalgamate the small freeholds, and bring them under one fence and one owner."

Whatever his correspondent may think on the subject, the editor will certainly agree with me, that the circumstance complained of, is any thing but an evil. As far as the small farmer is concerned, he is a gainer by it; for he is exchanging unprofitable for profitable labor. As far as agriculture is concerned, the larger the tract in possession of a single individual in a poor and exhausted country, the better. The whole will then have a fairer chance of feeling the beneficial effects of a milder husbandry. Thus have the evils of our agricultural system a tendency to correct themselves, without the intervention of legal assistance, and to bring about those very advantages, which in England have been produced by a singular but very fortunate combination of circumstances.

FENCEMORE.

Ruffin's response (Farmers' Register 3 [December 1835]: 458–59) *follows.*

Though the foregoing argument is addressed principally to the editor, we shall not trouble our readers with a reply—which indeed, if attempted, would necessarily be drawn from the same materials that have been already used in various parts of this journal. We shall be content with remarking on a single passage of the last paragraph, which shows, on the part of the writer, a misapprehension of our views.

Our correspondent is greatly mistaken in supposing that we agree with him in considering an amalgamation of small farms, forced in the manner stated, as "any thing but an evil." Both large and small farms have peculiar and important advantages, as well as disadvantages, and it is essential to the improvement and profit of agriculture, and to the interest of the people, that there should be farms of both classes—and of every size, except such as are either too large or too small to yield, under suitable and proper management, fair profits. But the great evil is, the frequent change of landmarks—the converting large farms to small, and small to large—the very thing which our correspondent welcomes as a boon bestowed by the operation of the law of enclosures. "Fencemore" had just stated, (and very correctly,) that the tide of emigration from Virginia is swelled by the law of descents, which "is constantly reducing estates to a size which *forbids a fair remuneration for agricultural employment.*" Then comes the alleged benefit of the law of enclosures, which by making it impossible for the owners to fence and till their little freeholds, compels them to be sold, to be united in some newly formed and newly arranged large farm. Both the changes are necessarily attended with prodigious losses, to the land owners and to the public—and the losses are such as can never be repaired. The small shares of a farm, divided among the heirs of the former owner, are necessarily in most cases of much less value than when united. The purchaser who unites various such shares of several ancient properties, certainly places them in a better situation for profitable culture—but even this object cannot be reached except by additional and great waste of labor and of capital, which would have been unnecessary but for the change of owners and of landmarks. The life of one farmer is spent in fixing an estate in the best form for one property—as a single well managed farm. He dies, and all the loss is sustained by his heirs and the country at large, which necessarily attends the cutting up of this large and well arranged farm into four or five pieces, neither of which is worth holding separately. One child is impoverished by having all the houses, and of course very little land—all the others are houseless. One or more have nothing but forest—others not a tree for fencing. They may possibly spend another generation in struggling under these evils, and in undoing all their father's arrangements. But soon or late, (with that aid of the law of enclosures which "Fencemore" considers in this respect so beneficial,) these reduced properties are, one after another, drawn into the adjacent large farms, and again all the arrangements of the last owners are useless, and are lost to them in the price, to the purchaser, and to the commonwealth. To continually *do and then undo*, is the operation of our governmental land policy—and whatever may be its benefits, they are purchased at an enormous sacrifice of the profits and capital of agriculture.

The agricultural prosperity of a country would be greatly promoted, if the dividing landmarks of farms could never be changed, except by sale or gift, both parties *being alive* at the time, and consenting to the transaction—when it may be supposed that the change would be advantageous to the individuals concerned, and consequently, to the public interest. But changes forced by the operation of law, whether they be in dividing a single farm, or consolidating the parts of the separated shares of several, must always be injurious to both private and public interests. We speak not of the political, moral, or social benefits of the divisions of farms, as caused by the law of descents. Let others make the most of these benefits, concerning which it is not our business to treat. But as they affect the interest and improvement of agriculture—as obstructions to reaping the full amount of product which the entire surface of the country would yield to labor judiciously directed—we regard the continual division of farms under that law, together with the consequent consolidations of various disjointed and unsuitable fragments to form large farms, caused by sales forced by the law of enclosures, as a combination of inflictions on agriculture which yield only in magnitude to the present exhausting drain, or rather flood, of emigration to the west.

5. Malaria—Against Mill Ponds
On the Sources of Malaria, or Autumnal Diseases, in Virginia, and the Means of Remedy and Prevention (1838)

Throughout the course of publication of the Farmers' Register, it has been one of the main objects of the editor to attract attention to the causes and effects of malaria, or unhealthy marsh-effluvia, and to enforce his views as to the means of restraining or preventing this greatest of the evils under which the eastern half of Virginia suffers. To forward this end, every fit opportunity has been availed of; and the subject has been treated, directly and at length, or incidentally and slightly, in various articles in these volumes. But there has been found but little if any encouragement to persevere in this course. The editor has, alone, and without any certain evidence of approval of his views and his course, and certainly without any practical adoption of his recommendations, labored in this cause, which, to his understanding, demands the support of all, on considerations of economy and agricultural improvement and profit, as well as on the more important grounds of the strength or frailty of the tenure by which the people of half of our entire territory possess and enjoy health, happiness, and even life. It is under such impressions of the high importance of the whole subject, that the readers of this journal are again invited to its consideration; and, probably, for the last time, by the present writer, if there continues to be no more interest excited, and action produced, in regard to the evils existing, and which are multiplied ten-fold in power by the ignorant and careless legislation of this commonwealth.

The views of the writer on this subject were presented generally, and at some length, in an editorial article (pp. 41 to 43) in Vol. V,[1] on the causes of, and means for preventing, the formation and the effects of malaria in eastern Virginia; and also in sundry shorter incidental passages in each of all the volumes, in connexion with articles on marshes, mill-ponds, and canals, &c. &c. But as it would be requiring too much of readers that they should either remember, or carefully refer to these various articles, a general, though slight view of the whole subject will be here presented, sustained by additional facts, which have been recently learned by personal inquiry and observation.

That the common autumnal or bilious diseases of eastern Virginia, and especially of the tide-water portion, which is most subjected to them, are principally caused by the effluvia rising from wet lands, is a matter in which all concur. The general difference between the presence of these disorders, in low, wet, or marshy countries, and their absence, or scarcity in mountainous and dry regions, is so great, that none can mistake, or differ about, the *general* causes and effects. But from this general opinion, which is true in the main, (though having numerous and important exceptions,) there is deduced the erroneous conclusion, that these opposite general effects produced on health, in extensive regions either generally low and wet, or generally hilly and dry, are produced by these opposite natural features, and cannot be very materially altered by art; as art cannot materially alter the natural character of the land. Or, in other words, that nature has made one great region low and sickly, and another high and healthy; and that man cannot do much to counteract the law of nature in either case. Perhaps none may maintain this position, in argument, without admitting partial exceptions in numerous particular cases and localities. Indeed, every man will say that care may lessen the causes and mitigate the operation of malaria, in a sickly region, or increase both in a healthy one. But, judging from the action of both the people and their laws, which speaks more strongly than words, it may be inferred that it is a general belief that such bendings of nature from her course can be but slight, in particular cases, and scarcely worth estimating on a broad scale, or through an extensive country. In entire conformity with this supposition, it is a notorious fact that very few individuals in Virginia have done any thing considerable, or on system, to protect their dwelling places from malaria; and the government has not only done nothing for general protection, but has actually caused the worst of the existing evils, and is encouraging their continued increase and aggravation, by the fixed legal policy of the country; which permits the raising of mill-ponds, which are productive of little else than malaria and disease; and indirectly, but effectually, forbids the drainage of extensive swamps. The production and deadly effects of malaria, in eastern Virginia, for the greater part, is to be charged, not to the laws of God, but to the laws of man; which, in this respect, operate to put away or sacrifice some of the most precious of God's blessings, offered to all, to gratify the whims, or the blind and often mistaken avarice of a few individuals. There are, doubtless, great *natural* differences as to the sickliness of differently situated regions; as between the low tide-water region of Virginia, the central or hilly, and also the mountainous region. But, in their natural state, before damaged by mill-ponds and other of man's miscalled improvements, the low-country was probably less afflicted by malaria, than the hilly parts now are, or may be rendered by the full extension of these injurious operations of man. This is a matter of mere supposition, and cannot possibly be subjected to the rigid test of proof by known facts. But, from reasoning, and inferences from such facts as are known, it seems most probable that some of the now most sickly counties on tide-water were, at the first settlement of the country, less sickly than the hilly and originally very healthy county of Brunswick, for example, has become in latter years.

Even the very important fact of increased and increasing sickliness in this country, is entirely without support from any known written authority; and the whole subject has been so little examined, or thought of, that to most readers the position here assumed may be entirely new. There are no statistics of health to which we can refer for proof. But general and historical facts, few as they are, if fairly considered, will suffice to place the question beyond dispute.

Before proceeding further in this part of the argument, let me remark, that I am opposed in the outset, and shall be opposed throughout, by the reluctance felt by every individual to believe, or if believing, to admit, that his particular property, or place of residence, is more sickly than in former times. This self-delusion, and consequent, though perhaps undesigned effort to deceive others, is almost universal. Each man claims for his own place more healthiness than in truth ought to be admitted; and the combined effect of all these individual claims, is to maintain that the whole country is more healthy than is true, and more so than each individual would have claimed for it, with the exception of his own farm and his own neighborhood. It is against this universal prejudice and obstruction that I have had to contend in seeking for facts, and shall have to contend in argument; and, with such opposition, there is but small hope of maintaining my ground, or producing conviction of the soundness of my views, in the minds of those who have so prejudged the case.

One of the strongest proofs of the greater former healthiness of the low country, was the settlement of our English ancestors having been made and continued at Jamestown. It was on May 13th, when they landed; and now, a residence on that spot, or in that region, continued for five months after that time of the year, would be fatal to half of the strangers from a northern climate, even though provided with all the comforts and necessaries which a long-settled country affords, and all of which the first settlers most deplorably needed. It is true, that for some years after the first settlement, there was much sickness, and numerous deaths; and that in fact the infant colony was more than once on the point of extinction. But these diseases and deaths do not seem, from the direct and the still stronger indirect testimony of history, to have been attributed by the sufferers to an unhealthy location; and there were sufficient other causes for all that was suffered, in the usual and unavoidable privations of the first colonists of a new and savage country, added to the extreme improvidence and mismanagement of these settlers, and their government, as detailed in history. Even after several years had passed, and though cultivating a very fertile soil, and aided by annual supplies of food from England, and with all the resources of trade with the savages, hunting, and fishing, still, want of food was one of the greatest causes of disease and death. Of course, there must have been, under any circumstances, more or less of disease caused by malaria; and although any predisposition to such disease, naturally induced, must have been violently urged to action, and aggravated to ten-fold malignity, by hunger, intemperance, exposure of every kind, depression of spirits, and every other painful emotion of the minds of men in such desperate straits— still, even with all these aids, the prevalence of autumnal diseases, the effect of malaria, was not so conspicuous as to stamp the character of sickliness on the location, nor to induce even the proposition to remove the colony, or afterwards its seat of government, to a much higher or more healthy situation.[2] The unavoidable inference seems to be, that the great sickliness of the early settlers was not attributed by themselves, to the climate. Yet, this was a question on which they could not possibly have been deceived. And even if most others had been deceived, by ignorance, and the want of experience of the effects of malaria, this could not have been the case with [Capt. John] Smith, the most efficient director, and the true founder of the colony; who would have known better, not only by his general intelligence, but also by his experience of such effects, gained in his campaigns against the Turks. It may be alledged, that fear of the savages, stronger than the dread of disease, caused the choice of, and after-continuance on, an unhealthy spot, because it was more easily guarded on the land-side, and perfectly accessible to ships. But spots equally favorable for defence, and on deep water, might have been selected at first, much higher up the river; and yet Jamestown and its immediate neighborhood continued to be the chief place in Virginia, after the power of the savages had been crushed, and settlements had been extended to distant and inland places. The proof of my position would be sufficiently proved by any attempt made now to settle Englishmen, just arrived, on the border of almost any of our tidewaters, and especially about the junction of the salt and fresh waters. Several such trials have been made with foreign laborers; but the first autumn was enough to put an end to each experiment, by inflicting so much disease and death as to prevent any of the foreigners remaining through another season, who could possibly move away.

There can be but little doubt also, but there was much less of autumnal diseases, or at least of violent and fatal diseases, before the revolutionary war than now. There was no such thing then, as the healthy residents leaving home in summer, as is so usual now, to spend the sickly season among the mountains,[3] or at the north; nor, does it appear, that there was much suffering for want of such resources, although, the climate must even then have become very far more unhealthy than in the early times of the colony.

Another striking proof of the increased tendency of the county to produce disease, even within the last sixty years, is presented by history, in the circumstances of the occupation of Yorktown by the British army in 1781, and the siege carried on by the American army; and especially in regard to the hastily-levied militia from the mountains, and other high and healthy parts of Virginia. Cornwallis chose his position first in Portsmouth, and afterwards in Yorktown, with a view to health, as well as defence, to await the arrival of reinforcements from New York. His army was concentrated at Yorktown, August 22. Washington reached Williamsburg, September 14, and the American army moved on thence to invest Yorktown, Sept. 30, and the surrender of the British army was made on Oct. 19th. Thus, both armies were exposed to the worst part of the malaria season, and the British army to the whole of it. Among the besiegers, were raw militia just raised for the occasion, from Rockbridge county, (of which portion I have been more particularly informed,) and probably from sundry others of the mountain-counties. There was certainly much sickness, and especially among the British troops; but not more than is usual in camps, and especially in besieged camps, suffering all the privations incidental to the confined situation. It does not appear, from the very slight notices in history, that there was more sickness than might have been expected if the same circumstances had occurred in the hilly middle region of Virginia. Yet, if the like circumstances could occur now, it can scarcely be doubted but that every soldier, not already acclimated, and accustomed to malaria, would be made sick; and that probably half of those just brought from breathing the pure mountain-air, would never return home.

Another indirect proof is presented in the great and deplorable decline of most of the lower counties of Virginia in wealth, and in the usual accompaniments of wealth, which formerly made a residence delightful in many neighborhoods in which there is nothing now left to invite any one to remain. It is true that other causes, political and economical, have concurred to produce this result. But the most potent of the several causes, was the slow and silent, but continual and increasing warfare on the health of body and mind, made by the action of malaria. By its operation, when scarcely amounting in effect to positive and known disease, the mind is sickened even more than the body. The buoyant spirits are tamed—energy is relaxed—the keen appetite for enjoyment, (which is the greater part of happiness,) is lost; and the victims of malaria cease to strive, or to enjoy; and either sink into apathy and listlessness, or urged by discontent, more than by any remains of energy, take the final step of emigration to the western wilderness.

But the upper country furnishes still stronger evidence, because of positive and unquestionable facts, to prove an increase of the product and effect of malaria. The hilly country between the falls of the rivers and the near[e]st mountain-range, with the exception of some comparatively small spots, on swamps and rivers, was formerly as free from this scourge as is now the mountain region. But the number and the extent of the unhealthy places have greatly increased, within the memory of those now living; and some large districts have been, in particular seasons, as subject to bilious diseases, and still more to violent ones, than the tide-water region. Indeed, in very many places, universally believed (unless by the mill-owners,) to be injuriously affected by the neighborhood of mill-ponds, these effects of malaria are of as regular recurrence in autumn, as on places near to any of the marshes of the low country; and are much more dangerous.

The third and highest region seems destined, notwithstanding its better defence in mountain sides and peaks, and the rarity of flat surface on which to form wide and shallow ponds, to take its turn next, as the victim of malaria. Already, in that part of the mountain-region in and about Frederick county, there have been particular autumns which seemed almost pestilential. And though such cases of general and virulent disease are rare, particular cases of autumnal diseases are now frequent in many such places where they were rarely heard of thirty years ago.

These statements may be considered by some as exaggerated or unfounded—and, by others, if admitted to be true, considered as showing the want of both patriotism and policy, in the writer thus exposing the enormous existing and still growing evils under which the country suffers. In regard to the former point, I admit, in advance, the scarcity of particular and positive facts, to serve as proofs, which is found throughout the whole subject; and that among the existing difficulties of obtaining such facts, (and still more by a single and unaided individual, who has had little opportunity to make proper researches,) I have to rely mostly upon general and loose opinions, and deductions from general facts. Hence, there is much liability of mistake. But if the public can in any way be driven to the examination of this subject, and numerous individuals be excited to search for facts, whether to sustain or to oppose my views, the arrangement and presentation of such facts will serve as materials, which are now almost totally wanting, and will enable this all-important question to be hereafter properly discussed, and correctly determined.

If there were no hope for relief, there would certainly be no use in exposing or dwelling upon these distresses of our people. But, though nothing yet has been done for relief, nor does it seem to have entered the imagination of our legislators—and though all they have yet done has been to add strength to the evil—still it is my confident opinion, that relief may be furnished for this sorest evil of the land, and furnished easily and profitably; and that it is perfectly within the power of man to dry up the most fruitful sources of malaria, and to bring the whole, or very nearly the whole of Virginia, to a state as healthy as that of any other country in the world. If such a result is indeed attainable, it is worth making every possible exertion for; and nothing will induce the smallest exertion, either by the people or the government, except a full exposure of the enormity of the evil which presses upon the country.

It is not my purpose to attempt to investigate the cause and trace the mode of operation of malaria. Though worthy of every care and labor, as a scientific question, it is one which as yet has entirely baffled every attempt at exposition. But though it is as yet unknown what is the chemical character of this subtle fluid, and what are the precise circumstances under which it is evolved, and what is the manner in which it exerts its baneful influence—still the main and most important points admit of no question. Thus, and in general, all persons, from the most ignorant to the most learned, agree that there is *something* which rises into the atmosphere, in hot weather, from marshy ground and stagnant waters, which tends to produce the common autumnal fevers in those who are much exposed to breathing the air contaminated by this admixture.

Though I speak of malaria as an aeriform fluid, or gaseous product, it is not designed to found my argument upon the truth of that opinion. Though, for convenience, as well as because inclining to the belief, malaria is here spoken of as a material aeriform product; yet, it may be also used as a term to designate the particular *condition of circumstances* produced by certain causes, which condition operates to produce and strengthen autumnal diseases. Still less do I mean to maintain that malaria, even if material, is of any one kind of gas, or any particular combination of several kinds. Besides these, there are many other common points on which the learned investigators of malaria totally disagree; and so much does each one insist upon deducing general principles from his own particular facts, (or supposed facts,) and so slightly and incorrectly have such facts been observed, that the general reader becomes lost in the contradictory positions of different instructers. Thus, judging from particular and isolated observations, with some writers, there is no condition of circumstances, which will not sometimes, in a warm climate produce malaria; and with others, upon equally partial and imperfect observation of other facts, the production is denied to be usually caused by any of the circumstances which are generally deemed the most certain and fruitful sources. One writer, perhaps, has known an exemption from disease in those who lived close to a stinking marsh, or a stagnant pond; and hence he denies that these are sources of malaria, and accordingly searches for them in other circumstances. Another has known the effects of malaria on troops encamped in a high defile in the mountains of Spain, where the soil was dry and stony, and no water except rapidly flowing rivulets, and the place some miles distant from the nearest marsh or lake. Hence he concludes, that even such a locality as this, in certain (unknown) circumstances, throws out abundance of malaria. Considering the circumstances under which most of the works on malaria have been written, it would be strange if they were not quite contradictory. The authors of most of them were army-surgeons and physicians, who observed the effects of malaria, in some deadly region, upon soldiers not at all acclimated. Perhaps the author was confined to a garrison, or at least limited in his observations to the line of march of an invading army; and in a country to which he was totally a stranger, and among a hostile people, whose opinions he could not learn, and whose language he probably did not understand. If a physician of Lord Cornwallis' army, who had merely accompanied his march through Virginia, and been cooped up in Yorktown during the siege, had written a treatise on the diseases of the country, he would have been better prepared to treat of them than most of those who have essayed such tasks; and he probably would have considered as a regular disease of the country the fatal "jail-fever," which swept off in numbers the absconding slaves who had joined the British army, and were crowded together in Yorktown, until the surrender, and which form of disease has never been known there, before, or since.

All agree that decaying and putrefying vegetable matter is one of the greatest, if not the only source of malaria. Of course, then, in addition to the sufficient abundance of the material, the circumstances most conducive to its putrefaction, must be the most favorable to the production of malaria.

The presence of moisture, a certain degree of heat, and the access of air, are circumstances *essential* to fermentation, and of course to the production of malaria; and neither can take place without the aid of all three of these things. Much moisture would be less favorable than a less quantity; and entire covering by water would, by excluding air, prevent fermentation, and its consequence the formation and escape of malaria.

It is also one of the few settled points, among scientific investigators, that malaria is very light, at least when warmed by the sun; and hence the fact known to many in this country, that those who live on the borders of marshes, and of mill-ponds, sometimes escape all injury from their exhalations, when others, who live on high hills, and at much greater distances from the sources, suffer greatly by the disease produced. Facts of this kind are numerous, and of regular annual occurrence, in Gloucester county. The whole of the wide and very level low-grounds furnish residences very healthy, compared to the tide-water region in general; though intersected in every direction by tide-waters, and though there still remains much swamp land unreclaimed, such as the whole body of low-ground was when in a state of nature. But the high, dry and hilly land, which forms the ridge of the county, is less healthy; and the highly elevated and beautiful sites of mansion-houses overlooking the low-grounds, are universally sickly in autumn.

If all the facts in regard to the action of malaria, were as regular and uniform as this one, just stated, as in Gloucester, there would be far less doubt on the subject. It is the uniform character of the country, in its high-land, low-ground, and also the water, and the long extent of each, which causes these effects to be so uniform there. Owing to causes stated in the description of the low-grounds of Gloucester (page 178,)[4] there is but little malaria evolved there; and if that, as supposed, rises by its greater levity, the regular daily sea-breeze must cause it to float towards the high-lands; and the long and regular line of ridge cannot fail to receive it, and in no very different proportions. But in most other situations, even though malaria should be produced in great quantity and with direful effects, yet these effects are so extremely irregular, in the places, the times, and the intensity of their operation, that they cannot be certainly traced to their true source; and therefore, that source may remain scarcely suspected, while it is dealing out death somewhere in almost every season. Away from the vicinity of the sea, nothing can be more irregular than the winds; yet, supposing a mill-pond to produce a regular and large supply of malaria every autumn, (though that supply is itself extremely irregular,) it depends upon the direction, force, and continuance of every change of wind, whether and where, and to what extent, the malaria will produce disease. It is therefore not at all strange, nor opposed, as is thought by some, to the regular annual production of malaria or causes of sickness, by each mill-pond, that the visitations of sickness, at any one place, should be very irregular, and the difference be often totally inexplicable from any known causes, or variation of circumstances.

According to the views presented, there must be more or less malaria (or the gaseous products which, under certain conditions, form malaria,) evolved in every country where there is much vegetable matter to ferment, and sufficient warmth of climate to carry on fermentation. But, in the small quantity which is unavoidably extricated in every such temperate and fertile country, these products seem to be harmless. Perhaps a small quantity is absorbed as food by growing plants, and this aids the production of the earth. If so, this beneficial operation is made easy by another quality of malaria, which is well established as true. This is, that though it is so expanded by the sun as to rise above the lower air, still it remains on the surface of the earth in the night, after being extricated, or perhaps descends again from above, when condensed by the cold night-air, and of course lies in contact, through the night, with growing plants. Hence it is, that sleeping on the ground, or in the lowest apartments, and being exposed to the night-air, invites the attacks, and increases the virulence of malaria; and hence also it is, that the keeping of fires at night, even in warm weather, has been found highly useful to health, in places much subject to autumnal fevers.

Though it may then be theoretically true that every good soil, in every agreeable climate, is throwing out malaria to a certain extent, it is only large quantities that are hurtful; and in practice, we have only, if possible, to avoid the formation of the hurtful excess of the products of fermentation. If, in lower Virginia, we can guard against the existing and increasing excess of malaria, our situation would be one of the healthiest in the world. For while we are comparatively free from the many and fatal disorders of the lungs to which the inhabitants of northern, and what are usually and improperly called *healthy* countries, are peculiarly subject, we have no source of disease peculiar to our location, save this one, which I fully believe, it is within our power to guard against.

Putrefying animal matter, alone, however offensive in scent, is supposed not to produce malaria. It cannot be doubted but that decomposing vegetable matter is its source, because there is no production of it where there is no such material. Still, vegetable matter, alone, or even when mixed with some putrescent animal matter, does not seem generally to produce malaria in great quantity, or with manifestly injurious effects on health. Thus, the gradual fermentation and rotting of the litter in cattle-yards, when left to stand through summer and autumn—or when the same was heaped and so left, (as was formerly the general practice in lower Virginia on all farms where manure was an object of care—) never was known to be certainly and highly injurious to the residents on the farm. Doubtless, malaria, and to an injurious extent, was always thus produced; but I have never known a sensible difference in regard to health, in years when either of the practices above-named were pursued, and when the material was carried out and applied to the fields in the spring, before fermenting. Yet, if judged by the test of some of the causes and effects as described and reasoned from by writers on malaria, on well-filled yard of litter, rotting through summer, ought to have produced enough malaria to kill half the inhabitants of the farm; and effects, in general, which would have been so disastrous, and so sure, as to leave no doubt of the cause of the evils, and of the absolute necessity of preventing the recurrence in future.

But the putrefaction of vegetable matter, mixed with other things, as earth and water, and under peculiar circumstances, (though neither the precise admixture nor the circumstances are known,) produces disease to such extent, that there is no doubting or mistaking the connexion of causes and effects. Such sure and abundant sources of malaria are the following materials. 1st, The putrid and stinking water of stagnant ponds, partially dried by the heat of summer. 2d, The mud bottoms of such ponds, or of streams reduced by drought, rich in decomposed vegetable matter, and left bare of water only in summer. 3d, Fresh-water marshes, of vegetable soil, frequently, but not regularly, covered by the tides. 4th, Fresh-water marshes, laid dry by embankments, and thereby permitted to rot away rapidly. 5th, The meeting of salt and fresh waters on land full of vegetable matter. Of these several and most important sources of malaria, I deem the third (fresh-water marshes in their natural state) to be the least hurtful; and that the sources numbered 1st, 2d, 4th, and 5th, increase in virulence in the order in which they are named. The greater evils produced by the last are universally admitted, but still by an erroneous deduction from the premises. The belt of the tide-water region of Virginia, in which the fresh water flowing down the rivers mingles with the refluent salt water from the ocean, is well known to be more subject to autumnal diseases, than any other extensive space in the country. The breadth of this belt varies much in different seasons. The parts of the rivers in which the fresh and salt waters meet, and where each alternately has possession, as the tide ebbs or flows, may be but a few miles wide, and even that space is not stationary. But if the limits of this belt be fixed by the highest points to which the rivers have been known to be brackish, in driest summers, and by the lowest points where they are fresh in winter, then this belt may be considered for the time as 40 or 50 miles wide, and, in length, stretching across all the tide-waters of the state. But in the much narrower space where this mingling of the salt and fresh waters usually takes place during the heat of summer, malaria acts with most intensity. Hence the general opinion, that it is simply the meeting and mingling of the fresh and salt waters which cause disease. This is not so, or but in a very slight degree. It is either the passage of fresh-water over salt-water marshes, or of salt-water over fresh-water marshes, that causes the great production of malaria, and disease. This is an important distinction, and the truth or error of the position deserves the most careful investigation. If the mere mingling of the waters were the cause of sickliness, any relief for this part of the evil would be hopeless, as the waters *must* meet and mix together, *somewhere*. But if it be as I suppose, the evil may be greatly restrained by works of art, or by simply preventing the unnatural accumulation of vast reservoirs of fresh water in mill-ponds, which when discharged, by breaches in the dams, or by opening the flood-gates, overflow salt-marshes, which the natural or unobstructed stream never could have covered.

Salt-water marshes, not touched by fresh-water streams, are not unhealthy to any considerable extent. This is susceptible of proof by innumerable examples in Virginia on the borders of the ocean, or of the waters of the Chesapeake bay. It is rare, however, to find a large salt-marsh attached to extensive high-land, which is not reached by some small stream; and every salt-marsh of course must sometimes be well washed and freshened by the heaviest falls of rain. Therefore all must, slightly and at some times, be prejudicial to health. These, however, are exceptions of but small practical or sensible operation.

The view here taken of the manner in which malaria is produced most certainly, and acts most injuriously, though not sustained by any known authority in this country, nor by any other precisely as stated here, is not therefore presented as original. I derived it, and thence deduced my application to this country in a modified form, from the interesting report on the malaria of Italy by Gaetano Georgini [*sic*], of which the substance was published in two different papers in the Farmers' Register, (p. 502 of Vol. IV, and 460 of Vol. V.)[5] In this report the author shows by the most conclusive argument and facts, that the irregular irruptions of sea-water over tracts of marshes, or other low-grounds, of fresh-water alluvial formation, caused the long continued and worst effects of malaria; and that by simply guarding against the entrance of sea-water, the country was restored permanently to healthiness. He says nothing of the reverse operation, the irregular floodings, with fresh-water, of salt-marshes. But what is produced by the one, can scarcely fail to be as well produced in the other case. The *mode* in which the effect is produced is not attempted to be explained by the learned author quoted above; nor does any explanation seem sufficient to my mind. The rapid and abundant production of malaria may perhaps be aided, if not entirely caused, by the luxuriant cover of fresh-water plants, in the one case, being partly killed, and made ready for putrefaction, by being covered by salt water; and in the other case, in this country, by a like injurious operation on the plants peculiar to salt marshes, produced by the overflowing of fresh water. We know that certain plants flourish best in salt and wet soil, as others do in wet soil entirely free from salt; and that respectively with these different growths, the salt and the fresh marshes are heavily covered. It must follow from a sudden change in the condition, from salt to fresh, or the reverse, that the health of the entire growth must be greatly injured, and much of it subjected to death and decay.

The next most fertile source of malaria, (or perhaps what is even of greater malignity, for the small space occupied,) is presented in what is entirely the work of man—the miscalled *improvements* made by embanking and partially or entirely drying tide-marshes. The soils of these marshes, as I have ascertained by careful analyses, are composed, for about half their weight of vegetable matter, and probably nine-tenths of their bulk is of that material, destructible by decomposition, when circumstances are favorable to that result; and drainage and cultivation produce precisely the condition which is most favorable. When covered twice every day by flood-tide, a marsh soil of this kind, though composed of the most putrescent materials, is but little subject to decomposition; because being always thoroughly water-soaked, even when not entirely covered, and by water continually changed, the air is too much excluded, and the wetness is too much in excess, to favor the progress of decomposition. When the marsh rises so high as not to be covered by the daily or frequent tides, then decomposition is more favored by the drier state of the surface, and to a greater extent, malaria is evolved, and health injured. Hence the inference, that the higher and the drier the marsh, the more it is injurious to health. But as soon as such a vegetable and putrescent soil is made nearly dry, and still more when cultivated and exposed to be penetrated by the air, decomposition proceeds under the most favorable circumstances. The soil sinks annually and rapidly, not so much by drying (as commonly supposed,) as by actually rotting away, and in a few years, it is reduced to so low a level, as again necessarily to pass under the dominion and shelter of the water. The more complete the drainage, and the more perfect the management as arable or tilled land, the more rapidly is that end reached. In the progress to this end, a layer of the whole soil, of from one to three feet in thickness, will have passed off into the air in the gaseous products of putrefaction, of which enormous products, a large proportion will be malaria, and the effects produced by it on the health of some of the neighboring population are generally so evident as to leave no doubt of the source of the evil. More full details on the effects of embankments of tide-marshes are to be found in previous articles in this work.[6]

The production of malaria by the last named operation, the embanking of marshes, however, is necessarily of very limited extent—and moreover of very limited duration. Nature soon asserts and enforces her rights; and the hopes of the improver, and the land so improved, are together overwhelmed by the reinstatement of the waters, and this source of disease is thereby cut off.

Tide-marshes, however extensive and injurious in their operation on health, still are limited to a comparatively small proportion of our broad territory. But there is another source which spreads disease over half the state, and which is entirely of artificial formation, and of which the evil effects have been becoming more and more extensive, and more and more virulent, from the early settlements of the country to this time. This widespread and generally operating source of disease and death is furnished by the numerous mill-ponds, of variable height of surface, which are now scattered over the whole face of eastern Virginia, and of which every individual case adds something to the general and enormous amount of injury to health and to life.

The law of Virginia in regard to the erection of mill-ponds, with perhaps the exception of the fence-law, is one of the most stupid, and most regardless both of private rights and general interests, of all in our code; and it is far more objectionable than the former, inasmuch as while the one merely robs private and destroys public wealth to an enormous amount, the mill-law permits and encourages also the destruction of health and of life throughout the whole land. It is true, unfortunately, that this opinion is not entertained by but few persons: and that even with those who admit that all such mill-ponds are injurious to some extent, their estimate of the amount of evil is much below mine. It is my object, to awaken the community to a sense of the enormity of the evil, and thereby to induce the commencement of measures of remedy and prevention. The universal acquiescence in this policy of our country, and the almost universal ignorance of the evils which it produces, requires strong language to enforce novel views in opposition to long established opinions. But it is confidently believed that my denunciations will be justified by reason and by facts, and by the magnitude of the existing evils.

There has long prevailed in Virginia a mania for building water-mills, which was not restrained by insufficient regular supplies of water to fill the ponds, nor by the insufficient prospect of business and of profit, even if there were no failure of water. In consequence, there have been not only erected mills on every stream barely sufficient to keep a common corn-mill in operation, but also on as many others where the water-power was either insufficient, or totally failed, during the driest season of every year. In the tide-water region, the mills for grinding wheat-flour, or any thing else for sale abroad, are limited to the falls of the large rivers. All the others, (and probably there is on average one for every square of five miles,) are merely designed to grind for toll the corn used for bread in the immediate neighborhood; and, considered merely in regard to money-cost and profit, it is most likely that half the mills in the country do not get enough toll-corn to pay for more than the costs of maintenance and repairs of their establishment. The more worthless the mill, on account of the insufficient supply of water, the more productive it necessarily is of malaria, diseases, and death. It will be difficult for me to make those who are unacquainted with our country believe that hundreds of mills have been built, and that most of them are still kept up, and many more new ones will probably yet be added to the number, which cannot yield any clear profit, above the entire cost, to the owners, independent of cost in property to the neighbors, and the cost (whatever that may be) of health and life to the country at large. Still the fact is notorious throughout lower Virginia, if it does not extend through the higher middle country. The only reason that I can conceive for so many unprofitable investments of this kind, is, that many residents of the country build mills, as many others raise race-horses, more for amusement and excitement, and to vary the monotony of their lives, than for profit. But this propensity of individuals could not have done much mischief to the country at large, but for the encouragement offered by the government. According to the law, and the long-established usage under the law, any man who desires to erect a mill, and for which it is necessary to pond the water on some of his neighbors' land, has nothing to do but to apply for an order of the county-court, by which the sheriff summons a jury to meet on the spot, to judge of, and assess the damages that will be sustained by the owners of the lands designed to be covered by the pond. The jury is generally composed principally of men as ignorant and unfit for such investigations and estimates as the neighborhood can furnish—and they decide by guess as to how much land will be covered, and what damage will be sustained in the loss of the use of the land. There is no question entertained as to whether a mill is at all required by the demand of the neighbors for meal; and if the question of the effect on health is even named, it is addressed to a body entirely unacquainted with, and regardless of the whole subject. In fact the question as to health has rarely been considered in any such cases; and never duly considered. If the land that will be covered by a pond, though very rich, is then in the state of swamp, and totally unproductive, such an uninformed jury, as the case is usually submitted to, will be very ready to decide that such land is worth nothing; and if $3 an acre is given as damages, for the land actually to be covered by the pond, it will be deemed a liberal allowance. The court will rarely refuse to sustain the verdict of the jury.

Though the use of the land thus covered is forever taken from the owner, or, for as long as the mill-owner may choose to keep up his pond, still the right of property is not changed. This small reservation of right, or feeble homage to justice, serves as a still further injury to the community, and is not of the least value to those to whom the right is reserved. It would be far better for all parties, if, when land was thus condemned to be covered by a mill-pond, that the damages assessed, however low and pitiful compared to the damages actually sustained, should have been deemed the purchase-money of the land, and the absolute right of property vested in the mill-owner. If this were the case now, there are many mill-ponds in Virginia which would be forthwith laid dry, even though the mills should necessarily go down; because the land covered by the ponds is now known to be worth more for cultivation than the mill is for toll. Hundreds of other mills, of greater profit and value, also, in that case, would be better supplied with water by canals than by their present ponds, by which their value as mills would be increased, to the owners and to the public, and the nuisances of the ponds be equally abated. But as the law now stands, if a mill, which will not bring in a net rent $50 a year, covers by its pond 500 acres of rich land belonging to other persons, the mill-owner has no interest whatever in draining the pond, because its drained bottom would belong to other persons. In any case approaching to this, and in which there would be a gain to all the individuals concerned, by draining the pond, still it is not done, and the nuisance continues long after it is well known to be such, because there is a contest between the several owners of the pond and of the land covered by it, in regard to their respective shares of profit to be gained by emptying the pond. Many such cases still exist in Virginia; although many of the most unprofitable ponds, from proper views of economy, have been drained, and either substituted by cheaper and more efficient canals, or the mills put down entirely. An old mill-pond in Dinwiddie county, which covered 1200 acres of land, has been drawn off, and thereby an indifferent mill exchanged for a large fertile farm. This would not have been done, even if the mill was worthless, but for the ownership of the mill, and the land covered by the pond, falling into the same hands. There is a mill-pond now kept up in Prince George county, which is supposed to cover nearly 400 acres of land; and there are many others not much smaller, on different branches of swamps in lower Virginia. The larger the pond, in general, the greater proportion of bottom is left dry in autumn, and the more disease is therefore produced; and though the draining of such large ponds would be so much the more an object of gain, there is the less chance for its being done, because of the separate ownerships and interests.

Almost all the mills throughout the lower part of Virginia, and also a large proportion of those in the more hilly middle country, are worked by streams which are inadequate to the daily supply of the mill, and evaporation from the pond, even if the grinding is not necessarily suspended or diminished at any time. To guard against the temporary failure in dry weather, the full "head" of the pond, (or the level of water for which damages were assessed, and to which the water may lawfully be raised,) is much higher than the lowest level that will work the mill. The land covered is also usually very nearly level, so that to raise the water 10 or 15 feet at the dam, will often back the water from one to two miles up the low-grounds. If the variation between a full head of water, and the lowest level, be 5 feet perpendicular, it will often cause the uncovering of many acres of the bottom of the pond to the hot sun, and thereby furnish a most fruitful source of malaria in every such case. Rich alluvial mud, as this always is, thus exposed, in hot weather cannot be otherwise than very injurious to health; and there is not a pond-mill in Virginia, with a variable head, which has not more or less of the pond every summer thus converted to a fruitful seed-bed and nursery of disease.

Besides this, there is the not rare occurrence of the pond being entirely drawn off in summer, by the breaking of the dam, and suffered so to remain for weeks or months, before being again repaired and filled. In this case, a double quantity of bottom is exposed to putrefaction, and fitted for the discharge of unhealthy miasma.

At all times, in ponds supplied by streams as feeble as most of those used for mills in Virginia, the water approaches to a stagnant state; and therefore, of itself is a producer of malaria. In dry seasons, when unusually low, the putridity of the water of such ponds is perceptible to the sense of smell; and it must be then far from harmless.

Another, and in certain situations, the greatest evil of mill-ponds, remains to be stated. The others above-mentioned are the effects of the scarcity of the supply of water; this is from the excess, which is found in all streams, at some times, even though the most deficient at others.

To guard as much as possible against the expected scarcity of water, the mill-owner aims to hold, when rains increase the usual supply, as "full a head" as he has a right to maintain. When this supply is exceeded, as it frequently is, and greatly, if the dam is not actually broken, and the whole emptied, in one prodigious flood, at least the flood-gates are opened widely, and a discharge made ten-fold greater than would have occurred, during equal time, if the stream had not been obstructed by a dam, and had discharged as regularly as the supply was increased. It will be evident, on considering these circumstances, that water from a mill-pond, whether discharged by flood-gates, or otherwise, must be far more variable in height, and in extent of inundation on the land below, than the natural stream, unobstructed by art; and still more than the stream opened and improved, and its course facilitated by art. An ordinary natural stream, which might have a very uniform discharge in dry weather, and would rarely overflow its banks in wet, if dammed across for a mill, would often have its bed, below the dam, left almost dry; and at rare and irregular times, would be converted to a tremendous flood, which would sweep over many hundreds of acres more than the floods of the natural steam could have reached. Besides the immense damage caused to cultivated land by these floods, (and which kind of damage is rarely estimated or thought of by juries, when mills are established above,) there are numerous hollows made, and filled with water, which, on the retreat of the flood, (as hasty as its inroad,) remain so many stagnant pools, until made dry by evaporation. The whole land, thus covered, is saturated with water; and, from the nature of the rich alluvial soil, is throughout, as it dries, made a producer of malaria.

But the worst part of this evil, by far, is when these artificial floods of fresh-water pass over salt-marshes—which happens in all the country in which the fresh and salt waters meet; and this combination of causes I consider the most efficient producer of disease in that part of the country, and the thing which ought most especially to be guarded against. According to the views before presented, the passage of fresh water over salt-marshes, no matter to what extent, is one of the most sure producers of malaria, and of a particularly malignant kind. The mill-ponds, alone, form other and far more extensive, if weaker sources of the poison; and by the union of the two, the mill-ponds exert all their usual bad influence above the dams, and spread ten-fold more pestilential effects below, by inundating the wide salt-marshes, which by natural streams would scarcely have been affected.

On Nansemond river there are lands already rich, and having inexhaustible supplies of the best marl, which have been sold at $10 the acre. There are hundreds of estates in the same belt of country, which cannot be sold for as much as the cost and present value of the buildings. And this otherwise fine country, so accursed by disease, owes its condition principally to the streams which flow into the salt tide-waters, and which are so numerous, and their sources so interlocked, that there is no spot safe, by remoteness of position, from these combined effects of mill-ponds and salt-marshes. It is therefore sufficiently evident why that otherwise finest part of the state, for agricultural improvement and profit, should stand among the lowest in both these respects. Yet this part of Virginia might be rendered both healthful and fruitful, and the delightful region which God has permitted it to be made, if man would accept and avail of his bounties, by merely using half the expense for improving, which has been lavished to inflict pestilence and poverty on the country.

These statements and expressions of opinion will be unpalatable, if not offensive; and perhaps may subject the writer to the charge of being willing to injure the residents of the region for whose relief in this respect he is most anxious, and of the facility and cheapness of obtaining confident. If the exposure and probing of the ulcer be never so painful, let it be remembered that it is done solely for the purpose of seeking for, and applying, a sure remedy.

There is still another source of malaria, which it is necessary to touch on in connexion with the above-mentioned, though it has been already treated more fully elsewhere, and therefore will be slightly mentioned here.[7]

From the vegetable matter upon the driest land, as it ferments and decays, there must be extricated more or less of the gaseous matter, which, when in excess, is injurious to health. According to this view, the whole surface of the country, and especially that most heavily covered with vegetable matter, may furnish malaria. The degree of hurtfulness of this product will depend on the power of growing vegetables to feed on, and of the soil to absorb and fix in it, this matter, which, according to its direction and quantity, may either enrich land, feed plants, or poison men. In earlier publications I have stated at large my reasons for believing that all the products of vegetable decomposition, on naturally poor lands, are lost to the land; and as the ultimate results of decomposition are gaseous, or aeriform, they must go off into the air. These products constitute or cause malaria, and its injurious effects on the health of the inhabitants. But calcareous matter serves effectually to fix there the enriching principles of decaying vegetable matter, until they become the food of growing plants. Hence the deduction that a naturally poor soil, made calcareous, will no longer throw off gaseous products, or malaria, into the air; but will store it up as fertilizing manure. The sure remedy for the irregular and generally slight degree of sickliness thus caused, is, to marl or lime all the land that requires calcareous earth. But that remedy would not be sufficient, if mill-ponds or marshes in the neighborhood continued to send out large additional supplies of the aeriform poison.[8]

The correctness of my deductions as to the very injurious effects of mill-ponds on health, will be denied on several grounds, which, so far as expected, I will anticipate as objections, and state with answers, as follows:

Objection 1. Admitting generally, and to some extent, the ill effects of mill-ponds in producing noxious exhalations, and autumnal diseases, it does not appear, that these effects can be either so great, or so sure, as is charged above. The residents on the farms nearest to mill-ponds are not always, and often not at all, more sickly than those who reside several miles distant. The house of the slave who acts as miller, is usually near the mill, and close to the pond; yet families so situated are generally as healthy as any others, and sometimes are healthy in a remarkable degree, compared to the neighborhood generally.

Answer. Near the mill-dam, or the lower end of the pond, may well be less affected by the exhalations from it, than places a mile or two more distant. That part is the deepest of the pond, and of which also the banks are steepest; and perhaps half a mile in length of the bottom of the upper and shallowest part of the pond, and of alluvial mud, might be left naked in drought, before a margin of steep hill-side, of three feet width, could be exposed near the mill. Further—from the greater lightness of the malaria, it will rise high in the air, and would soon be carried far away by a moderate breeze. If the wind be moderate, and steady to one direction, and still more if its course be confined to an opening by or between woods, or to a narrow valley between high hills, it may well be imagined that the poisonous air might injuriously affect persons perhaps five miles from the pond, and who would not suspect the operation of so distant a source; while others, close to its border, but in a different direction, or on a different level, might escape its influence.

Objection 2.—There is not enough difference in the usual or average healthiness of families the most exposed, and others the least exposed to mill-ponds, to attribute much of the effects to these causes. Whole neighborhoods, in some autumns, are very healthy, and in others very sickly, without either condition seeming to be connected with any certain and known state of the nearest mill-ponds.

Answer.—The extreme lightness of the poisonous air, and great and frequent variations in the direction, force, and continuance of the winds on which it is borne, make it generally impossible for it to be known from which particular pond or ponds the malaria rises, or where it is carried. It is most probable that the exhalations of twenty ponds, of which the most remote may be thirty miles apart, may be mingled together, even by the winds of a single day, and thus combine and average the effects of all. Further—if all the mill-ponds of a country furnish one half of the active and injurious malaria, and the other half is thrown off, nearly equally, by the whole surface of the land, (though some parts would receive the strongest doses, and others escape with having only the weakest,) it would be impossible to understand the mode, and estimate the intensity, of operation of the known general causes; or to refer, with certainty, any one effect to its special or principal cause. Thus, a farm, relieved from all malaria of its own product, by marling and by drying its mill-pond, though evidently showing the benefit in increased general healthiness, might still be sorely visited by the seeds of disease from other and remote sources, directed and concentrated by a steady wind.

Having presented these views of the origin, action and effects of malaria in this country, I can better exhibit the progress of the causes which I believe to have operated, and which are still continuing to operate, to produce the change from a healthy, to an unhealthy state.

When our ancestors first reached this shore, nearly the whole country was in a state of nature. The savages had cleared for cultivation but a few fertile spots on the banks of the rivers; all the rest of the land was under one great forest. The streams had not been obstructed by the cutting down of trees across their beds, (by which, in many cases, streams have since been choked, and swamps thereby formed, or greatly extended.) No dams had obstructed the free and regular course of the streams, and therefore no great artificial floods were formed. The soil not having been cultivated, was not exposed to be washed away by the rains into the rivers. The waters therefore were generally clear, instead of being generally muddy, as since all these circumstances have been changed. In this former state of things there could have been existing but few sources of malaria. The first sources formed by the civilized settlers, was in making ponds to supply water-mills. But while these were yet few in number, the constructers of course chose the best and most unfailing streams; and the ponds were also, for a long time, surrounded by dense and tall forests. Such hilly land as the margins of the ponds would certainly not be brought into cultivation, while so much that was far better, and easier to till, remained unoccupied. Hence, such ponds produced but little malaria, and that little was warded off from the settlers, or taken up, by the forest growth. The general wooded state of the country, also, for a long time, rendered the supplies of water more regular, and prevented the severe droughts, which would have altered greatly, as is usual now, the levels of the ponds.

The clearing, cultivation, and consequent washing of the lands of the upper country, greatly increased the muddiness, and quantity of alluvial deposite of the rivers, and thereby increased the marshes both in breadth and in height. More mills continued to be built, and on streams worse and worse for water-power, as the choice became less open, and the mill-mania began to grow; and, in the general, each successive construction of a pond was less productive of profit, and more productive of disease, than its predecessors. The number of mills not only continued to increase, and is increasing to this day, and in the oldest settled parts of this state, as well as the newest, but gradual changes also took place in the condition of the old mills which greatly increased their fitness to produce disease. By the long continued deposite of mud from the streams, and the washing of the now cleared and tilled hill-sides, the ponds became more shallow, and the waste of water by evaporation therefore became greater; while the supply was lessened, in consequence of the extended clearings of the great forest which had before covered the whole country. To remedy the increasing deficiency of water, the owners of old mills, who were not prohibited by circumstances, raised the level of their contents, still more increased the daily evaporation, and also the violence of floods, and the variable height and surface of the water; all of which again combined to increase, still more than before, the product of malaria. The consideration of the progress of all these circumstances, and their bearing on each other, will serve to explain why a particular neighborhood might formerly have been healthy, though having two or three mill-ponds within or around it, and why it might gradually have become very unhealthy, in the course of time, by the malaria produced by the ponds of the same mills, or perhaps by the addition of one or more new pond, to the former number. But, in such cases, so gradual would be the general change, and so irregular and variable the attacks and virulence of the autumnal diseases, that the sufferers would not attribute the change, (even if they admitted it to have taken place,) in their average degree of health, to causes which had so long existed, without being charged with doing mischief; and in which causes, no change of condition had been observed. Add to this, that self-love makes every man reluctant to believe, and to confess, that his own farm, or his own neighborhood, has become more sickly; and the change for the worse is attributed to transient causes, until the former state of things is almost forgotten, and the present is received as if it had always been the usual condition of circumstances.

During all this time, other causes were working to produce other nurseries of disease, and impediments to agricultural products and improvement. The wet alluvial bottom-lands, bordering on small rivers and still smaller streams, were for a long time neglected, and deemed of little value, except for their fine white-oak, cypress, and other noble timber trees. These were cut down so as to fall into or across the streams, when in reach, more often than otherwise; and in consequence of such obstructions, continually increased in number for more than a century, the before open streams were choked, and the bordering low-grounds converted to swamps; and those which had been swampy at first, were made still more so, by obstructing the sluggish streams and spreading them over the whole surface, and causing that surface continually to rise, by fallen trees and alluvion. But wet as are such swamps for the greater part of the year, most of the surface is dry in autumn; and the scanty water is then stagnant in numerous pools, until added to by the first heavy rain, or a flood from a mill-pond discharged above. Of course all these circumstances added enormously to the previous annual decomposition of vegetable matter, and consequent production of malaria. Such swamps as these, formed by nature and increased by art, are those on the Chickahominy, Blackwater, and many other long but gentle streams. To form or increase their evil qualities and tendencies the law has given full permission, and no small aid; but it positively, though indirectly, *forbids the drainage* of all such extensive swamps, and preserves them still as mere nurseries of disease. A general law for permitting and facilitating, under proper regulations, the draining of these great swamps, would be a measure which would be most beneficial, not only for improving the healthiness, but for increasing the agricultural products of the country.

But though the tendency of the general changes in the physical condition of the country was to increase the causes of autumnal diseases, there were numerous particular exceptions, in works serving to promote health. Of this kind were the opening and straightening of the choked channels of small rivers, and many large streams, in the hilly country, where there was enough descent to enable each individual proprietor of flooded low-ground, to relieve it by operations confined to his own land. The effectual drainage of much land of this kind has produced so much benefit to health, as, in many cases, to balance, and even exceed, the increasing pestiferous effects of the neighboring mill-ponds. Such facts would be taken, by most persons, as proofs that the increase of mill-ponds had not increased disease.

Such benefits have been produced by the gradual draining of the extensive low-ground of Gloucester, which in its former and natural swampy state, must, necessarily, have been an abundant source of malaria. This change, together with other circumstances stated in the recent description of that part of the country, has operated to render Gloucester as free from bilious disorders as any part of the tide-water region—save the adjoining county of Matthews.[9] The remarkable general state of healthiness of all these very low lands, at present, as well as the exceptions, and evident causes of the exceptions, furnish the most clear and important evidence of the truth of the position, that mill-ponds, and floods of fresh-water discharged over salt-marshes, are the great sources of malaria in Virginia. As stated formerly,[10] there are but few fresh-water streams discharged on salt-marshes in these two counties, and not a pond-mill on the low-grounds, nor indeed in the whole county of Matthews, save one on its border nearest the highland. The facts presented here, alone, will prove the great and certain benefit to be obtained by even a partial and imperfect avoidance of the action, separate and combined, of these two great sources of malaria.

The most important part of this subject is the consideration of the remedies for the evils described. But although the means available for this end, in my opinion, are ready, cheap, and sure, still it is needless at present to argue in their favor at great length. Unless the people are aroused to a proper sense of the evils under which the country suffers, no regard will be paid to the consideration of proper remedies; and if the former object can be gained, the latter will then necessarily follow.

The most important of these remedies, and of which the proper use, I maintain, will remove nearly all the existing sources of malaria, and make lower and middle Virginia in general as healthy as any region of the earth—will be merely here stated concisely but distinctly.

1st. To prevent the continuance of any mill-ponds of very uncertain supply, and variable "head," or height of water.

2d. To furnish to the land-floods, of streams swollen by rains, or by any mill-ponds still left, the quickest and best possible discharge to tide-water by open canals, so as to prevent the fresh-waters passing over any salt-marshes.

3d. To drain the great flat swamps; all of which require a continued canal to be extended from the lowest out-let, up to the head of the supply of water, in the most effective course, and on a general plan, through the lands of many different proprietors. The drainage of lands, so situated, is effectually forbidden by the existing laws; as there is no power to act, unless all the proprietors concur in every particular of the execution and expense of the drainage; which is obviously impossible.

4th. To refrain from embanking from the tide any marshes of the usual putrescent and perishable soil.

5th. To apply marl or lime to all lands needing calcareous manures, and on which they could be furnished at not too great cost for even such great improvement of soil and product, as would certainly be obtained in such cases.

The two last means of prevention are altogether within the province of individuals, and will be used, or not, according to the views of different individuals, as to the agricultural profit to be expected from such operations.

The three first-named means of remedy would each require the action of the legislature, to enable them to be used to any extent.

The necessity for a general plan being authorized by law for inducing and compelling combined operations to drain swamps on long and sluggish streams, though merely for agricultural improvement and profit, is already evident to most intelligent farmers; and perhaps nothing is now wanting to procure such legislation, but the proper exertion of some of the individuals who are most interested on the subject.[11]

The giving free vent to land-floods, also, by wide and straight canals, and preventing them, by dikes, from overflowing the salt-marshes, though a kind of work requiring public money, as well as legal authority, still may be hoped for, when the necessity of the measure shall have been made evident.

But there is no such prospect of success as to the most important reform needed, in the putting down of all fever-breeding mill-ponds; and he who will venture to advocate this general measure, will be regarded by most of those whom he aims to serve, as more an enemy than friend to their interests, and more deserving to be treated as a lunatic, than to be respected as a judicious advocate for valuable public improvements. It is not in the vain hope of now enforcing my views by extended argument, but to offer explanations, and thereby prevent misconstruction, on some particular points, that some further remarks will now be offered.

Even if the public mind had been prepared for a full legal reformation of the police of mill-ponds, and for the laying dry all such as are nuisances to health, there would be no accompanying necessity for injuring the private interests of mill-owners, nor of causing material loss or inconvenience to the customers of the mills. In the first place, in justice to the vested rights of the millers, (however unjust to others, and injurious to the public may have been the original creation of their rights,) I would advocate full compensation being made for every sacrifice of value in their ponds, which should be required and compelled for the general benefit. But *not more* than full compensation for all value thus destroyed should be granted; and many of the fever-breeding ponds have really no pecuniary value to their owners or to the public; and most others may, to greater advantage, be supplied with water by canals, instead of by ponds. Even if one-third of all the mills should be thus put down entirely, these would be such as now always fail in dry seasons; and the more permanent and regular supplies of water, which all the remaining mills would receive from the canals substituted for ponds, would render them able to furnish the whole country with meal, with regularity, certainty, and in abundance, and therefore more suitably and conveniently to the consumers, than all the mills, good and bad, now in operation. By an important innovation in the law in regard to mills, (enacted March 2d, 1826,) every owner of a mill is authorized to cut a canal through the lands of other persons, if required by the nature of the locality, so as to substitute the pond by a canal. Before this amendment of the old law, no mill-owner could effect any such improvement, unless in the rare case of his own land extending under the whole course of the desired canal. The privileges offered by this new provision have already been availed of in many cases, in Charlotte, and the neighboring counties, and to great advantage in regard to health as well as to increased power to the mills, and great value gained in the rich drained bottoms of the ponds being put under cultivation. Slow as such lessons are usually learned, and slow as new agricultural improvements are brought into extended use, this highly beneficial and profitable improvement, cannot fail to be adopted generally in the course of time.[12] The main obstacle to the early and general substitution of canals for ponds, wherever the change is practicable, is the absurd legal distribution of rights in the mill-ponds and the land which they cover, as stated on a preceding page (223); one person being vested with the perpetual right to keep the land overflowed and worthless, while others have the right of property in that land, to be exercised only in the never-expected event of the owner of the pond drawing it off and draining the rich bottom; and that for the gain of others, more than himself. Now I would get rid of this absurd conflict of rights, by vesting the full property in the land covered, in every mill-owner who would draw off the pond; or if he did not avail of the privilege offered, the land should be given up to its former owners, or to any one else, who would construct a canal, and thereby secure to the use of the mill an equally good supply of water-power.

Each of the several remedies proposed and stated above, would alone furnish a fruitful subject for investigation and discussion. But more extended remark[s] from this source, is as yet uncalled for. Other persons, having better practical information, and thereby prepared to confirm or to disprove the positions here assumed, are invited to aid in the discussion. Let the truth be made known, on whichever side it may be found; and should all facts and deductions presented serve to show that the present system greatly needs reformation, and to awaken the public to the importance of the object, then will be the suitable and propitious time to ask attention to remedies proposed for the then acknowledged evil, inflicted by the action or permission of the government. Whenever the legislature is prepared to act decisively on this whole question, there will be before them a subject for the internal improvement of Virginia, far more important in beneficial results, than the roads and canals which have cost millions of dollars to the treasury; and yet which will be cheaper, compared to the profit to be certainly counted on, than the most humble or contemptible job yet carried through by public expenditure and as a public improvement.

But according to the existing law, any single individual who clearly sustains injury to health from any particular mill-pond, has even now the legal power to have that particular nuisance abated, by means of suit for damages for the injury thereby sustained. It has been judicially settled, that such ground of suit for damages is not prevented by any previous assessment by the first jury, nor by any lapse of time during which the mill has been standing; nor is the ground removed by the new damages awarded for injury already sustained and sued for. No matter how often damages may have been given to the plaintiff by successive verdicts, and paid by the defendant, there will continue ground to sue, and recover, as long as the pond remains, and is hurtful. It is surprising that the law, so favorable to the interests of mill-owners, and regardless of all conflicting interests and rights of other persons, should have permitted, in this particular, so much of remedy for the previous injustice and injury inflicted by the law. And it is still more surprising, that after legal decisions have so clearly shown the remedy, that of so many thousands of individuals who are unquestionably suffering every autumn from the neighborhood of stagnant mill-ponds, so few should have availed themselves of the offered means of relief.

If the importance of this general subject were duly appreciated, its investigation would become an object of the care, and be conducted at the expense of government. If the legislature of Virginia (for example) would institute a "General Board of Health," or "Commission of Sanitary Police," for the purpose of investigating the subject of malaria thoroughly, and of reporting the sources and proper remedies, the body of evidence which would be collected, and the after-results, might be made worth many millions of increased pecuniary value to the state, besides the far greater benefit to be produced to the health, the physical and moral qualities, and the general happiness of the people. At any possible cost of such an investigation, and of the system of measures founded thereon, the public improvement and benefit produced thereby would exceed the expenses an hundred-fold.

6

Public Health and Recycling
Desultory Observations on the Police of Health, in Virginia—As It Is, and as It Ought to Be (1837)

Introductory Remarks

The greater and more important part of the first of the following papers (marked No. 1 of the series,) has been already twice laid before the public. The first publication consisted principally of theoretical views, then but little sustained by known facts, and not at all by any known authority, and was presented in the first volume of the Farmers' Register (page 76) as a "Supplementary Chapter to the *Essay on Calcareous Manures*."[1] The second publication was embodied in the second edition of the *Essay*, much extended, especially in the testimony of facts—and which will here be given unchanged, except by the addition of some new passages, of argument and illustration. It may require apology for thus embracing matter that is already in the possession of many of the most inquiring and intelligent readers. But on the other hand, very many persons have not seen, and cannot readily have access to either publication—and moreover, as presenting general principles, this portion is most suitable to precede and introduce the practical applications that will follow. General reasoning and propositions, or instructions in a didactic form, are always less impressive, and less likely to be useful, than when applied to actual and well know[n] facts and circumstances. It is to supply the latter deficiency, and to endeavor to attract the attention of those most interested in this subject, the residents of towns, that the succeeding portion is now offered. Of this numerous class, few have given any attention to either of the previous publications, merely because they were *agricultural*, and therefore erroneously supposed to be of no value to mere townsmen. But though the matters treated are of high importance to agriculture, as offering rich sources of fertility to the country, they also are of not less importance to towns *directly, by preserving cleanliness* and *guarding health*, as well as *indirectly*, in fertilizing all the lands of their vicinity. If the views maintained should ever be acted on extensively, and to much public benefit, it must be by townsmen, and especially by town magistracies and councils leading the way in the work. It is therefore, that I most earnestly ask the attention especially of the residents of towns; and if attention is given, I hope to prove that they can greatly aid their own individual comforts, and pecuniary interests, by adopting a general system of economical police that will also improve agriculture, and enrich a large portion of the neighboring country.

Under the general head of the "Police of health in Virginia," I propose to treat 1st of the general action and effects of calcareous matter in preventing the wasteful and injurious decomposition of animal and vegetable matters; 2ndly, of applications of these principles to the police of towns especially; and 3rdly, of the causes of autumnal diseases in the country generally, and the means for removing them. The last proposed part may be delayed for the gathering of more numerous facts; and the whole subject will be treated in the desultory and irregular manner admissible in presenting much more investigation and discussion, before it can be given a methodical and well digested form.

No. I. The Action and Effects of Calcareous Earth in Preserving Putrescent Matters and Thereby Promoting Cleanliness and Health

This first number, as stated above, being merely a portion (chap. xix,) somewhat enlarged, of the second edition of the *Essay on Calcareous Manures*, the language is in some passages better suited to its former than its present position. The general theory of the action of calcareous earth, in producing fertilization of soil, is considered as established by the earlier and larger portion of the *Essay*, and therefore will be taken as established grounds throughout these observations, and not as a subject yet to be proved, or even discussed. It will merely be here stated generally, that the most important positions maintained in that work are these: that calcareous earth (carbonate of lime) has the property of combining with the products of all putrefying animal and vegetable matters, and thus preserving them from waste, and securing them for enriching the soil and feeding vegetation; and that other earths (sand, clay, &c.) have no such chemical action on putrescent matters, and cannot retain them, or profit fully by their value, for enriching soil, except by the addition and aid of calcareous earth.

The operation of calcareous earth in enriching barren soils, has been traced, in a former part of this essay, to the chemical power possessed by that earth of combining with putrescent matters, or with the products of their fermentation—and in that manner, preserving them from waste, for the use of the soil, and for the food of growing plants. That power was exemplified by the details of an experiment, (page 31,) in which the carcass of an animal was so acted on, and its enriching properties secured. That trial of the putrefaction of animal matter in contact with calcareous earth, was commenced with a view to results very different from those which were obtained. Darwin says that *nitrous acid* is produced in the process of fermentation, and he supposes the *nitrate of lime* to be very serviceable to vegetation.[2] As the nitrous acid is a gas, it must pass off into the air, under ordinary circumstances, as fast as it is formed, and be entirely lost. But as it is strongly attracted by lime, it was supposed that a cover of calcareous earth would arrest it, and form a new combination, which, if not precisely nitrate of lime, would at least be composed of the same elements, though in different proportions. To ascertain whether any such combination had taken place, when the manure was used, a handful of the marl was taken, which had been in immediate contact with the carcass, and thrown into a glass of hot water. After remaining half an hour, the fluid was poured off, filtered, and evaporated, and left a considerable proportion of a white soluble salt (supposed eight or ten grains). I could not ascertain it[s] kind—but it was not deliquescent, and therefore could not have been the nitrate of lime. The spot on which the carcass lay, was so strongly impregnated by this salt, that it remained bare of vegetation for several years, and until the field was ploughed up for cultivation. But whatever were the products of fermentation saved by this experiment, the absence of all offensive effluvia throughout the process sufficiently proved that little or nothing was lost—as every atom must be, when flesh putrefies in the open air: and I presume that a cover of equal thickness of clay, or sand, or any mixture of both, without calcareous earth, would have had very little effect in arresting and retaining the aeriform products of putrefication. All the circumstances of this experiment, and particularly the good effect exhibited by the manure when put to use, prove the propriety of extending a similar practice. In the neighborhood of towns, or where-ever else the carcasses of animals, or any other animal substances subject to rapid and wasteful fermentation, can be obtained in great quantity, all their enriching powers might be secured, by depositing then between layers of marl, or calcareous earth in any other form. On the borders of the Chowan [River, in northeastern North Carolina], immense quantities of herrings are often used as manure, when purchasers cannot take off the myriads supplied by the seines. A herring is buried under each corn-hill, and fine crops are thus made, as far as this singular mode of manuring is extended. But whatever benefits may have been thus derived, the sense of smelling, as well as the known chemical products of the process of animal putrefication, make it certain that nine-tenths of all this rich manure, when so applied, must be wasted in the air. If those who fortunately possess, this supply of animal manure, would cause the fermentation to take place, and be completed, mixed with and enclosed by marl, in pits of suitable size, they would increase prodigiously both the amount and permanency of their acting animal manure, besides obtaining the benefit of the calcareous earth mixed with it.

But without regarding such uncommon, or abundant sources for supplying animal matter, every farmer may considerably increase his stock of putrescent manure, by using the preservative power of marl; and all the substances that might be so saved, are not only now lost to the land, but serve to contaminate the air while putrefying, and perhaps to engender diseases. The last consideration is of most importance to towns, though worthy of attention every where. Whoever will make the trial, will be surprised to find how much putrescent matter may be collected from the dwelling house, kitchen, and laundry of a family; and which if accumulated (without mixture with calcareous earth,) will soon become so offensive as to prove the necessity of putting an end to the practice. Yet it must be admitted that when all such matters are scattered about, (as is usual both in town and country,) over an extended surface, the same putrefaction must ensue, and the same noxious effluvia be evolved, though not enough concentrated to be very offensive, or even always perceptible. The same amount is inhaled—but in a very diluted state, and in small, though incessantly repeated doses. But if mild calcareous earth in any form (and fossil shells or marl present much the cheapest,) is used to cover and mix with the putrescent matters so collected, they will be prevented from discharging offensive effluvia, and preserved to enrich the soil. A malignant and ever acting enemy will be converted to a friend and benefactor.

The usual dispersion and waste of such putrescent and excrementitious matters about a farm-house, though a considerable loss to agriculture, may take place without being very offensive to the senses, or certainly injurious to health. But the case is widely different in towns. There, unless great care is continually used to remove or destroy filth of every kind, it soon becomes offensive, if not pestilential. During the last summer (1832) when that most horrible scourge of the human race, the Asiatic cholera, was desolating some of the towns of the United States, and all expected to be visited by its fatal ravages, great and unusual exertions were every where used to remove and prevent the accumulation of filth, which if allowed to remain, it was supposed would invite the approach, and aid the effects of the pestilence. The efforts made for that purpose served to show what a vast amount of putrescent matter existed in every town, and which was so rapidly reproduced, that its complete riddance was impossible. Immense quantities of the richest manures, or materials for them, were washed away into the rivers—caustic lime was used to destroy them—and the chloride of lime to decompose the offensive products of their fermentation, when that process had already occurred. All this amount of labor and expense was directed to the complete destruction of what might have given fertility to many adjacent fields—and yet it served to cleanse the towns but imperfectly, and for a very short time. Yet the object in view might have been better attained by the previous adoption of the proper means for preserving these putrescent matters, then by destroying them.—These means would be to mix or cover all accumulations of such matters with rich marl, (which would be the better for the purpose if its shells were in small particles,) and in such quantity as the effect would show to be sufficient. But much the greater part of the filth of a town is not, and cannot be accumulated; and from being dispersed, is the most difficult to remove, and is probably the most noxious in its usual course of fermentation. This would be guarded against by covering thickly with marl the floor of every cellar and stable, back yard, and stable lot. Every other vacant space should be lightly covered. The same course pursued on the gardens and other cultivated grounds, would be sufficiently compensated by the increased products that would be obtained: but independent of that consideration, the manures there applied would be prevented from escaping into the air—and being wholly retained by the soil, much smaller applications would serve. The level streets ought also to be sprinkled with marl, and as often as circumstances might require. The various putrescent matters usually left in the streets of a town, alone serve to make the mud scraped from them a valuable manure; for the principal part of the bulk of street mud is composed merely of the barren clay brought in upon the wheels of wagons from the country. Such a cover of calcareous earth would be the most effectual absorbent and preserver of putrescent matter, as well as the cheapest mode of keeping a town always clean. There would be less noxious or offensive effluvia, than is generated in spite of all the ordinary means of prevention; and by scraping up and removing the marl after it had combined with and secured enough of putrescent matter, a compost would be obtained for the use of the surrounding country, so rich and so abundant, that its use would repay a large part, if not the whole of the expense incurred in its production. Probably one covering of marl for each year would serve for most yards, &c. but if required oftener, it would only prove the necessity for the operation, and show the greater value in the results. The compost that might be obtained from spaces equal to five hundred acres in a populous town, would durably enrich thrice as many acres of the adjacent country: and after twenty years of such a course, the surrounding farms might be capable of returning to the town a ten-fold increased surplus product. After the qualities and value of the manure so formed were properly estimated, it would be used for farms that would be out of reach of all other calcareous manures. Carts bringing country produce to market, might, with profit, carry back loads of this compost eight or ten miles. The annual supply that the country might be furnished with, would produce very different effects from the putrescent and fleeting manure now obtained from the town stables. Of the little durable benefit heretofore derived from such means, the appearance of the country offers sufficient testimony. At three miles distance from some of the principal towns in Virginia, more than half the cultivated land is too poor to yield any farming profit. The surplus grain sent to market is very inconsiderable—and the coarse hay from the wet meadows can only be sold to those who feed horses belonging to other persons—and to whom that kind of hay is the most desirable, that is least likely to be eaten.

But even if the waste and destruction of manure in towns were counted as nothing, and the preservation of health, by keeping the air pure, were the only object sought, still calcareous earth, as presented by rich marl, would serve the purpose far better than quick lime. It is true, that the latter substance acts powerfully in decomposing putrescent animal matter, and destroys its texture and qualities so completely, that the operation is commonly and expressively called "burning" the substance acted on. But to use a sufficient quantity of quicklime to meet and decompose all putrescent animal matter in a town, would be intolerably expensive, and still mere objectionable in other respects. If a cover of dry quicklime in powder was spread over all the surfaces requiring it for this purpose, the town would be unfit to live in; and the nuisance would be scarcely less, when rain had changed the suffocating dust to an adhesive mortar. Woollen clothing, carpets, and even living flesh would be continually sustaining injury from the contact. No such objection would attend the use of mild calcareous earth; and this could be obtained probably for less than one-fifth of the cost of quicklime, supposing an equal quantity of pure calcareous matter to be obtained in each case. At this time the richest marl on James River may be obtained at merely the cost of digging, and its carriage by water, which if undertaken on a large scale, could not exceed, and probably would not equal three cents the bushel.

The putrescent animal matters that would be preserved and rendered innoxious by the general marling of the site of a town, would be mostly such as are so dispersed and imperceptible that they would otherwise be entirely lost. But all such as are usually saved in part, would be doubled in quantity and value, and deprived of their offensive and noxious qualities, by being kept mixed with calcareous earth. The importance of this plan being adopted with the products of privies, &c. is still greater in town than country. The various matters so collected and combined should never be applied to the soil alone, as the salt derived from the kitchen, and the potash and soap from the laundry, might be injurious in so concentrated a form. When the pit for receiving this compound is emptied, the contents should be spread over other and weaker manure, before being applied to the field.

Towns might furnish many other kinds of rich manure, which are now lost entirely. Some of these particularly require the aid of calcareous earth to be secured from destruction by putrefaction, and others, though not putrescent, are equally wasted. The blood of slaughtered animals, and the waste and rejected articles of wool, hair, feathers, skin, horn and bones, all are manures of great richness. We not only give the flesh of dead animals to infect the air, instead of using it to fertilize the land, but their bones, which might be so easily saved, are completely thrown away. Bones are composed of phosphate of lime and gelatinous animal matter, and when crushed, form one of the richest and most convenient manures in the world. They are shipped in quantities from the continent of Europe, to be sold for manure in England. The fields of battle have been gleaned, and their shallow graves emptied, for this purpose: and the bones of the ten thousand British heroes who fell on the field of Waterloo, are now performing the less glorious, but more useful purpose of producing, as manure, bread for their brothers at home.[3]

There prevails a vulgar but useful superstition, that there is "bad luck" in throwing into the fire any thing, however small may be its amount or value, that can serve for the food of any living animal. It is a pity that the same belief does not extend to every thing that, as manure, can serve to feed growing plants—and that even the parings of nails and clippings of beards are not used (as in China) in aid of this object. However small each particular source might be, the amount of all the manures that might be saved, and which are now wasted, would add incalculably to the usual means for fertilization. Human excrement, which is scarcely used at all in this country, is stated to be even richer than that of birds; and if all the enriching matters were preserved that are derived not only from the food, but from all the habits of man, there can be no question but that a town of ten thousand inhabitants, from those sources alone, might enrich more land than could be done from as many cattle.

The opinions here presented are principally founded on the theory of the operation of calcareous manures, as maintained in the foregoing part of this essay: but they are also sustained to considerable extent by facts and experience. The most undeniable practical proof of one of my positions, is the power of a cover of marl to prevent the escape of all offensive effluvia from the most putrescent animal matters. Of this power I have made continued use for about eighteen months, and know it to be more effectual than quicklime, even if the destructive action of the latter was not objectionable. Quicklime forms new combinations with putrescent substances, and in thus combining, throws off effluvia, which though different from the products of putrescent matter alone, are still disagreeable and offensive. Mild lime, on the contrary, absorbs and preserves every thing—or at least prevents the escape of any offensive odor being perceived. Whether putrescent vegetable matter is acted on in like manner by calcareous earth, cannot be as well tested by our senses, and therefore the proof is less satisfactory. But if it is true that calcareous earth acts by combining putrescent matters with the soil, and thus preventing their loss, (as I have endeavored to prove in chapter viii.) it must follow, that to the extent of such combination, the formation and escape of all volatile products of putrefaction will also be prevented.

But it will be considered that the most important inquiry remains to be answered. Has the application of calcareous manures been found, in practice, decidedly beneficial to the health of the residents on the land? I answer, that long experience, and the collection and comparison of numerous facts, derived from various sources, will be required to remove all doubts from this question; and it would be presumptuous in any individual to offer, as sufficient proof, the experience of only ten or twelve years on any one farm. But while admitting the insufficiency of such testimony, I assert that, so far, my experience decidedly supports my position. My principal farm, until within some four or five years, was subject in a remarkable degree to the common mild autumnal diseases of our low country. Whether it is owing to marling or other unknown causes, these bilious diseases have since become comparatively very rare. Neither does my opinion in this respect, nor the facts that have occurred on my farm, stand alone. Some other persons are equally convinced of this change on other land as well as on mine. But in most cases where I have made inquiries as to such results, nothing decisive had been observed. The hope that other persons may be induced to observe and report facts bearing on this important point, has in part caused the appearance of these crude and perhaps premature views. (1833)[4]

Even if my opinions and reasoning should appear sound, I am aware that the practical application is not to be looked for soon; and that the scheme of using marl in towns is more likely to be met by ridicule, than to receive a serious and attentive examination. Notwithstanding this anticipation, and however hopeless of making converts either of individuals or of corporate bodies, I will offer a few concluding remarks on the most obvious objections to, and benefits of the plan. The objections will all be resolved into one—namely, the expense to be encountered. The expense certainly would be considerable; but it would be amply compensated by the gains and benefits. In the first place the general use of marl as proposed for towns, would serve to insure cleanliness, and purity of the air, more than all the labor of their Boards of Health and their scavengers, even when acting under the dread of approaching pestilence. Secondly, the putrescent manures produced in towns, by being merely preserved from waste, would be increased ten-fold in quantity and value. Thirdly, all existing nuisances and abominations of filth would be at end, and the beautiful city of Richmond (for example) would not give offence to our nostrils, almost as often as it offers gratification to our eyes. Lastly, the marl after being used until saturated with putrescent matter, would retain all its first value as calcareous earth, and be well worth purchasing and removing to the adjacent farms, independent of the enriching manure with which it would be loaded. If these advantages can indeed be obtained, they would be cheaply bought at any price necessary to be encountered for the purpose.

Most of the foregoing part of this chapter was first published in the Farmers' Register, (for July 1833) and as supplementary to this Essay. That publication drew some attention from others to the subject, and served to elicit many important facts, of which I had been before altogether ignorant, in support of the operation of calcareous earth in arresting the effects of *malaria*, and the usual autumnal diseases of the southern states, and other similar regions. These facts, together with the results of my own personal experience, extended through two more autumns (or sickly seasons as commonly called here and farther south,) since the first publication of these views, will now be submitted. Most of the facts derived from other persons relate to one region—the "rott[e]n limestone lands" of southern Alabama: but that region is extensive, of remarkable and well known character and peculiarities, and the evidence comes from various sources, and is full, and consistent in purport. The facts will be here embodied, and the more important statements from which they are drawn, will be presented more fully in the Appendix. (See N.)

The first fact brought out, was, that in the town of Mobile, near the Gulf of Mexico, the streets actually had been paved with shells—thus presenting precisely such a case as I recommended; though it had not been done with any view to promoting cleanliness or health. The shells had been used merely as a substitute for stones, which [may] not be so cheaply obtained. Nor had the greatly improved healthliness of Mobile since the streets were so covered, (of which there is the most ample and undoubted testimony,) been attributed to that cause, until the publication of the foregoing opinions served to connect them, as cause and effect. This can scarcely be doubted by those who will admit the theory of the action of calcareous earth; and the remarkable change from unhealthiness in Mobile, to comparative healthiness, is a very strong exemplification of the truth of the theory. But it is not strange, that when so many other causes might (and probably did) operate to arrest diseases, that none should have considered the chemical operation of the shell pavement as one of them, and still less as the one by far the most important. The paving of streets, (with any material,) draining and filling up wet places, substituting for rotting wooden buildings new ones of brick and stone—and especially fires—all operate we know, (and particularly the last,) to improve the healthiness of towns; and all these operated at Mobile, as well as shelling the streets. Neither was the shelling so ordered as to produce its best effect for health. The streets, alleys, and many yards and small vacant lots were covered, and so far the formation and evolving of pestilential effluvia were lessened. But as this was not the object in view, and indeed the chemical action of shells was not thought of, the process was incomplete, and must necessarily be less effectual than it might have been made. The shelling ought to have been extended to every open spot where filth could accumulate—to every back yard, to every cellar, and made the material of the floor of every stable, and of all other buildings of which the floor would otherwise be of common earth. In addition, after a sufficient lapse of time to saturate with putrescent matters the upper part of the calcareous layer, and thus to make it a very rich compound, there should be a partial or total removal of the mass, and a new coating of shells laid down. The value of the old material, as manure, would probably go far towards paying for this renewal: and if it is not so renewed, the calcareous matter cannot combine with more than a certain amount of putrescent matters—and after being so saturated, can have no farther effect in saving such matters for use, or preventing them from having their usual evil course.

The burning of towns is well known to be a cause of the healthiness of the places being greatly improved, and that that effect continues after as many buildings, or more, have replaced those destroyed by fire. Indeed this improvement is considered so permanent, as well as considerable, that the most sweeping and destructive conflagrations of some of our southern towns, have been afterwards acknowledged to have proved a gain, and a blessing. The principal and immediate mode of operation of this universally acknowledged cause, is usually supposed to be the total destruction, by the fire, of all filth and putrescent matters—and in a less degree, and more gradually, by afterwards substituting brick and stone for wooden buildings, which are always in a more or less decaying state. But though these reasons have served heretofore to satisfy all, as to the beneficial consequences of fires, surely they are altogether inadequate as causes for such great and durable effects. The mere destruction of all putrescent matters in a town, at any one time, would certainly leave a clear atmosphere, and give strong assurance of health being improved for a short time afterwards: but these matters would be replaced, probably in the course of few months, by the residence of as many inhabitants, and the continuance of the same general habits—and most certainly this cause would lose all its operation by the time the town was rebuilt. But there is one operation produced by the burning of a town, which is far more powerful—which in fact is indirectly the very practice which has been advocated—and the effect of which, if given its due weight, furnishes proof of the theory set forth, by the experience of every unhealthy town which has suffered much from fire. If any estimate is made of the immense quantity of mild calcareous earth which is contained in the plastering and brick-work of even the wooden dwelling houses of a town, (and still more of those built of masonry) it must be admitted that all that material being separated, broken down, and (soon or late,) spread by the burning of the houses and pulling down of their ruins, is enough to give a very heavy cover of calcareous earth to the whole space of land burnt over. It is to this operation, in a far greater degree than to all others, that I attribute the beneficial effects to health of the burning of towns.

I proceed to the facts derived from the extensive body of prairie lands in Alabama which rest on a substratum of soft limestone, or rich indurated clay marl. It was from these remarkable soils that the specimens were obtained which were described at page 22. Some of these, indeed all that have been examined by chemical tests of the high and dry prairie lands, contain calcareous earth in larger proportions than any soils of considerable extent in the United States that I have seen, or tested. The specimens not containing free calcareous earth are of the class of neutral soils; and the calcareous earth, which doubtless they formerly contained, and from which they derived their peculiar and valuable qualities, may be supposed only to be concealed by the accumulation of vegetable matter, according to the general views submitted in chapter vii. The more full descriptions of the soils of this remarkable and extensive region which will be placed in the Appendix (at N) render it unnecessary to enlarge much here. It will be sufficient to sum up concisely the facts there exhibited—and which agree with various other private accounts which have been received from undoubted sources of information. The deductions from these facts, and their accordance with the theory of the operation of calcareous matter, are matters of reasoning, and as such, are submitted to the consideration and judgment of readers.

The soil of these prairie lands is very rich, except the spots where the soft limestone rises to the surface, and makes the calcareous ingredient excessive: in the specimen formerly mentioned, the pure calcareous matter formed 59 parts in the hundred of this "bald prairie" land. The soil generally has so little of sand, that nothing but the calcareous matter, which enters so largely into its composition, prevents it being so stiff and intractable, that its tillage would be almost impracticable; yet it is friable and light when dry, and easy to till. But the superfluous rain water cannot sink and pass off, as in sandy or other pervious lands, but is held in this close and highly absorbent soil, which throughout winter is thereby made a deep mire, unfit to prepare for tillage, and scarcely practicable to travel over. This water-holding quality of the soil, and the nearness to the surface of the hard marly substratum, deprive the country of natural springs and running streams: and before the important discovery was made that pure water might be obtained by boring from 300 to 700 feet through the solid calcareous rock, the inhabitants used the stagnant water collected in pits, which was very far from pure, or palatable. Under all these circumstances, added to the rank herbage of millions of acres annually dying and decomposing under a southern sun, it might have been counted on, as almost certain, that such a country would have proved very unhealthy: yet the reverse is the fact, and in a remarkable degree. The healthiness of this region is so connected with, and limited by, the calcareous substratum and soil, that it could not escape observation: and they have been considered as cause and effect by those who had no theory to support, and who did not spend a thought upon the mode in which was produced the important result which they so readily admitted. Their testimony therefore is in this respect the more valuable, because it cannot be suspected. The intelligent author of the extract from the *Southern Agriculturist*, which will be given in the Appendix (N) is altogether unknown to me—and it is presumed that he had never heard of this Essay, nor of these views of the action of calcareous earth.

After adducing the foregoing mass of evidence, for which I am indebted to others, it will appear very unimportant to add what will follow from my personal observation—especially, as the opinion has been expressed above, that the experience of any one individual, on any one farm, or in any one location, though continued for ten or twelve years must be very insufficient as proof of a permanent change of healthiness, and of the actual causes of such changes. But, as in the absence of more striking facts, and of practical proofs, my own limited experience was formerly brought forward—it is proper here to add, that the two autumns that have since passed, have brought no circumstances to weaken the opinions advanced, and many that have served, on the contrary, to strengthen them. (1835)[5]

On my principal farm, Coggin's Point, the position of the homestead was always most inconveniently situated, and became the more so as the clearing and improvement of the poorer and more remote parts of the land were extended. For this reason, in addition to others, the farm buildings, and negroes' dwellings had been gradually removed, as the expense could be best encountered, until the old homestead was entirely abandoned in

1831, for a more eligible location. This would prevent the different degree of healthiness found here, before and since marling, from presenting a fair statement of proof. But still, there is no doubt of the general results showing a great and decided improvement in respect to health—and this was evident, before as well as since the removal of the dwelling place of the slaves. The greater number of these had been moved to an intermediate location, (with a view to health) before these benefits of marling were either felt, or anticipated—where a portion of them remained until within the last few years: and the circumstances attending this location, furnish ground for the opinion maintained, which is not liable to the objection referred to.

The poorer farm (Shellbanks) which was made a summer residence for my family in 1828 and the two succeeding years, and a permanent dwelling place since 1831, was marled to the extent of 120 acres, including all the land around the houses, in 1828; and in a few succeeding years, the space marled amounted to more than 300 acres. During this time, the yard was covered heavily with marl—and in 1832, when the approach of Asiatic cholera caused such alarm, the floor of the cellar of the house, (which is very damp,) the stable floor, and stable yard, were also covered, and every other vacant spot. In addition, the plan of collecting, for manure, all putrescent animal matters in a pit, and covering or mixing them frequently with marl, has been pursued for several years, though not with as much care and economy as ought to be used. In this pit, for experiment as much as for profit, the carcasses of animals have been several times placed, and preserved (as before) from giving out any offensive odor, merely by the covering of marl, until their very slow decomposition was at an end. The health of the family, during the first two or three autumns, was about as good as on what are considered healthy places in the tide-water region of Virginia— all of which are more or less subject to bilious disorders in autumn, though deserving well (as indeed does the whole country) to be considered more than usually exempt from all other diseases. We had among the members of a large family, some intermittents, and some more severe bilious fevers during that time. But there has been a still greater and unlooked for improvement since—and for the last two years, I believe that all residing permanently at this place, have enjoyed as good health, as could be hoped for in any situation in the United States. Among the domestic servants and their young children, last autumn, there were a few slight agues, (which were attributed to some of those acts of imprudence to which negroes are so notoriously addicted, even if not necessarily exposed,) and which were scarcely worth notice, but as exceptions to the general healthiness. The land not being then tilled, there were no field laborers. Among my own large family, and other white persons who were permanent residents, there was not a single ague, or the slightest disease to be counted as one of climate, or proceeding from *malaria*. But I repeat, that many such facts are necessary, and much time, and the testimony of many different persons from various places to be brought together, before the causes can be fully admitted of such mysterious effects, as disease and its removal. It is to be hoped that the facts and deductions here presented, however defective, may at least serve to attract the attention of many other, and more competent investigators, to this highly important subject.

To the time when this last publication is made, (June, 1837,) there has been no reason to doubt the actual facts of autumnal diseases (the effects of *malaria*,) being lessened by even the partial use of marling—nor the inference that they would almost cease to occur (where no mill-ponds and undrained lands existed,) if all the surface of a considerable extent of country were made calcareous, and all rapidly putrescent and otherwise offensive matter were preserved and kept harmless by being combined with marl, applied from time to time, as required. But it should be remembered, that, as yet, there has been no instance of the greater part of any whole neighborhood, of so much as a few miles in extent, being marled; nor even of all the surface of any one farm; and that, therefore, we have no means of judging by experience of the full measure of benefit to be derived from such a general change of the character of the soil. The most that has yet been done, any where, is the marling of all the cultivated and arable land—leaving unmarled, and as much as ever the abundant sources of vegetable decomposition and of disease, all the woodland, hill-sides, and the wet bottoms. Now, as the remaining woodlands are generally among the poorest of our soils, that is, (according to the theory maintained,) soils incapable of combining with and retaining the products of decomposition—and as they are covered annually with leaves, which in time all rot and finally pass off into the air— it follows, that the lands so left are among the most fruitful of malaria. It is obvious that the remedy is but partially and inefficiently in operation, while from one third to one half of every farm is left unmarled, and free as ever to evolve the cause of disease. So sure does this opinion seem to me, that I have commenced acting on it, by marling the woodland that is not designed to be cleared for cultivation—and shall continue, as more necessary labors permit, to do so, until not an acre of the farm is left without being changed in character by calcareous earth.

It is proper to add, as an opinion founded on but limited experience, as yet, that though the cases of sickness on Coggin's Point Farm, have certainly diminished very greatly—there not being one case recently, where there formerly were ten, or perhaps twenty—still that the diseases seemed to have changed in kind, and to have increased in severity and danger. Formerly, there was almost no sickness except from ague and fever (or very rarely, a case of mild bilious fever,) which, though few persons escaped from through the autumn, and some suffered several relapses, the attacks, were never dangerous, and required little skill, and but a few days, to cure, for that time. Bad as was this state of things, it seemed that the ague and fever acted as a safety-valve to the system, and while it seldom permitted the enjoyment of long continued robust health, it prevented the occurrence of more dangerous or fatal diseases, such as are the most common among the fewer diseases of what are deemed healthy regions. The diseases of my negroes for the last six or eight years have been of a more inflammatory kind, and are not confined to autumn: and there have been certainly more severe and fatal diseases, and more that have had medical aid, than formerly, when there was so much more of sickness of one kind, and at one season. In short, it seems that the diseases are no longer (or but in few cases,) those of the low country and of a bilious climate, but are like those of the upper country, which, though occurring but rarely, are generally of a serious nature. The facts on which this particular opinion has been formed, are yet too few, and of too short continuance, to attach to them much importance; and even if they were less doubtful, I have not the medical knowledge to trace these new effects back to their causes. Still, it is deemed due to candor, and to the desire for a fair and full investigation of the subject, even if making against my own views, that these opinions should be stated. There is no other subject, th[a]n this, taken in general, which more deserves and requires investigation—and in the present inchoate state of the discussion, the expression of even erroneous opinions will not be useless, if it should serve to elicit more full or correct ones from other sources.

Since November, 1835, I have ceased to reside on Shellbanks farm,[6] and therefore have no later personal experience of the continued effect of marling in preventing the formation of the seeds of autumnal diseases. But it is understood from a physician, Dr. John S. Epes, who has since rented and resided on the farm, that it has maintained its reputation for a remarkable exemption from those diseases. The last summer and autumn (1836,) were unusually sickly throughout most of the surrounding neighborhood, and the residents of Shellbanks did not escape the visitation; but nearly all of their cases were entirely different from the diseases which were general in autumn, before marling the land. Nor can entire and continued exemption, even from the diseases caused by *malaria*, be reasonably counted on any where, (according to the views submitted above,) while the greater part of the surrounding and adjacent lands remain unmarled, and in their original state of unfruitfulness in every thing, save the poisonous products of vegetable decomposition.

When my opinions of the beneficial operation of calcareous earth, in soil, or mixed with putrescent matter, in destroying or disarming the sources of disease, were first published, and until after the last publication of the same in the *Essay on Calcareous Manures*, I had no knowledge that similar grounds had been taken by any other person. But since, in the recent publications of a French writer, M. Puvis,[7] I have found the same general opinion expressed, and many important facts given in confirmation. These views are presented at length in the several articles translated from Puvis in the 3rd and 4th volumes of the Farmers' Register—to which it is enough here to refer, and to request the recurrence or attention of those readers who desire more extended statements and proofs of this highly important effect of calcareous earth as an ingredient of soil.[8]

No. II. The Police of Filth in Towns, and Its Bearing on Comfort, Decency and Health

The delightful season of opening summer has arrived, and the face of the earth, as formed by nature, and not deformed by man, is seen in its fairest aspect and brightest colors. Every thing shows life, in youth and beauty, and nothing yet exhibits indications of decay. Every feature of the natural landscape, in every region, however varied, is beautiful to the eye. The most barren and worthless of our lands, though the most wretched in appearance after cultivation, before being touched by man, are covered with magnificent forests. Nature has not made a scene that is displeasing to the eye; and even this granite region, barren and unsightly as much of it now is, was once wide scene of universal beauty. It is man that wastes the beauties and blessings of nature, and deforms and defiles whatever he touches.

The opening of summer in our towns, presents a very different aspect, and is accompanied with very different associations. It is true, that some beautiful gardens are seen, in which the hand of man (or more generally of woman) has improved on nature, by bringing together, in numbers, nature's choicest ornaments. But these are exceptions to the general appearance. The broad sloppy flats, receptacles of collected rain water and oozes from hill-sides, which during winter and spring merely barred the way of walkers, or, at worst, gave them wet feet, and colds and pleurisies, now are drying up, without the Corporation being put to the cost of the small amount of ditching that would have kept the ground dry at all times. A "green mantle" overspreads the standing pools—and all will soon become a naked, ugly, and foul-scented mud. The thickly settled and commercial parts of towns may, perhaps, have nothing visible worse than men, and merchandize, brick houses, and paved streets: but all the out-skirts and vacant places are full of abominations to cleanliness and health, and of offence to the nostrils as well as to the eyes. The commencement of warm weather gives activity to decomposition, and the soft air is redolent of its products: and in sundry different spots of every town, the effluvia arising from filthily kept yards, of stables and hog-styes, of privies, and sometimes the breezes tainted by a dead cat, or, if without the suburbs, by carrion of larger kind, are offered to our sense of smelling, in doses of various degrees of intensity, and in every variety of combination. We become accustomed by the habit of endurance to these, as to all other evils, and in time, are scarcely conscious of the magnitude of the nuisance. But its offensiveness is estimated at the true value, by visitors fresh from the pure air of the country.

Now approaches the time when the Police, and the Board of Health will begin to bestir themselves to abate nuisances of this kind, but in such a way as to effect no manner of benefit. Their operations merely consist in moving decomposing matter, or its sources, from one spot to another, there to proceed as before—and by thus moving and dispersing filth, to hasten its decomposition still more, though rendering its products less evident, by their being more widely diffused. But the total amount of the production of such effluvia is not the less in quantity, nor the less hurtful, because, by being more wide-spread, and diluted, and by contaminating more of the atmosphere, the scent is less concentrated and offensive. All the operations of the most industrious and zealous Board of Health do not lessen the amount of decomposition within the limits of a town, unless the putrescent matter is actually thrown into and floated away by a rapid river, or otherwise conveyed away to poison the air some where else, where there may be fewer people to breathe of it. Every removal, and exposure of new surfaces, serves only to quicken the progress of decomposition.

It is not a little remarkable that this general state of filthiness is caused and maintained, in a great degree, by the fastidious or squeamish nicety of our people. It is almost universally considered that it is quite too dirty a business, too offensive to the imagination, as well as the senses, to use carrion and human excrement for manure. If this silly prejudice did not operate, and if proper economy were used to collect, preserve, and apply these rich and most decomposable substances, the profit which they would bring as manure, would far more than pay for the expense of the proper procedure to preserve the matters, and at the same time to maintain cleanliness. But it is not only that the contents of privies are suffered to accumulate, because of their being no profitable demand for them, (as exists in countries where the worth of manure is better understood,) but there is that want of accommodation in the number and situation of privies, which operates to the injury of comfort, of decency, and in many cases, directly as well as indirectly, to the injury of health. We are so exceedingly nice, or proud, that we desire to conceal the existence of such humiliating necessities of our nature; and no conveniences for the purpose are provided, and kept in proper order for public use: and the privation is a matter of extreme inconvenience to all decent visitors to a town, who have not acquired a knowledge of, and a right to use, some such places. The same morbid feeling of shame that prevents on the one side the accommodations being afforded, also prevents on the other any complaint of the want of them. But the ground for complaint does not the less exist—as every countryman can testify, and even every townsman, when visiting another town than his own. So nice and squeamish are our people on such subjects, that to treat of it by word or writing, would be considered by very many as both ridiculous and offensive; and when one ventures still farther, as I shall do, to recommend modes of removing the nuisance, and converting it to profit, there is much ground to expect that nothing will be excited, except a sense of the ridiculous in some, and a feeling of disgust in others. But I have never been deterred from urging what was deemed highly expedient, by the dread of being laughed at, and as to exciting disgust, it is just what is desired, provided it can be directed against the habits which are held up to condemnation.

In large cities, necessity has compelled the adoption of means to get rid of excrementitious and other filth, by a general system of sewers, or subterranean passages, into which all such matters are thrown, and by the flowing of water through, in abundance, they are washed into the adjacent river. The sewers of some great cities have been constructed on a plan so vast, and at so much expense, and were so excellent in their operation, that they have been considered as not less worthy of admiration than the magnificent temples and palaces. If only the object was to cleanse a town, and there was sufficient command of water, and of money, there could be nothing to object to this plan. Certainly the expense of constructing the sewers would be an objection no[t] worth notice, when compared to the value of their intended effect. But if the system were not perfect, and the supply of water always abundant, the evil would be made so much the greater by being concealed from observation. There is another objection to this plan, in its contaminating and corrupting the waters of the rivers into which the sewers empty; and it may well be doubted whether water so defiled, does not itself throw off deleterious effluvia, and is not rendered more liable to cause decomposition in whatever decomposable matter it may reach: and thus that the waters are not only made to stink, but also to poison those who have destroyed their purity. But the greatest objection to this plan, is the utter destruction of so enormous an amount of rich manure, which if properly preserved and applied, would soon make rich and fruitful the poorest surrounding country. And to properly accumulate and preserve all this manure, and prevent its being offensive to the senses, or injurious to health, might in most cases, (and certainly on all the eastern coast of the southern states,) be effected not only at less cost than by a proper system of sewers, but at less than the present wasteful and expensive system of employing laborers, under direction of the town police, and boards of health, so to stir up and move about the excrement, as to produce its most speedy decomposition, and total passing off into the air, and thereby to give the full benefit of its evolving effluvia to the nostrils of the towns-people.

The remedy is that which has been proposed in general terms in the preceding part of these observations; to provide *calcareous earth* (either marl, or whatever other form may be cheapest,) enough to cover every spot in the town, in which decomposable filth can accumulate; and this to be renewed from time to time, as needed. The calcareous matter would form a chemical compound with the putrescent, so as to preserve the latter from all waste, and from giving out any offensive odor; and once a year, (when in situations not convenient at all times,) and in cold weather, the accumulations might be removed to the country to be used as manure; and the richest as well as the most permanent manure in the world, this compound of animal and calcareous matter would be.

The object would be to *accumulate*, as much as possible, instead of *dispersing*, the most putrescent matters. And for this purpose, as well as to afford the general accommodations now so much required for comfort and for decency, and also for health, there should be large and well constructed privies erected in suitable situations, and at convenient distances apart, throughout the town, free for the use of all males without exception. The pits should be large and sufficiently deep, but accessible to carts, to bring marl, and to remove the contents. At the expense of the town (as the whole system ought to be,) there should always be kept a heap of rich marl near to each pit, and a sprinkling, once or twice a day over the excrement, would effectually secure it from wasting, or being offensive. By such places of accommodation being furnished, and kept in the neatest condition by regular attendants, there might be, and would be abated many of the small private receptacles, which necessarily (as now managed) are more or less filthy nuisances. And the *buckets* which now are at night emptied on all vacant and forbidden spots, (and requiring the unceasing activity of the Police and Board of Health to attempt to prevent,) would be then emptied into these pits, with certainty, simply because they would offer the nearest and most convenient places of deposite. There would then be no inducement remaining for the defiling of every spot of vacant ground; and such places, instead of being abominations to the senses and the minds of all decent observers—and absolutely forbidden to the footsteps, and even to the distant view of modest women—would be clean and lovely grass plots, serving to refresh and relieve the eyes tired of seeing brick walls and stone pavements. I will touch but gently on the *moral* nuisance that exists in so many cases in every town, where these vacant spots, the only public places "of ease," are overlooked by the back windows of the houses of respectable families, the members of which, though at considerable distances, are nevertheless unavoidably subjected to witness indecent exposures, still more offensive to the mind than to the eye.

In addition to the public and general accommodations proposed, there should be a certain and sufficient quantity of marl carried at certain intervals of time, to every private lot, (unless the occupant took measures to provide himself with it,) to be used as wanted, for similar purposes. This would prevent, what is almost impossible now to avoid, there being offensive accumulations, or still more offensive removals and dispersions, of foecal matter on private lots.

It would be impossible to approach the truth in estimating what would be the expense of such a system in any particular town, until it shall have been tried. But there can be no doubt but that the benefits would far overbalance the cost. Many expenses and evils, much worse to bear, and now continually encountered, would be, by these means, avoided. Such of these as bear on private individuals, I pass over without notice. For one item, the public would save all that part of the labors of their police, which is now most unprofitably devoted to this object.

But even if it is admitted that the means proposed would be as effectual as I imagine, in preserving cleanliness, and cutting off sources of disease—and that the compound formed is of all the supposed value as manure, still it may be objected that it would be long before prejudice and incredulity will be so removed as to make this manure an article of sale—and consequently, that all expectations of returns from sales must be visionary. Even if there should be no sales for two or three years, and if the manure should be merely taken for the trouble of carrying it away, the expense would be well afforded as a mere matter of police. But two years' use would make manifest the value of this compound manure, and the demand and the price would afterwards gradually increase, until it would nearly or quite defray the whole expense of the plan.

But the town of Petersburg has at once the best possible customer for all that the plan would supply for some years, in the farm of the Poor-House, belonging to, and cultivated at the expense of the town. To this land, now, much putrescent manure is carried, removed by the Police from the town. But except in winter, or at the rare and short other periods when manure can be (or is) at once advantageously laid on the field, these supplies are heaped up for future use, and of course, rot away as rapidly as possible, and give ten times as much of their products to the air as to the soil. Besides—even if there was not necessarily this great waste from the decomposition of manure altogether putrescent, when moved and heaped in warm weather, there would be very little profit from its application. The lands lying over the belt of granite which passes through Virginia, and which forms the falls of the rivers flowing to the Atlantic, are, naturally, among the most destitute of lime, and consequently are among the poorest and the least capable of retaining putrescent manures when applied to them. Such are the lands surrounding and within a few miles of Richmond and Petersburg, and probably all the other towns at the falls of our rivers. Particular individuals, by lavish use of the cheap and rich manure of the public stables, have highly, though but for a short time, improved some of these lands, and reaped heavy crops, and, possibly, made great profits. But still the demand for such manure by the hungry, yet wasteful soil, is continual, and if it is not frequently repeated, the original poverty soon returns. But few persons have used these means, to much extent, and most neighboring residents are satisfied that the town manure is too costly to be carted to their farms. Yet though the richest stable manure, (richest because it is principally of animal matter,) may be bought from the tavern and livery stables at $12\frac{1}{2}$ cents for the largest single horse loads' (20 to 25 bushels,) it mostly rots away in bulks in stable yards, for want of regular purchasers even at that low price. So it is however—from the little town manure carried to neighboring farms, the little permanency of effect of what is used, and the general impression that it is not worth using—it results that most of the lands, lying even within the short distance of a mile from the towns, are wretchedly poor, and yield but little for the support of the town, either in grain, grass, or garden vegetables for market. Indeed it may well be doubted, whether a large proportion of the population of the vicinity do not buy (or obtain otherwise) from the town, as much provision as they sell to it. This state of things has continued, with but little actual improvement, as long as these towns have stood; and it may safely be predicted, that unless calcareous manures are used to fix the otherwise fleeting value of the putrescent matters, that the general condition of things will never be much better. It is not then strange, that with the neighboring farms so poverty-stricken, the town markets should be badly supplied, and at high prices, with all the small articles of daily purchase and consumption, which, though small, make up the greater part of the comfort, and (at usual prices) cause the greater part of the expense of living. Just let the reader imagine what would be the difference in these respects, if the lands surrounding each town, for as much as six miles distance, were as rich as they well could be, and produced in abundance, clover and other grasses, a full supply of garden vegetables and other small articles for the daily markets, besides their large crops of grain and other staple products. The comforts of all the persons living in town, so far as they depend on food, would be greatly increased, while the expenses of living would be made less than at present—and yet the suppliers of the market would be better rewarded than by the present miserable system, because rich land and good farming can always undersell the poor and unproductive; and a market generally well supplied is a more sure, and therefore a better place of sale, than where demand is irregular and, of course, prices irregular, though often very high. It will be under such a state of improvement that market gardens and market farms will be profitably kept—and the towns will be abundantly supplied, and from their neighborhood, with milk, cream, butter, eggs, fowls, and fresh meats of fat young animals, as well as with vegetables. The surplus product of hay, grain, and other field crops, of such highly enriched districts, would make no small addition to the sales and the export trade of the towns, and would serve to increase their population, and thus furnish a still increased demand for the products of the neighboring lands. It is also probable, that if the fish, of the rivers which flow by towns, were not driven away by the filthiness of the water, that their numbers would be the greater on account of the neighborhood of a town, (and the abundance of food thrown into the water,) instead of being reduced almost to nothing, as is notoriously the case. Even the shad, and other fish of passage, whose instinct strongly impels them to seek the higher waters of rivers, to deposite their spawn, are mostly deterred from passing through the flood of filthy water that a town supplies; and the people on the upper waters suffer thereby a privation, as do the townsmen by the driving to a distance the more fixed residents of our fresh water rivers.

It may however be reasonably objected, by those who have not studied the qualities of soils and manures, that too much value is counted on from the use of this proposed compound matter. It would be unnecessary here to repeat at length all the grounds on which that estimate is founded. For the amount of early and annual increase to be expected from marl on naturally poor soils, and for the permanency of its effects, I refer to the reasoning and the facts presented in the *Essay on Calcareous Manures*, and also to the opinions of the hundreds of farmers in lower Virginia who are now thus improving their lands. For the chemical power of calcareous earth in combining with, and preserving from waste, putrescent matters, I refer to the general reasoning on this head in the Essay, and the statements made in the first of these communications. As to the enriching value of human excrements, it is known, in Europe and in China, that they are the richest of all. In England; it is stated in agricultural books, that two wagon loads is a sufficient dressing for an acre—probably because more at once would be hurtful to the crop. In France, there are in operation regular establishments set up by private adventurers, for *desiccating*, and thus preparing for use, the products of the privies and public sewers of large cities; and sufficient profits are made to support these establishments, by selling the dried manure (*poudrette*) to the farmers. Its great richness, in small weight and bulk, makes it well suited for distant transportation, and extensive sale. From the accounts that I have read of these establishments, it may be inferred that much previous decomposition, waste of value, and extrication of offensive effluvia, must take place in the material, before it is brought to the desiccating establishment—and that both the previous and subsequent manual operations must be highly disagreeable and disgusting. Besides, the desiccation seems to be sought more by mechanical than by chemical means—and any dry pulverized earthy matter is used to absorb the fluid and to make the mixture dry. There does not seem to be much choice in the earthy substances. Thus they propose gypsum, and burnt earth, and quicklime, as well as chalk, rubbish of demolished buildings, and coal and wood ashes. The first two of these substances, according to my views would be but of little effect, acting as they do only mechanically; the quicklime, (which it seems is preferred,) would be decidedly injurious; and the mild calcareous character of the latter substances would render them, only, proper for the desired results. The profit of this business in France, alone, would be sufficient proof of the greater value of the far more simple, economical and effectual and cleanly plan which I recommend, and which is also perfectly in accordance with the chemical properties and action of the substances used in the compound. I annex the only known description of the French process, below, in the application for a patent by the inventor, Donat, and which was communicated, with the introductory comments, by the Board of Health of Philadelphia, to the Agricultural Society of Pennsylvania.[9]

French method of preparing poudrette and urate for manure.
To the Pennsylvania Agricultural Society.

The attention of the Philadelphia Board of Health has been earnestly directed towards discovering some mode of disposing of the contents of privies, which would remove from the precincts of our city, where the deposites are made, a nuisance at present of a very formidable character, and which must necessarily increase. In pursuance of this object, the board has concluded, that an effectual remedy for the evil is only to be sought in the conversion of the offensive substance into *inodorous manures*, after the methods now successfully practised in many parts of Europe, and especially in the cities of Paris and London.

The principle by which this object is effected, is simple, and consists in the drying, or desiccation of the urinary and foecal matters, either apart or together, by the addition of certain absorbent substances, such as plaster, lime, chalk, ashes, &c. It is probable that the ashes of the Lehigh and Schuylkill coal may be thus usefully disposed of. The manure prepared from the foecal or more solid contents of privies, has long been known and highly esteemed by the gardeners and agriculturalists of France, under the name of *poudrette*. That prepared from the urinous portion is comparatively of modern invention, and is called *urate*.

Aware that such a plan is not to be carried into effect under the special direction of either your society or their own body, the board lays the subject before you, in the hope that its advantages will be properly investigated and made known, so as to lead to useful results; for, surely, nothing can be more worthy of general and special encouragement, than a plan not only calculated to promote the health and comfort of our large community, but to render essential assistance to the most important of the useful arts, insuring at the same time liberal profits to those actually engaged in its execution.

That your society may be placed in possession of more particular information relative to the subject under consideration, the board would refer you to numerous highly favorable reports and interesting proceedings of the most respectable associations established in Europe for the encouragement of agricultural and useful arts, among which we would especially call your attention to those of the French "Royal and Central Agricultural Society," and the "Society for the Encouragement of National Industry," during the years 1818–19–20.

The following translation of a French document, furnishes an accurate detail of the process by which the *urate* is manufactured, and throws much important light upon the subject generally:

Certificate granted upon the application for a brevet (patent) of invention to M. Donat, (Joseph Etienne-Victor-Gabriel,) residing at Paris, department of the Seine.

The Ministerial Secretary of the state, for the department of the interior, considering the Memoir of M. Donat, proprietory, residing in Paris, Rue des Bons-Enfans, No. 28, in which he states his desire to enjoy the proper rights secured by law of the 7th of January, 1791, to the authors of inventions and discoveries in all kinds of industry, and to obtain in consequence, a brevet of invention for fifteen years, for the sudden drying of the urinary portion, and manipulation of the contents of privies, within the twenty four hours succeeding their removal; all by particular means and processes, of which he declares himself the author, as it appears from the verbal process addressed at the time, to the depot of documents attached to the secretaryship of the department of the Seine, the 19th of January, 1819.

Considering the designs of the apparatus, and the descriptive memoir of which the following is a copy.[10]

I have contrived a plan which affords me the means of extracting from urinary and foecal matters, a manure very superior to those hitherto known. Desirous of securing to myself the exclusive enjoyment of my invention, I have made application to the prefecture, department of the Seine, conformably to the laws of the 7th of January and 25th of May, 1791, for a brevet of 15 years, for the complete and immediate desiccation of foecal and urinary matters together, or separately, by means of absorbents which I add, such as lime, plaster, chalk, marl, ashes either natural or mineral, such as are taken from the different ash mines. Substances having calcareous bases may be calcined for the absorption of a greater quality of liquid, at least when the high price of the combustible, or the low price of the absorbent, do not offer greater advantages in using it directly from the quarry.

This variety of absorbent substances, assures to every country the means of manufacturing a very abundant and active manure with human dejections. The product of my operations is inodorous, for two reasons: The first is, that when urine is employed, it gives out no odor after the absorption of its moisture: The second is, when the foecal matters are sufficiently mixed with the absorbent, I bury them at least 18 inches deep, to prevent the disengagement of the odor during the fermentation necessary to the good quality of the manure.

I give to the manure made with pure urine and one of the aforesaid substances, the name of *urate*. I believe that this composition, mixed or combined with that resulting from the combination of foecal matters with a certain quantity of one of the aforesaid absorbent matters, produces a manure of great activity. The only difficult point is, to ascertain the proper proportions for the admixture.

For the manufacture of the urate on a large scale, it is necessary to construct at least six basins, in form of a watch glass, inverted. They should hold about 12 hectolitres, (about 300 gallons,) of which there will be six of urine and six of the absorbent matter of one of the kinds formerly designated, freshly calcined.

The cask or vessel holding the urine, is to be so placed that it will empty itself through its bung into the basin. During this operation, one workman is employed in pouring in the plaister, another in mixing it in the basin with a rake or scraper.

When the mixture is finished, the operators pass to another, and so on to the sixth. Then the first is emptied of the purpose of commencing operations anew. The mixture is finished by further drying in the air.

At the end of the day, the quantity of urate which has been made since morning, is to be broken down by means of a cast iron cylinder rolled over it; after which it is sifted, (*passée à la double claie,*) and then immediately stored or packed up, to prevent the absorption of moisture.

By this combination, the urine being dried by its union with the absorbent matter, which is itself a manure, unites all the vegetative powers of its two component parts, and will constitute the most productive of all our manures, in consequence of the very small quantity that is necessary to employ to procure the best results.

I have designated six substances as being proper to absorb the superabundant water of urine, and I have only mentioned them without pointing out any particular one, as I thought that no country is without some one of them. But in case I am mistaken in this opinion, very great advantages may still be derived from urines, by mixing them with burnt earth, (that of heath soil is to be preferred,) or with natural ashes, (*tendres naturelles*). I only estimate the value of this mixture as a means of obtaining all the salts of urine in a solid state, which will facilitate its transportation and employment in agriculture.

The ancients considered urine as the most powerful of manures. This is not therefore the end of my invention, which consists alone in its sudden desiccation and solidification, and the draining or drying up of those infectio[u]s depositories of this substance to be met with in the environs of large cities, where they furnish inexhaustible sources of unhealthy exhalations.

I leave it to the learned societies to express their judgment upon the qualities of the urate, and restrict myself to the application for a brevet of invention, for a method of preparing it immediately, so as to destroy at the same time the odor of the urine.

(Signed) DONAT.

Paris, January 19th, 1819.

(Here follows the certificate granting the brevet or patent to M. Donat, for 15 years, signed by the Ministerial Secretary of the interior department, Count Decazes.)

Signed for and on behalf of the board of health

SAMUEL J. ROBBINS.

President.

Attest,

THOS. H. RITCHIE, SEC.

Philadelphia, January 16th, 1826.

However beneficial may be this plan, both for cleansing a city and for forming rich manure, it is evident that the process is not conducted upon uniform, and rarely upon correct principles—which I consider are only conformed to when the absorbing matter used is some form of *mild* calcareous earth. By calcining this material, as the inventor recommends, a destructive, instead of a preserving ingredient is formed—and one which never should be used to mix with foecal or other animal matter, if the value of the product, as manure, is of any account. Far better than this French method is that which has been in use time out of mind among the Chinese—a people, who however unenlightened in science, are in advance of most other nations in the means for preserving and increasing the fertility of the earth. Travellers have informed us, that in that country, human excrements are mixed with clay marl (doubtless rich in calcareous earth). The mixture is made up in the form of cakes, which after being dried, are free from all offensive odor, and indeed give to the senses no indication of their composition; and they are exposed in quantity in the streets for sale, as manure for the neighboring lands.

There is an obvious objection to, or ground to doubt, the effects imputed to decomposing filth as causing disease, in this well known fact, that in spite of this and all other sources of disease, our towns are more free from autumnal sickness, (the effects of *malaria*,) than much of the neighboring and surrounding country; and that the towns have all become more healthy, as they have increased in age and population. These general facts are admitted to be correct, but the inference from them is denied, on the following grounds.

In the first place—bad as may be the effects of the gaseous or aeriform products of animal putrefaction, it is well known that they are much less productive of malaria than are those of vegetable putrefaction. This I believe is a well established and universally received medical fact. And as our *police of health* in the country is at least as bad as in the towns, (though the nuisances are of a different character,) it may well happen, that the vast quantity of decomposing vegetable matter in the woods and in the fields, where there is no calcareous ingredient in the soil to combine with the products of decomposition, and to fix them there—together with the pestilential effluvia from the numerous mill ponds, which more or less affect injuriously half the places of residence in lower and middle Virginia—may produce more malaria and disease, than the decomposition of animal filth in the towns. Besides, there are counteracting agencies always operating to lessen the ill effects of decomposition of filth in towns, though such operation is neither intended, nor understood, by those who profit by it. From various sources, the calcareous earth in towns is always accumulating. The ashes of all the wood consumed as fuel, furnish a large and rich supply—and though these are sometimes conveyed away for manure, still the far greater part is scattered about the town. Coal ashes, in a much less degree as to strength, also add to the stock. The waste of lime, and the old cement of buildings repaired or demolished, all furnish calcareous matter, and all, though without its being designed, are in time spread every where. But the burning of a town, or a large portion of it, as stated in the first part of these papers, furnishes the great supply of calcareous matter—enough indeed to give a very heavy dressing to the whole space burnt over, and much more—

and to serve to combine with all the animal matter for a number of years, and to give permanently to the soil of the town, that valuable quality which is entirely wanting in that of the surrounding poor country.

There is one still more foul abomination in our present system, which has grown out of the want of proper public accommodations, and the extreme difficulty (not to say indecency) of daily removals of uncombined and unchanged excrements from private houses. The practice alluded to belongs to the most crowded parts of cities, and has proceeded from them, and from Europe, to this country, where as yet it is but little used. Where space is very costly, deep pits are dug beneath privies, from which the contents are not removed for years together, and more probably *never*. They do not become *full* (or at least very slowly,) and thereby *compel* their being emptied—because after a certain bulk of the highly putrescent matter has been accumulated, the waste by decomposition goes on nearly or quite as fast as the increase from the daily additions of material. If quicklime is added, this decomposition is hastened, and a different, though but little less offensive odor is substituted. But whether these depositories are cleaned out at long intervals, or not, there can be no question but that nineteen-twentieths of the whole mass goes off by decomposition, and is mixed with the atmosphere; and however diluted, or however altered by mixture, helps to form the air breathed by the inhabitants of towns—who are too delicate, and too fastidious, to have all such nuisances prevented by proper, general, and public regulations. We have not yet been enough crowded in our towns for the last mentioned practice to have gone to much extent. But as it is the result of (supposed) necessity, it will increase with the growth of the towns; and as such receptacles will be of course concealed as much as possible from observation, their existence will not be known, nor the extent of the evil estimated, and scarcely even suspected.

In France, in past times, when there was neither the refinement of manners, nor the knowledge of the evils produced, that now would forbid the introduction of such a usage, large and deep covered pits, or vaults, to privies, were common in the smaller towns, and which were by no means kept for private use. In such public places (*fosses d'aisance*)[11] the rapid accumulation made it absolutely necessary to remove the matter sometimes, though very rarely; and a description of the state of things at such times, and the effects produced, on health, and even life, and the necessity of guarding against them, will serve to show to our citizens, who have never thought of any evils except that of offensiveness to the senses and to decency, that effluvia, always hurtful and sometimes deadly in effect, are actually evolved.

And it should be borne in mind, that the same effluvia must be extricated from smaller accumulations also, though the effects are diminished according to the smaller amount and more gradual extrication, or more diluted state of the doses inhaled by the surrounding population.

There is another and still more disgusting, and still more evident effect of accumulations of putrescent animal matters in towns, presented in the infiltration of the fluid parts through previous strata of earth, and the consequent admixture with the water supplying springs and wells. This part of the subject may be resumed, and treated more at length in a future number of these observations, so as not now to interrupt the consideration of the effects of gaseous or aeriform products of accumulations of putrefying animal matters.

No. III. The Police of Filth, in Towns, Continued. Facts and Opinions on the Subject from French Authorities

The following article, which I have translated from Rozier's "*Cours Complet d'Agriculture*," etc. (Paris edition, of 1815,)[12] will serve to present in a stronger point of view the dangers to health caused by accumulations of foecal matters in towns. The reader may be instructed by its facts and reasoning, as to the importance of the subject to health—and he will also be amused by the display of technical terms, and form of scientific arrangement and classification, applied to such a subject. But this manner of the French author, nevertheless, furnishes additional evidence that the subject has been long studied in his country, and therefore, that the results obtained, and the opinions derived, are the more entitled to respect.

It is proper to premise, that in France, and elsewhere in Europe, the poorer cultivators and inhabitants of the country do not generally have separate and isolated dwellings, as in the United States, but are collected in villages, or hamlets which are surrounded by the fields which the inhabitants cultivate, and the pastures on which their cattle graze. This state of things, which was originally required for mutual security, and which old habits still retain in use, has no existence in the United States, except among some cultivators of French descent, on the Mississippi, and the laborers in cotton or other large factories. It is to villages, and perhaps crowded villages, that the author refers, in the following piece, when speaking of the "*country*," and the "*fosses d'aisance*" of which he treats were common to the use of many persons. The injurious effects described, like the circumstances which produced them, are also (as yet) without parallel in this country. But we have no right thence to suppose that our different habits lead to no danger, or to much less evil than the deep and large vaulted *"fosses d'aisance."* In the latter, as described in the French account, the accumulation of foecal matter, and the concentration of the energy of its poisonous products, serve to exhibit its worst virulence bearing upon a small space, and upon the very few persons most exposed by nearness, or actual contact. But if the matter was diffused, as by the practices and habits of our population, the same kind of fermentation would proceed, the same products be exhaled, and as much deadly aeriform poison be evolved and breathed, but rendered scarcely sensible, in effect, by being widely diffused over much space, greatly diluted, and thus divided—among a much greater number of persons.

(TRANSLATION.)

Fosse d'Aisance. This subject relates directly to agriculture, as furnishing one of the most excellent manures, at the same time that it interests the health, and even the life of the cultivator; for, how many casualties occur in the country (villages) by the emptying of these pits, for want of knowing the means of preventing them!

We shall not speak of their construction, which makes an essential part of the art of building; but we are going to enter upon some details in regard to the various substances which compose the matter, or contents, of the pits. We have not to fear exciting the *disgust* of the cultivator; accustomed as he is to excrementitious matters, he will consent that we shall instruct him concerning that of which he is less informed.

These substances are distinguished by the names of the *crust*, (*croute*) *hècate*, *vanne*, and *scrapings* (*gratin*). The *crust* has often sufficient thickness and firmness to sustain the weight of the laborers walking on its surface! The *hècate* is the pyramidal heaps. The *vanne* is the liquid part, usually of a green color, and is corrupted (*infecte*). The *scrapings* are the parts adhering to the walls, and to the bottom of the pit.

The crust is sometimes pushed up (from the mass below) by a sufficiently voluminous layer of mephitic gas, so as to induce the belief that the pit is full. In this case, the emptying of the pit may be put off to a future time, by merely opening into, and facilitating the escape of the intermediary layer of gas, and thereby lowering the crust.

We proceed now to the accidents occasioned by the emptying, and often even by the mere opening of a (covered) pit. Frequent as such accidents are in the cities, they are much more so in the country (villages,) in consequence of the little experience of this operation. The two only means of preventing them are quicklime and fire.

This article, in Rozier, occupies much extent,[13] and contains a *Memoire sur les Fosses d'Aisance*, which I had prepared when I was engaged in investigating the subject, in concert with Laborie and M. Parmentier.[14] The employment of quicklime consists in slaking it to put it in a state of powder, or to make a fluid by mixing it with very little water, and to introduce it into the contents of the pit, by stirring it with a pole; then the mephitic exhalations are destroyed or confined. The proportion of lime required, depends on the mass of matters, and the cessation of the existence of mephitic gas, of which we may be assured by letting down a lighted candle to the surface. If the flame is extinguished, or even burns dimly, there is still mephitic air undecomposed, and more quicklime is to be added.

As to fire, there are many modes of applying it. Either a chafing-dish of burning coals is placed in the pit, and left there to burn out completely, or dry straw is lighted on it. It is useful to make (in the walls) air holes; they may serve for the escape of the lightest gases; but how little do they draw, when the atmosphere weighs on their orifices! For it is of the *fosse d'aisance* as of the barometer—or rather it is one of the most faithful of barometers. The weather will continue clear, as long as, from the holes over the pit, there arises *ammonia*, that pungent odor so sensible to the eyes, and to the smell.

The crust is sometimes firm enough, I have said, for the workman, without inconvenience, to walk upon its surface. This surface is commonly covered with sulphur, as is also often the vault (or arched covering) of the pit to which the sulphur sublimes, and fixes on. Sulphur is a very abundant product of the fermentation of animal substances; under these circumstances it is formed in the humid mode. When charged with the excavation of the half-moon of the gate of Saint-Antoine, which, from a very remote period, had been made a common receptacle of filth, (*voirie*,) I was struck with the enormous quantity of sulphur with which the earth was impregnated. It is to this sulphur, or rather to the sulphuretted hydrogen gas, which is formed in the pits, and is the most mephitic of known gases, that are principally owing the accidents produced by the operation of emptying these receptacles. M. Dupuytrein[15] has thrown much light upon the different gases held by the matter of the *fosses d'aisance*. At the time when I was occupied in this investigation, Lavoisier,[16] the Abbè Fontanes,[17] whom I had invited to repeat or examine the experiments, could not pronounce upon the diversity of emanating gases; of which one kind, the *mitte*, (as vulgarly termed,) which limits its effects to causing to the workman a momentary blindness, seeming to affect merely the system of optic nerves—whilst the other occasions the painter's colic, and conducts its victim to the state of paralysis, to asphyxia, and finally to death. The experiments of M. Dupuytrein, though interesting in their relation to science, have changed nothing in the preservative means which we have stated, to wit, quicklime and fire, as the destructive or ventilating agents. It is to the negligence of these means, I repeat, to which is to be attributed the accidents that occur from time to time.

Lime is the most energetic disinfecting agent. Throw it into the putrid fluid (*vanne*) of a *fosse*, and it becomes instantly inodorous; it fixes, (*enchaine*,) it decomposes all the mephitic gases. It is thus, that when thrown into a *fosse*, it suspends the extrication of infectious emanations, at the same time that it arrests the tumescence and fermentation of the matter, which is lowered, and the space of time before the emptying becomes absolutely necessary, is thereby prolonged. It is recommended by many to throw snow into the pits, on the pretence of its economizing the emptying, because, as it is said, the snow consumes the excrementitious matter. If the soil in which a pit is sunk is so pervious as to permit the infiltration of the liquid portion of its contents, then the adding of snow (that is, water,) by giving more fluidity, will facilitate the imbibing by the soil. But if the pit is well and solidly constructed, and loses nothing by filtration, the addition of snow does but augment the mass, and hasten the time of emptying. It is thus that people assert every thing, because they believe every thing—and ignorance loves best that which is the most improbable.

There is a phenomenon which it is suitable to mention in this article, as sometimes causing accidents. It often happens that children throw lighted papers down through the orifices of the seats over the *fosses*. When our *barometer-fosse*, in place of ammonia, exhales sulphuretted hydrogen, this gas, the most combustible of all, takes fire, with explosion; and if a sufficient quantity of sulphur is formed upon the crust, or upon the arched roof, the bursting of the roof of the pit may be the result. This gas of *fosses* is also met with in mines, and takes fire there from the lamps of the workmen. But in the mines, the explosion is seldom attended with injury, because there is communication by galleries, or by the ceiling, with the atmosphere: the workman throws himself flat on the floor, and it is as much if his hair is singed by the meteor-like fire, which burns but little, and very rapidly.

But why should *fosses* be permitted to exist, when their contents occasion so many accidents? Their cleaning out is the profession the most abject and most disgusting; and it is difficult to conceive how men can devote themselves to it voluntarily. Certainly, humanity would not permit that a legislator should inscribe such a punishment on the penal code. The workman employed in this wretched business, raises the stone that serves to close the entrance to the vault, and often there immediately exhales a *mofette*, or gas, dangerous or mortal. Another *mofette* is found under the crust into which he cuts, which escapes at the first stroke of the hoe. He puts down his ladder, and descends into this gulf; he makes the [s]ign of the cross, asking the protection of Heaven. He draws out the putrid fluid; at the end of some minutes, it is the *mitte* which reaches him; he is struck with blindness; he is drawn out, if not blind, at least deprived of sight for more or less time. Or perhaps it is the *plomb*;[18] his knees fail, he staggers, and has a universal trembling; cold seizes him, he breathes with difficulty, and he is conducted to the hospital to await the coming of convulsions, violent colic, and other pains, and paralysis which often becomes permanent. Another workman succeeds the first; he has a rope fastened around his breast and beneath his arms, while the other end is held by one of his comrades above, who follows him with his eye, and is ready to draw him up, if he plunges into the *vanne*, or falls extended upon the more solid mass, struck by asphyxia, (fainting) if not by death. It was doubtless, the existence of *fosses d'aisance* among the Greeks, which has furnished grounds for the fable of the mouths of Styx and of Cocytus. The noted Grotto del Cane (of Italy,) does nothing but produce asphyxia—that is to deprive of the signs of life, which are restored immediately by the subject being plunged in the water of the neighboring lake, Agnano. But it is not the same with the asphyxia occasioned by the emptying of *fosses d'aisance*. The sulfuretted hydrogen gas is quite another thing to the carbonic acid gas.

The numerous accidents occasioned by the emptying of *fosses d'aisance* were among the first objects which exercised my zeal in the career of public utility to which I have consecrated my labors. In consequence, I enlisted the solicitude of government, and proposed to it to unite myself with Laborie and M. Parmentier, for continuing the researches which were alike interesting to humanity, to science, and to agriculture. Chemistry had analyzed the excrementitious matters; it had commenced to analyze the gases; but it had not penetrated into the interior of the *fosses d'aisance*, the only laboratory in which to examine the phenomena which the most putrescent substance presents. From these researches, it has resulted, that no workman who will take the precaution which we have proposed in the use of quicklime, and of fire, ought to perish in cleaning out *fosses* or wells; or in the excavation of mephitic soils, to which I have applied, with no less efficacy, these cautionary means against death and asphyxia.

Soft stone (*pierre tendre*) should be used for the construction of *fosses d'aisance*; hard stone has not sufficient resistance. The gases, the most active, the most solvent, exhaling from the excrementious matters which are undergoing an uninterrupted process of fermentation, tend to soften the stone, which they penetrate to a great thickness. I have seen walls of extreme solidity, of which the surface might be crumbled by the fingers—not only the wall of the *fosses*, but those of the body of the (upper) building forming the privies; whilst the soft stone permits the penetration of the viscous fluid, which thus forms a coating that prevents infiltration.

The circular form is so much the more necessary, as I have seen (square) *fosses* of which the cleaning caused no accidents to the moment when, the centre being emptied, the corners were commenced upon. Nothing is more dangerous than to meet with bunches of straw or hay, which have been thrown into the pit; it is rare that they do not conceal a *mofette*, or mephitic gas. In general, all foreign substances add much to the dangers of emptying; it is thus that soap waters (which have been used for washing,) may cause a *fosse* to be fatal to the workmen engaged in emptying it.[19]

I will observe that pits for farm-yard dung, ought to be considered as true *fosses d'aisance*, in regard to their putrid fluid part, the mephitic gas which they evolve, and consequently, of the accidents, which are of familiar character, which attend the emptying of *fosses* of farm-yard dung; so that the means indicated as safeguards in the one case, suit also for the other.

No. IV. The Waste and Destruction of Town-Made Stable Manure, and of Other Rich Materials

The almost universal mismanagement of the manure made in tavern and livery stables, where many horses are usually kept, and the gross ignorance or total disregard of the principles, shown in every particular, are still more strange, and less excusable, than any of the errors and neglects in the matter treated in the last two numbers. In regard to that, it must be confessed that there is much to excite prejudice and disgust—and therefore, there are strong objections to economizing the value and applying it to proper use.

But in regard to the manure furnished by domestic animals, no such objections exist; and no owner of a stable, or cultivator of a farm, has any doubt, (even when there ought to be doubt,) of the manure obtained from horses, (more especially,) being rich, and beneficial in use. Yet, notwithstanding this opinion, the nearly pure animal manure of stables, in Richmond, (where all manures are more in demand for reason of their being more market farms) sells, for 25 cents the cart load (of 20 heaped bushels)—and in Petersburg, at half that price the sale has been always slow, until very recently. Even at these low prices, the buyer sometimes complains of paying too much—and with truth, inasmuch as the half of the bulk, and nine-tenths of the strength of the manure, are destroyed before the small remainder of value is applied to his field.

The dung and urine of corn-fed horses make a manure so rich, and therefore so inclined to run into violent and destructive fermentation, that it is very difficult to avoid great loss of value from that result, even in cool weather, and when there is mixed with the animal matter ten times its quantity of vegetable matter, used for littering the stalls. Few farmers, of even the careful and judicious, who put stable manure in heaps, escape some loss from destructive, *consuming* fermentation, which produces the "*fire fanging,*" whose presence is so plainly marked by *white mouldiness*. When the fermentation is very violent, from too little moisture, and this *fire-fang* is dry, and found throughout the whole mass; then nearly all the rich parts of the bulk have been already decomposed and driven off in gases—and the mouldy and scorched bulk left, though still deemed rich, because it is *stable manure*, is perhaps dear at the usual miserable price of about half a cent the bushel.

Straw, the usual material for litter, or any other kind of litter for stables, is a costly article of purchase in towns; and under the present bad system, this dearness must continue, as even the carriage, from the places of supply in the country, must make the article costly. The owners of public stables cannot afford to pay the prices for litter which the present scanty demand would make necessary to be asked by the seller, so long as there is no better price offered for the manure that is made. Still less, under the existing circumstances and present opinions, can the owners of private stables, or persons who keep only one or two horses, or cows, in the town, afford to pay for a sufficiency of litter—as their trouble and cost would be greater, and their sales of manure less profitable, than of the large stable owner—because every small business is less profitable than large operations. Therefore it is in vain to expect a change from the owners of town stables, unless it is made their interest, by the owners of neighboring farms presenting a sufficient demand for the manure, in its recent and most valuable condition. Therefore, while so little country demand exists for the manure—and while travellers and other horse owners will submit to the present abuse of the animals in stables—there will be no beneficial change in the management. I shall hereafter propose advantageous means for supplying the deficiency of litter; but at present, shall continue the consideration of matters as they now are.

According then to the present system of stable management—if it were not for the bad quality of the hay, which causes much of it to be pulled out of the racks without being eaten, and thus to be trodden under—the horses in most public stables would have nothing beneath their feet when standing, or their bodies, when excessive fatigue absolutely compels them to lie down, except the earth, wet and saturated, if not actually made miry, by their urine.

The rack is kept filled with hay, and the horse perhaps receives a very full allowance of grain. But a dry and clean bed to stand and to lie on, with half as much corn, and a fourth as much of *good hay*, would probably keep the animal in more comfort, and as good health and flesh—while the materials for manure, or the product in the stable, would be increased ten-fold.

The urine of animals, considering the very large proportion it makes of their excretions, is lit[tle] inferior in value, as manure, to the more solid part; and this is lost, on an earthen floor, not well covered by absorbing litter. It merely serves to make the floor filthy, and very uncomfortable, if not absolutely injurious to the health of the horses. When a plank floor is used, the horse is at least kept out of the mire—but the urine leaking through the planks, settles into the earth below, and is there continually decomposing, and throwing up ammonia, and other gases. A *better plan* than either of these is seen in one of the livery stables in Richmond. It has a plank floor, and Shockoe creek passes under the house—and by a few auger holes in every stall, all the fluid, semi-fluid, and much of the solid matter, are gotten rid of with less trouble than by the usual modes, though not more effectually—and the horses at least are free from the inconvenience of standing in filth.

But as every stable owner has not the convenience of a creek to receive the richest parts of his manure, let us proceed to consider the more usual modes. The stable is cleaned out every morning, by sweeping up the dung, and the little wet hay (or other litter if any,) mixed with it, and carrying it in wheelbarrows to be thrown on a *heap* in the yard. The mud formed by the urine with the earthen floor, is swept over very lightly, lest any or it should be removed. Of course, so far as the end aimed at is attained, the urine is all lost to the manure; but acts, in decomposing, to do all the harm it is capable of in exhaling gases. What may be that amount of harm, under these circumstances, or whether there is any produced, I do not pretend to say. It is enough that, as manure, the urine is totally lost. The richness of the materials of the heap, and its want of continued and sufficient moisture, induce speedy, rapid and violent fermentation, which is soon destructive of all the richest and most soluble parts of the vegetable as well as the animal ingredients; and this fermentation continues as long as there is enough remaining moisture to support it. The slow and gradual additions to the manure heap, made every day, just serve to maintain a continual and fierce heat. These daily supplies of new materials are like fuel added to fire. The residuum, after this scorching operation, is the stable manure that is usually applied to the soil. It may be true that the destructive process is not often so effectually performed, nor carried to its final end—and therefore that all the possible damage described above is not produced. But there is scarcely a doubt, but that, on a general average, three-fourths of the value of the new manure, actually produced daily in the stables, is totally dissipated, and lost in the air. Let it be remembered, that the richer the mass, and the larger the heap, the more violent and destructive will be the fermentation, and the greater the loss.

In addition to the theoretical views presented, and the statements of general results, which go to show the little value of manure that has passed through this burning process, I will mention a particular fact just stated to me by a highly respectable gentleman and practical farmer, near Petersburg. When he took possession of the farm on which he resides, he found on it a considerable heap of stable manure which had been brought from the town stables, and put in a heap, *for preservation*, as it is usually considered, to remain until it was time to apply it to the next year's crop. The succeeding occupant, (my informant,) made the application—and obtained from it scarcely any benefit. The manure was much *fire-fanged:* it probably had been thrown into a new and violent fermentation by its removal, and second heaping, in addition to all that is usually suffered in the stable yard.

If, instead of saving the manure according to any of the usual modes, the cleanings of every day were regularly thrown into the river, (as done in part, at the Richmond livery stable named above,) or into the fire, it would be not much loss of value, in the end, compared to the results now obtained—and there would be great benefits gained in other respects. Besides all other possible evils of keeping up a constantly hot fermenting mass in every stable yard, it serves, in town or country, as a breeding place for myriads of stable and house flies. These insects are prompted by instinct to deposite their eggs in the hot mass; and I have seen in the bulk when opened, the maggots which the eggs had produced in such prodigious numbers, that it alone satisfied me of the impropriety of thus providing hot beds and the best possible breeding establishments for these pests. It probably requires a certain combination of heat and moisture to effect the hatching so uniformly and in such great numbers. But whether hatched gradually, or nearly at the same time, very slight observation of a livery stable in summer, will show that the horses, though standing in the stalls, are far from idle; for it requires incessant movements of their heads, feet, and tails, to partially relieve themselves from the continued attacks of legions of sharp-biting flies.

Equally ignorant and destructive management is found in regard to many other smaller sources of rich manures in towns—which are either given up entirely to corrupt the air and the waters, or but a small part of their value is used to nourish vegetables and improve the soil. I should extend these remarks beyond the reader's patience, or sufferance, if every department of this copious subject was examined. I shall therefore pass by all these miscellaneous matters, and confine my remaining observations on town manures, to the means of avoiding the worst of the present losses caused by stable management.

In proposing other modes of economizing the value of stable manure, (and which are applicable also to street droppings, and to sundry other rich materials, not specially named,) I shall keep in view the existing obstacles to every such improvement; and therefore, even if all the changes recommended, were adopted, they would not constitute the *best* system that the most enlightened economy would fix upon. My directions and suggestions, moreover, will be made as concise as possible—and perhaps, may be altogether superfluous to those readers who will keep in mind the general principles that have been advanced and enforced, and who will simply aim, in practice, to avoid the manifest and great evils and losses of the present system of management.

First—supposing the supply of litter to continue, as now, very inadequate for making manure in proper quantities, and for sufficiently moderating the fermentation of the parts. In this case, the manure should never be heaped, but be kept thinly spread until removed to the land, or to be composted in the country with enough of other and poorer materials; and the removals ought to take place every day, or at not much longer intervals. Until so removed, the manure should be protected as much as possible from sun, wind, and especially from rain, or other water so abundant as wash away, or to carry into the earth, the juices of the manure. But all these circumstances, injurious as they may be, are not so wasteful and destructive as the usual hot and violent fermentation. A cellar, or basement floor, under the plank floor of a public stable, if not easily accessible to carts, offers the easiest and best means for cleaning out, and disposing of the manure as made; but heaping there, also, should be avoided, if to remain even 24 hours; and thorough and equal soaking of the urine through the solid parts of the manure, should be secured. The earthen floor of this cellar, or of the stable itself, if there is no cellar beneath, should be covered once or twice a year with a layer of rich marl, several inches thick. This would save (by combining with and fixing) much of the rich fluid which otherwise would sink in the earth for want of absorbing litter, and ultimately be decomposed, and lost. When necessary to remove this temporary calcareous floor, it would furnish of itself one of the most powerful of manures, and be an excellent ingredient to mix with the putrescent matter.

It is an important object at present for town stable owners to procure such kinds of litter as are the *least* disposed to rot, because it will last so much the longer, and a less quantity will serve the purpose of keeping the horses comfortable; at the same time, the least putrescent of vegetable matters, and which on that account are now scarcely thought of as materials for manure, would be effectually decomposed, and brought to the proper state to feed growing vegetables, by the contact and chemical action of so large a quantity of rich animal matter. Thus it well deserves consideration whether very great benefits might not be gained by using as litter such new materials, as spent tanners' bark, saw-dust, the empty hulls of cottonseed, (where hulled previous to expressing the oil,) and the waste of cotton factories. Very large quantities of one or more of these materials are furnished in most of our towns—and which now are thrown away into the rivers, or elsewhere, as altogether worthless. If these, and all other vegetable litter still were deficient, and rich and dry marl were cheap enough, *that* might substitute litter in part; and the daily removal of the wet portion, and the replacing it with from half a bushel to a bushel of the clean and dry marl, would keep the horses comfortable, and secure the liquid manure that would otherwise be lost.

Some of these suggestions might be rendered worthless, and other[s] would be carried into use far more beneficially if the main deficiency of materials were supplied, in a ten-fold greater amount of straw, leaves, or other vegetable litter from the country. This would be best commenced if some individual, who is the owner of a town stable and also of a neighboring farm, would profit by these hints, and furnish both the demand and the supply. He would have both parts of the trade, and would make all the profit that it offered, and have no one to bargain with, or consult, but himself. But he would gain nothing more than would any two persons, the one a stable owner, and the other a farmer, who knew the values to be saved, and would so agree to act that nothing should go to waste. If the country man would supply the stable with good litter of any kind, (leaves, or cornstalks, as well as straw,) and for remuneration carry back a load of rich manure for every load of litter, he would be making a most profitable exchange; and on the other hand, the stable owner after paying away this manure, would have left for sale more in quantity, and there would be more of value saved in it, than in his entire stock as now managed.

The cultivators of farms within short distances of our towns, have very great profits offered to their acceptance, by their using properly the quantity of manure to be obtained from the town stables, (independent of all other, and untried sources)—but, they have availed but little of their advantageous position, and cannot do much better, without adopting some steps which are now almost totally overlooked and neglected. It is not my present purpose to give directions for applying manures, or for farm management—but considering the general deficiencies, in these respects, it would not be proper to omit stating that I consider it essential to the profiting in the best manner by town manures, that the following requisitions should be fulfilled: 1st That the land is to be either first made calcareous, or the putrescent matter previously mixed or composted with enough marl to combine with, and secure it; 2ndly, that the rich town manure, be not suffered to pass through *violent and destructive* fermentation; and 3dly, as far as may be effected, that such manures be applied to crops as fast as brought to the farms, and, in preference to grain crops, to be given as top dressings to young clover, or other grasses.[20]

7 Reclamation—Against Erosion
On Draining: Addressed to Young Farmers
(1833–1834)

There is no one branch of practical farming which is more generally misunderstood or neglected in Virginia, than draining. The dryness of our climate, and the small proportion of our soils that are both level and of a retentive nature, cause this evil to be limited in its injurious effects, and to attract but little notice, compared to other defects in our husbandry. It may, and probably is the case, that many of our good farmers understand correctly the principles of draining wet lands, and execute the different requisite processes in an efficient and economical manner. But such cases are very rare in comparison to the many who are deficient both in theory and practice. It therefore seems to me, though not claiming to possess much knowledge on this subject, that even my imperfect views and experience may be serviceable to young farmers, and bad farmers—and let it be understood that my observations are designed for no others. Good farmers are requested to pass over the[se] pages, of the Farmers' Register, which my desultory remarks may occupy, and to pardon the writer for so occupying them to their loss. Another apology may be due to my readers, for the unpolished form and manner of my writing: if so, I will make it now, and then be done with apologizing. In the first place, I have barely time to write even hastily and carelessly; secondly, my matter will not be of a kind to deserve much labor in embellishing; and lastly, perhaps with all the pains and care I could bestow, I should not make it appear much better.

Drains or ditches are required for three different purposes, as follows: 1st. For collecting and discharging su[r]plus rain water on land which is generally dry. 2nd. For conveying streams. 3d. For collecting springs oozing from the hills, and diverting their course from the land below.

Ditches of the first kind, of which I shall now speak, are the easiest to make and keep in order, and are generally either omitted altogether, or made at double expense, to serve but half their purpose. A drain is required wherever there is a narrow depression of the surface of any land nearly level, in which the water of heavy rains collects and remains until it slowly passes off at the lower extremity, or soaks into the earth. Unless the loss of crop from such a cause is almost certain, and the space of an extent too great to lose, it is generally left to take its chance for a dry season, or gentle rains—by favor of which, the sink may sometimes remain dry enough through the summer. But usually, from excessive wetness, it costs double labor to till, and produces either a scanty crop, or none. When a ditch cannot be dispensed with it is commonly cut by the spade through the middle of the sink to its outlet, and the earth thrown on one or both sides of the ditch, in little separate hillocks, to let the water pass between them into he ditch. Every cleaning out of the ditch helps to convert these separate hillocks into a continued bank on both sides—and that end is still faster reached by the soil being turned towards the ditch by every ploughing, as horses do not (and cannot safely) cross such ditches with the plough. The trouble of stopping and turning the ploughs on reaching the ditch, and the margin thus lost or damaged on each side, amount to a serious disadvantage, even while the ditch serves properly as a drain to the adjacent ground: but in a few years that good is nearly or quite forfeited, by the margins of the ditch being so raised as to bank out the water, unless other means are used to prevent.

Nearly all the trouble and loss caused by this slovenly mode of ditching, may be avoided by using the *plough* in a proper manner to make and repair such drains. Mark off the middle of the sink, through its whole length, and with whatever crook its course may have. Then plough a "land," the sides of which shall be parallel to and equidistant from the middle of the sink, and of course on that line will fall the water furrow made by the finishing of the ploughing. The width of the "land" so ploughed may be from 10 to 30 yards wide as a shallow or deeper drain is wanting—and very often, a single deep ploughing, with a careful running of the last furrows, will serve to drain the sink as effectually as a new ditch cut by the spade. In this case, the work costs almost nothing—as the ploughing should be given when the field is in the course of being broken up for a crop. If the mode of cultivation is in ridges crossing this drain, an additional ploughing of the same land should be given immediately, which will doubly deepen the water furrow for the drain. But if the field is kept under flat cultivation, or in wide beds, that additional ploughing will scarcely be needed, as the drain may be easily and conveniently kept open, by ploughing out a similar "land" wherever the field is broken up. Whatever may be the mode of cultivation, the ploughs will cross this drain without the least difficulty, and there will be no land lost to cultivation. The ditch will scarcely be observed (being merely a water furrow,) but in fact, the land on each side, after a few years, slopes gently towards it for 10 yards perhaps, so that it is actually a drain of 20 yards width. The earth carried into the furrow by ploughs running across will scarcely fill it too much in the tillage of a crop of corn: but, if necessary, the earth so carried in is easily thrown out by shovels, and may be scattered over the widely sloped margins, without fear of raising a bank.

The poor level ridge lands, below the falls of our rivers, are full of shallow basins, which, though often dry in summer, are ponds of rain water all the winter and spring. These ponds are usually in a line along a wide shallow depression, descending towards one of its extremities. As the wetness of the earth, and the roots of trees (when the land is first cleared,) would forbid the effectual use of the plough in such places, a narrow ditch must be cut with spades, and brought from the lower outlet, through the middle of the line of ponds, so as to draw off all the standing water. But as soon as the land is fit to receive good ploughing, (which will be by the beginning of the next course of crops,) a broad land as directed before should be marked off, taking the ditch as its middle, and ploughed out. The closing furrows will be probably as low as the bottom of the old ditch, and sweep away all appearance of it, and leave it passable by ploughs and carts, though more serviceable than when it was a barrier to the passage of both. In short, in all situations of this kind, the plough seems to efface the ditches, while it renders them most efficient. The superiority of these drains, in cheapness and efficiency, to those cut in the best manner by the spade, may be easily conceived, by supposing a piece of flat and wet soil to be thrown into wide and high beds with clean deep water furrows, in the usual manner, by the plough—and compared with similar lands ploughed level, and then divided into beds by narrow trenches being dug between with spades. Every piece of well bedded flat land has in every water furrow such a drain as I have recommended.

In bedded level land, there will be many slight depressions, which even when so shallow as to be scarcely perceptible, will hold water after heavy rains, and destroy the growth of winter crops. If the beds prevent the opening of drains across entirely by the plough, at least it may commence and forward the spade work for these places:—These *grips* (as such temporary drains are called) should be opened only a little deeper than the water furrows which cross them, as soon as the field is sowed in wheat. They may be quite effectual as drains, without being wide or deep enough to obstruct the future ploughing of the field.

The next kind of ditches are carriers of streams, and serve to drain the adjacent land by sinking the level of the streams in ordinary times, and more or less preventing its overflowing its margins, when swollen by rains. These ditches are required in almost every alluvial bottom, formed by, and subject to the inundations of streams passing through: unless the body of water is too great to be manageable by such means.

The streams of lower Virginia may be divided into two kinds: 1st. Such as have so little fall in their course, as to form *swamps*, by overflowing, or at least saturating with water all the lowgrounds during the winter and wet seasons, and thus making the land a worthless quagmire at all times. 2nd. Such as have enough fall to leave the lowground firm and even dry, in ordinary times, except where injured by springs, or other water than that conveyed by the main stream.

The first class of streams are much the most important, on account of the many extensive bodies of swamp land which remain not only worthless, but nuisances, in several respects, and particularly as nurseries of disease; though no lands are richer, or could be brought into profitable use and cultivation, so easily and cheaply, compared to the great gain that would be obtained. Still, it will be unnecessary for me to treat on this branch of draining at length. In this point only, there is nothing to object to as faulty in the practice of individuals—for our laws (indirectly but effectually) forbid all such extensive drainings—and thus, our government shows a degree of negligence or stupidity—(it deserves no milder name—) which surpasses all of which evidence can be found in individual operations. When our country was first settled, it seems probable that these swamps were comparatively dry, and the streams unobstructed, except by the dams constructed by beavers. But every operation of our *civilized* population has served still more to raise, obstruct, and stagnate the waters. The only profit yet drawn from the swamps, has been by getting lumber from the large cypresses and other timber trees. In cutting down these trees, their tops are very often thrown into the course of the stream, where each serves to catch all the leaves and other floating rubbish, until it forms a dam, and raises, and often diverts the stream, to a new bed. The current is at no time sufficiently strong to remove such obstructions, although it may be spread over a flat of half a mile in width—and every one remains, until covered over with a deposite of mud. The law permits any land owner to add to these obstructions at his will—but (in effect) refuses the right to use the only means for bringing into profitable culture these great tracts of rich land, and of restoring health to the neighboring farms, which they now infect with bilious diseases.

Notwithstanding the great extent of overflowing waters on these flat swamps, the supply is much smaller than it appears, and they could be removed and kept within safe bounds by opening a canal from the outlet below, through the whole course of the swamp, as straight as the form and inclination of the land would permit: Level as such swamps are, there is plenty of fall for this purpose—and a ditch of 10 or 12 feet wide and 3 or 4 feet deep, would drain away the water which as now obstructed, inundates many thousands of acres. The expense of this central main carrier would be very inconsiderable, divided among all the owners of a large swamp; and when finished, nothing more would be wanting to make the land dry, except the small side drains to intercept the springs coming out of the highland, which each proprietor would dig for himself. The central canal being so nearly level might possibly be made also serviceable for winter navigation, by having temporary floodgates.

But cheap and profitable as such drainings would be, they are rendered impossible under our existing laws, because the concurrence of every individual owner of the swamp is necessary for the execution of the work. Blackwater Swamp (for example) is more than 60 miles long, including all its branches, and perhaps belongs to more than thrice as many individuals—and it is manifest, that from such a number, no such concurrence can possibl[y] be expected, even if there were among them no minors, or life estate holders, neither of whom can legally concur. If by possibility, only a single proprietor opposed the scheme, while all the others were in favor of it, he alone might obstruct the execution. Nor is there any remedy to be soon expected. If three out of every four of the proprietors of any of these swamps were to be awakened to the importance and profit of such a general plan of drainage, (and I am sorry to confess that such is far from being the case,) and were to petition the assembly for powers to make it [a canal], and to compel all to bear their share of the cost, the proposition would excite violent objections, and perhaps intolerant and unappeasable enmity to the scheme. Every small lawyer, in and out of the legislature, would be furnished with a most convenient theme. We should hear the plan denounced as an invasion of the "sacred rights of property," and the denunciation maintained by so many arguments (or what would pass for arguments) that the advocates would be glad to retreat from the wordy inundation. But as plausible as such arguments may be, precisely such might be urged against opening the existing, or any roads, through private property, if we can suppose such a case possible as the country being settled and cultivated, without having a public road within its limits. Roads are cut through private property without asking leave of the owner, and he is also taxed according to his property, to pay his share of the expense of construction. Sometimes it happens that the road for which a proprietor is so taxed in his landed rights, and on his purse, though beneficial to the public, is to him individually a source of inconvenience and of loss. Still these exceptions are properly considered as no objection to the general regulation, for the general good—and the lawyers raise no objections, because the policy is already sanctioned by *law*. But if all of lower Virginia had been one great swamp, held by thousands of individual proprietors, and which could be drained as easily as Chickahominy and Blackwater swamps now could be—according to our laws and to the arguments of lawyers, there would be no possible means, consistent with justice and the principles of our legal policy, by which this beneficial improvement could be effected.

But I have already said too much upon a branch of draining which was only intended at first to be named as a matter to be omitted. My purpose was to advise practical operations which each individual may perform—and I have allowed myself to digress (uselessly I fear,) upon what individual efforts are altogether forbidden.

Streams of the second class, having sufficient fall, are generally such as flow through a hilly country. The lowgrounds, or bottom lands, lying on the borders of such streams, form a large proportion of the best natural soils of lower Virginia. Indeed but few other soils are richer than these have been, or would be more productive, if they had been properly managed: but the general treatment of such lands has been so injudicious, that they have yielded but little net product, and in many cases have become nuisances, and a source of loss instead of profit. I allude especially to lowgrounds on small streams, not exceeding the size sufficient for an ordinary mill. Some of the usual and barbarous practices will be pointed out for avoidance, and also because their effects now present some of the worst obstructions to a proper plan for drainage and cultivation.

The bottoms through which the streams run, have been entirely formed during past ages by the earth washed from the higher lands by heavy floods from heavy rains, and deposited so as to form a nearly level surface. Of course the greater part of this deposite has been made from the main stream, and at the times when it overflowed the whole lowground. But it is not only during such floods that the operation is going on. At all times a shallow running stream is bringing down earth, and thus raising its own bed, until it leaves it for another and lower place, or when a flood comes, throws the accumulated sand out of its choked channel, over every place low enough to receive an accession. Thus, by the tendency of the overflowing water to cover mostly the lowest land, and from the greater subsidence of the suspended earth, where the water is most deep and still, nature works continually to keep such lands level from side to side. Before the adjacent hills were cleared and subjected to the plough, there could have been no great supply of earth, except from the richest soil on the surface—and that was furnished slowly and gradually.

Rich as these narrow bottoms were, our fathers did not readily undertake to drain and cultivate them. Before this was done, the adjacent highlands had in most cases been cleared, cultivated, and washed into gullies—and had served to throw upon the lowground more of barren subsoil in a year, than it had before received of rich mould in ten. Nor was this injurious deposite brought down by the principal stream, and spread over the whole surface. It mostly was brought by torrents of rain water, which for a little time swelled the rivulet to a flood, and by which the sand or gravel was carried out on the rich bottom soil, in points projecting from the ravine through which the torrent rushed. These points of sand, by their thickness and poverty, now form one of the greatest difficulties in draining and cultivating the lowgrounds.

The management of the neighboring highland, so far as its washing is promoted or prevented, is one of the most important things bearing on the alluvial bottom below. If no mischief had been already produced from this cause, the instructions that will be offered would be more simple, and yet far more serviceable. Prevention is always better than cure—and in these cases, the perfect cure is impossible. We have destroyed the greater part of the value of our low grounds, before we knew their productiveness.

Next come the injuries inflicted directly when such lands are under cultivation.

Bottoms of the kind under consideration are generally from 150 yards wide, to the narrowest size worth draining—meandering continually in their course,—and having sufficient fall or inclination to give a rapid course to the natural stream, and to allow the land to be effectually drained for cultivation. The stream is still more crooked than the valley through which it runs, and is often twice as long as would be the straightest course that might be given. Nor is the course of the stream always through the lowest part of the land—for the margins of the stream are often the highest parts, owing to the more plentiful deposite of sediment when the overflowing waters first rise over their banks. The first error usually committed is to leave the stream (if a large one) in its natural crooked bed, instead of giving it a shorter course: the next is, to impede still more the course of the water by allowing thickets of briers and shrubs to stand on the edge of the stream, and every kind of rubbish to be thrown into it. The crooked and choked channel causes the stream to overflow with a rain that would not swell the current injuriously, in a clean and straightened bed, though of no more average width, and occupying not one fourth as much land. The land lost is not only the bed and banks of the stream. A very crooked course makes it impossible for the plough to run in the same direction: and many points of land are formed too narrow to be worth cultivation. Hence a wide margin is left to grow up in thickets, and to harbor muskrats and other vermin—or to be kept cleared at more than twice the cost of proper cultivation. If the stream separates the lands of two proprietors, (as is one of our common follies,) the evil is far worse. If either cleans his margin effectually, and opens the stream, it will avail but little, while his neighbor uses no such care. In the meantime, with every heavy rain the obstructed current sweeps across the land, tearing away the ploughed mould, or covering it with its load of sand. The soil thus swept off serves to fill and render useless the cross drains made to convey the smaller streams. Every such flood saturates the soil with water to the great damage of the crop, and leaves the whole a picture of desolation.

Next let us examine the side, and cross ditches. In hilly and sandy lands, small springs ooze out of the hills so frequently along the side of the lowground, that there is almost always a necessity for a ditch on each side, to intercept them. The side ditches (unless they are covered or hollow drains, which will be treated of hereafter,) if bordered by cultivated land, are perpetually filling from the washing hillside soil; and the difficulty of keeping them open is still more increased, where they cross the points of sand brought down by former currents, and which are still increasing from every torrent of rain water. It is not strange, therefore, that the side ditches are seldom clean; and though they obstruct, they seldom entirely prevent the oozing waters finding their way to the lowground. Where the valley is narrow, or the main stream small, one of the side ditches is often made also the main carrier, or channel of the stream. This is a great saving, if the ditch is straight enough, and so situated in other respects as to be kept open, and deep: but otherwise, the side of the lowground is the worst location for the main stream.

Cross ditches are used to bring the water of a side ditch to the main stream, when it can no longer be continued down the side; or they lead from low places which though not exposed to spring water, would be made ponds by receiving the surface water from rains and floods, and would remain so until the water was soaked up by the absorbent earth. The beds in which the land is cultivated usually lead to these cross ditches, and their water furrows there vent the water that collects in them. As the cross ditches fill rapidly from the causes already stated, their repeated cleanings soon make a dike not only on *one* side, but on *both*, unless more judgment is exercised than is common. Then the ditch, when well open, may draw water from its head to its outlet: but not a drop can enter it along its whole course, so well is it defended by the bank of earth on each side.

Now for the rotation and culture.

Lands of this kind seldom form but a small portion of a whole field, and are therefore not often put under a different rotation. Say that it is the common three shift rotation, or 1. corn—2. wheat—3. at rest. When the winter comes preceding the year for corn, the land has been two years without a ditch being cleaned out: and if grazed, the treading of cattle and rooting of hogs have been aiding greatly to fill them up. If not grazed, the richness and wet state of the land have made it a wilderness of weeds and rubbish growth of all kinds. In the latter case, ditching in autumn would be almost impossible; but even if kept bare enough by grazing, no farmer has leisure for a heavy job of ditching before winter. At all events, it is never done. The land treated as I have stated is almost as wet as if no draining had ever been done—nay, it is often much the wetter for the work miscalled draining. But little of wet ditching can be done in cold weather: so it is in March, before the old drains are opened. Still the land is very wet from having remained so long water soaked—and it is ploughed before it is dry enough, because the season is too much advanced to wait longer. Under such circumstances the land cannot produce near a full crop, even if the draining was then perfect, and continued effectual for the remainder of the year. During the following crop of wheat, the drains are filling, and seldom opened, and during the year of rest afterwards, the former water soaked condition of the land is completely brought back.

This picture does not in every particular apply to all such lands, even when most badly managed: but some of the traits will suit all, not excepting some in the hands of the best farmers: for in the management of lowground especially, we often want the means to perform what our judgment directs should be done. For example—who is there who does not pronounce, when ditching in March, that he ought to have done the work in September? and who is there who profits by his own opinion and experience, so as to avoid the same error in future?

I now proceed to propose plans for draining and cultivating soils of this kind, all of which I have tried with success to such extent as was permitted by the situation of the land in my possession, and the circumstances under which it was placed.

It will perhaps be more plain to apply instructions to a particular case of common occurrence, than to attempt to embrace every variety of circumstance and difficulty. For this purpose, suppose the land under consideration to vary from 100 to 150 yards wide, the alluvial formation of a stream strong enough to turn a common mill, and which flows through in a very crooked channel in ordinary, and commonly overflows the whole bottom with every very heavy rain, or perhaps two or three times a year. The average descent of the stream and the land, from six to fifteen feet in a mile. The soil a rich sandy loam. Subsoil various: sometimes layers of sand within reach of deep ploughing, sometimes of clay, and sometimes (though rarely) the rich black surface soil shows no change for several feet in depth. We suppose farther that the land has been cleared and cultivated long enough to give the plough generally a free passage.

Lowgrounds of the kind under consideration are in general more exposed to water from numerous springs oozing out of the adjoining highland, than from the main stream—and therefore the removal of the former first demands the farmer's care. But neither the side, central, or cross drains can be finished, before the other kinds are in progress—and it will suit my arrangement best to speak first of the ditch or carrier of the main stream. The side drains require most skill and care, and their consideration will hereafter be undertaken. For the present, let it be understood that the side and cross drains are in the usual imperfect state of operation, serving to permit the imperfect cultivation of the lowground.

The great object is to give the main stream the shortest and best course through all the extent of low ground to be drained. With this view, the shape of the ground, and the force and size of the floods should be well considered, and the new course for the stream determined accordingly. In general, it will be cheapest to adhere nearly to the straightest course—which, in a crooked bottom, will cause the line to touch the projecting points of highland, first on one side, and then on the other. But desirable as are long straight stretches, we must take care to change their direction very gradually when a change is necessary. In a long straight course, with sufficient descent, the water acquires a force which enables it to keep its direction, in spite of considerable obstacles—and will rush across, and fill up with its deposite, any part of its channel which turns off at an angle, or with a short curve. To avoid this danger, it will be sometimes proper to begin a gentle curve before reaching the point where it would necessarily be made. There is another case in which straight courses should be departed from— that is, when with no great variation of direction, or increase of distance, the main carrier may be made to keep along the side of the low ground for a considerable distance, which will so far serve to avoid the trouble of another side drain, and also preserve the low ground in one unbroken body, at that place. But desirable as it certainly is to have the stream kept at the side of the low ground, it should not be done unless the location is good with a view to perfect drainage. A ditch at the junction of the low and highland, is far more subject to be filled with rubbish and earth brought by rains, than if in the body of low grounds—and, therefore, if so situated, its course must not be too crooked, nor the force of the current too small, to guard against that danger.

When the line from the ditch has been fixed, it should be marked off by stakes, wherever not plainly enough exhibited by some existing marks. It will be generally found that the line will divide the low ground into large pieces, shaped something like segments of circles, the straight sides of which will be the new line for the stream, and the curved sides made by the hollow bends of the inclosing high land. The new line will probably cross in many places the serpentine bed of the stream. The work should be commenced in the dry season, and on the driest parts of the land, if any are too wet for the operations required.

On a part of the new line, say from 50 to 200 yards in length, and extending from the old stream at one place of crossing, to another, lay off with a plough, well and deeply, a *land* of about 12 feet wide, the closing water furrow of which will be the centre of the intended canal. This width of ploughing will be sufficient, if a passage for the water six feet wide and 2½ feet deep will serve: but the larger the canal is desired, the wider should be the ploughing. Such a stretch as is here spoken of, is supposed to pass through the body of low ground. As soon as the plough has cut a few furrows, laborers with broad hoes begin to draw out the loosened earth, and to deposite it, with very little regard to accuracy, on the land outside of the ploughing. When the plough has closed its work and formed a deep water furrow, it begins again and goes over the same land, whether the hoes have finished ahead or not. A third time the same operations may be repeated, or until the ditch is either nearly deep enough, or the bottom has become too miry for the horse to walk on. In this manner the greater part of the digging and removing of the earth may be done at a very small cost, compared to spade work. Still there remains something for the spades to finish. After the last ploughed earth has been drawn out, the ditch of the desired width (say 5 to 7 feet) should be accurately laid off by a line, and the stakes first set up to mark the course. A single spade's depth will generally give sufficient depth, and the work will be very easy to perform. There is no need, generally, of digging low enough to divert at once the stream to the new course. It will be sure to take the new and straighter course at every rise of water, and will naturally deepen the new, and at the same time be filling up the old channel. This operation may be hastened by opening well the upper end of the new channel at each crossing place, and obstructing somewhat the old passage just below, by the top of a tree or other rubbish, which, though serving to impede the floods, will not prevent the passage of the stream in common times. It would be improper to stop the water entirely from its own channel, as that would prevent its being filled up, and it would remain in the way of cultivation. But if a current has choice of two channels, united above and below, the one straight, and the other crooked and twice as long, the effect will certainly be, sooner or later, to deepen and enlarge the first, and to deposit its mud and sand in its slower passage through the second, until it is entirely filled. It is much cheaper to let nature thus aid your draining operations, than to dig the carrier at once as deep as desirable.

When the first rough part of the excavation, by ploughs and hoes, is finished through one stretch, it may be begun on some other—either adjoining, or distant, as may be most convenient. As the old channel for a long time will continue to convey the stream, it serves to keep the new work in different dry sections, to be opened as may be convenient. Adjoining sections should be connected as soon as possible (and by the spade if necessary) so as to have the benefit of any flood of rain that may occur.

When the main carrier is intended to be made for some distance along the margin of the highland, the earth must be thrown by the plough altogether towards the lowland. For this purpose it will be cheapest to use a hillside plough, which, by shifting the mould-board, turns the furrow slice to the same side, whether going up or down. If a common plough is used for such places, it must cut only when driven down the course of the valley, and be dragged back empty, to begin another furrow, at the upper end of the stretch.

If the owners of low grounds would act according to their true interest, this plan would be extended as far as the nature of the land required it, without regard to whom might be the owner of any particular spot. Then each proprietor would be benefited by the drainage of the land below, serving as an outlet or vent for his own. But that is not now to be counted on, and each person must expect his drains to end with the termination of his land. If there is much fall in the stream at that place, the injury from this stoppage will not be considerable, except perhaps to the next land below. There the water increased trebly in velocity by its clear passage above, and finding no straight or sufficient channel below, will probably rush over the land, and expose it to all the damage which the owner will well deserve to sustain. If, on the contrary, the fall is inconsiderable, as in the swamps before described, the lower landholder could render ineffectual the draining of the land just above. In such cases a good vent to the water below is highly important, and the want of it may destroy half the benefit which might be derived from the whole drainage.

When streams are thus straightened, and their sides kept clean and smooth, they will carry off quantities of water that could not be kept within the former stream, even if four or five times the superficial extent. But I do not mean that inundations will be altogether avoided, though they will be comparatively rare—and, when they occur will be of short continuance.

But there is an objection (and unfortunately a very general one) to all such schemes of drainage. Streams are generally made to serve as dividing lines between different properties, and that circumstance alone is sufficient in most cases to prohibit any rational scheme of drainage. When lands were first taken up under the old patents, and sold out without accurate surveying, a stream was a very convenient land mark, because it could not easily be changed or mistaken. But for the drainage and proper cultivation of the lowground, the stream is the worst dividing line that could be fixed upon. All my foregoing directions on this subject must rest on the supposition that one person owns both sides of the stream—or that the different owners are willing to concur in the best general plan of drainage, and in the exchange of points of land cut off by the new carrier of the water. Either of these cases is so rare, that I must agree that the directions I have written are almost useless, and an unprofitable waste of the time of my readers, as well as my own. However, should I find that any value is attached to my suggestions, I may resume and finish my observations, as at first designed. It remains to treat of side drains, open or covered, intended to intercept springs having their sources in the highlands.

R. N.[1]

Part Three

Travels and Expositions

8 Farming the Great Dismal Swamp
Hasty Observations on the Agriculture of the County of Nansemond (1836)

The county of Nansemond, exclusive of the considerable portion embraced in the Dismal Swamp, and other smaller swamps and the tide-marshes, is generally of sandy, and naturally poor soil, moderately high and undulating on and very near Nansemond River and the tributary streams, and very level in the larger remaining part of the county. The smaller portion of river lands, and the larger body of the black lands, or "piney woods," as called in distinction from the river lands, are very different in several important respects, besides that of the nature of the soil.

From the very short and limited opportunity which I had for making observations on the agriculture of Nansemond, it seemed, with a few exceptions of spirited and profitable improvements, to be in a very low state: lower indeed, considering the great natural resources for making profitable improvements of the soil, and products from its cultivation, than any other part of Virginia, that I have yet seen. It is reported, however, that there is a still more rude and degraded state of agriculture, a still greater neglect of the advantages offered by the soil and its locality, in the adjacent counties of Norfolk and Princess Anne: as if the more that nature offers to man of her bounties, the less he is disposed to avail of their benefits, or indeed of any portion of them in a proper and profitable manner. It is indeed passing strange—and not sufficiently explained by all the reasons stated, that the country immediately surrounding the oldest, and long the most important commercial town in Virginia[1]—having every advantage of markets—with much productive soil, and the most abundant resources for enriching the balance—with waters offering nearly as rich products as the lands—and in addition, with the long continued, and long to be continued annual expenditure of vast sums from the national treasury on the great public works in this region—that notwithstanding all these and other advantages, that this oldest settled and longest cultivated part of Virginia, and of the United States, should be now the least removed from the wild state of nature, that the inhabitants generally seem to be losing rather than gaining, in the general progress of improvement, knowledge, and refinement. If, in addition, the mania for emigration to the west should rage hereabout as it is doing in the middle region of Virginia, the time may come when the howling of wolves will be heard from the suburbs of Norfolk.

The most general rotation of crops in Nansemond, (if any deserve the name of rotation,) is 1st, Indian corn, 2nd, grazing—and so on, as long as the land will bring a crop that will pay half the cost of tillage. A still more scourging course is not unfrequent, which is corn every year, as long as the land can bear it, or until the more extended clearings of the cultivator permit him to adopt the improvement of the two-shift rotation above described, and to let his fields have "rest" every other year, under the close grazing of such natural herbage as may attempt to grow. Upon the naturally high fertile shelly lands on Nansemond River, almost continual tillage under corn, as a matter of choice, for many years together, and with little or no manure, was the general practice: and such was the astonishing durability of the lands, that to the last they gave pretty good crops, and exhibited evidences of great power of resuscitation in the soil, if allowed any chance by proper resting, or putrescent manure. On the poor back lands, where under corn every year, field peas are also raised in every intermediate row. Of these but part are gathered for seed, sale, or home consumption, and the hogs are turned in to eat the balance, as soon as the corn is fit to be gathered. This mixture of the pea crop is general, and in some measure mitigates the severity of both the two-shift and one-shift rotations. But this kind of cultivation is by no means confined to this county. It has been very general through lower Virginia, and is still to be found in nearly all the lower counties. Nor is the improvement of soil less frequent in Nansemond than in some other of these counties—it is only the more remarkable and condemnable there, because the marl and other manures in that county offer such abundant means for improving the soil and increasing the products. In travelling on the railway, from Portsmouth to Roanoke,[2] I saw not an acre of the numerous and extensive cornfields that seemed to have produced more than 15 bushels—and much the greater part would not have averaged more than 10 bushels. It is true, that the manured spots and best grounds were generally in cotton—but even of these, very few fields, if any, would have brought 20 bushels of corn to the acre. I do not estimate the cost of tillage (at least in Virginia,) half as high as does the editor of the *Cultivator*,[3] who has stated that 30 bushels of corn per acre is the smallest crop that will give any profit to the farmer: but nevertheless, I think that not half of all the fields seen on this route can leave any clear profit on the cultivation.

In Nansemond River are caught immense quantities of fine oysters—and the Indian banks of shells all along the river, furnish testimony that the river has for ages yielded, in abundance this valuable product. On every elevated point on the river, and indeed generally along its banks on both sides, are found spread the prodigious deposites of decayed oyster shells which give the high and well deserved reputation to these lands for fertility, and durability of productiveness under long continued grain tillage of the most scourging kind. I could not learn any certain facts as to the exact length of time during which any particular shelly field had been kept in corn, without rest or manure, and what was its rate of productiveness after such hard treatment. But I found existing a general belief that but few of such lands had escaped 20 or more years of such continued cropping—and on some, the cultivation for 200 years has been never lighter than two-shift rotation above mentioned, with no more manure than the proprietor could scarcely help making. After all such treatment, these lands, however greatly reduced from their first exuberant fertility, were still what was deemed good and productive soil—and yielding crops that would have given the character of *rich* to any land producing such, in the south-eastern counties, away from tide water.

Having no leisure for more extended observations, I visited only Stockley, the farm and residence of Wills Cowper, Esq. which was the site of the principal town of the Nansemond tribe of Indians, and of whose separate huts, the ground still affords testimony of the localities, in the greater abundance of the shells, swelling above the general level at each spot where the surface had not been moved, otherwise than by ordinary tillage. Just opposite in the river is Duplin (now called Dumplin) Island, the former residence of the kings of the Nansemond tribe of Indians. This island contains about 14 acres of high lands, and a considerable body of marsh.[4] These collections of shells, all along the river, are from 2 to 3 feet deep nearest the river banks, and spread back, covering the whole surface, and gradually thinning, so as to form a large portion of all the fields on both sides of the Nansemond: and these original artificial deposites make the celebrated fertile shelly lands. It is but within a few years back that Mr. Cowper and a few others have greatly extended the surface of this kind of soil, by digging down some of the thickest parts of the bed, and carrying the shells, and the rich black earth with which they are always intermixed, to other lands, for manure. This is giving, at the same time, both calcareous and putrescent (or alimentary) manures—and the efforts have been as good as might have been anticipated from such applications, where both kinds of manure were wanting.

It was this shelly soil, and others similar elsewhere, though but little known by personal examination, that furnished to me long ago one of the strongest and earliest proofs of the value and manner of operation of calcareous manures. In investigating the subject at a later period, I obtained from Mr. Cowper specimens of the soil now first seen by me—all of which showed a remarkably large proportion of calcareous matter, though most of it (fortunately) is still in the state of coarse shells, or their fragments. The following statement of the composition of one of them, from the *Essay on Calcareous Manures*, (p. 20, 2nd Ed.) will serve as a specimen of some of the best soil of the Indian Banks.

"Oyster shell soil of the best quality from the farm of Wills Cowper, Esq. on Nansemond River—never manured, and supposed to have been cultivated in corn as often as three years in four, since the first settlement of the country—now yields (by actual measurement) thirty bushels of corn to the acre—but is very unproductive in wheat. A specimen was taken from the surface to the depth of six inches, weighing altogether

242 dwt., which consisted of

126—of shells and their fragments, separated by the sieve,

116—remaining finely divided soil.

Of the finely divided part, 500 grains consisted of

18 grains of carbonate of lime,

330—silicious sand—none very coarse,

94—impalpable aluminous and silicious earth,

35—putrescent vegetable matter—none coarse or unrotted,

23—loss.

500."

I refrain from stating more particulars of these interesting and valuable shelly lands, because it has been promised that better information on this general subject, and more full details, shall be furnished from a highly respectable source. It is enough here to say that the value of the Indian shell banks as manure, as well as of the marl found in various places, have been used to great advantage on Stockley, as well as on some other river farms.

Wheat is scarcely grown at all in Nansemond, and the shell lands are decidedly unfavorable to that crop, but very friendly to clover—though this grass has but lately been tried, or even supposed capable of being here made. On clover on this shell land, Mr. C. had tried gypsum without any profitable effect. This result is very different from what I had counted on. But there was also another difference in result—and one may serve to explain the strangeness of the other. Though perfectly satisfied that lime in soil, is essential to the sure and profitable growth of clover—and that shells or marl will insure its healthy growth on even sandy and poor land—I have not learned from experience, that calcareous manure alone would sustain *luxuriant* clover, on a sandy soil, even when aided by putrescent manure: though supplying gypsum in such cases, another necessary or specific manure for clover, will generally produce a heavy crop of this grass. But this made by Mr. C. was a very heavy and luxuriant crop, without gypsum—and was no better where it was applied. I infer that nature had somehow supplied the soil with gypsum in sufficient quantity—which supposition will serve to explain all the results.

The farms on the banks of the Nansemond River are in every respect, save one, most desirable as places of residence, as well as for agricultural profit. That exception is their being very subject to the only disease that has peculiar or much power in lower Virginia—autumnal fevers, the effects of malaria. I am loath to believe the cause of these diseases to be beyond human control. It is not probable that the neighboring salt marshes are the source of this evil, because in other parts of lower Virginia, in the neighborhood of other marshes apparently precisely similar, the country is comparatively healthy. So far as malaria emanates from the dry land, (as no doubt it does wherever circumstances are favorable to its formation,) there is little doubt but that this source may be closed up by the general and sufficient application of marl, or other calcareous manures. The great sources of autumnal diseases in lower and middle Virginia, is more likely to be found in the mill ponds—and these are abundant along, at least, part of the sickly borders of this river. If the meeting of salt and fresh water is a fruitful source of malaria, or the spreading of salt water over fresh marshes, (as stated in an article in the last number of the Farmers' Register,) then the discharge of floods of accumulated water, after heavy rains from these mill-ponds, may, by overflowing the adjacent salt marshes, produce much of these deplorable results, in addition to the universal extrication of malaria from the ponds themselves, when lowered by summer droughts. The pestilential emanations thus produced from mill-ponds, and from the discharge of their floods upon salt marshes, may thus serve to produce much the greater part of the autumnal diseases so prevalent on the banks of this river. The remedy for this great evil will only be found in the general abatement of the greatest physical nuisance in our country, viz. mill-ponds on small or insufficient streams. As to the smaller amount of such disease justly attributable to emanations from the general face of the country, it would be greatly limited, if not annihilated, by that kind of manuring which would most redound to the profit of the farmers, in other respects as well as in preserving health. Such has been the undoubted result on farms and in neighborhoods in Prince George, where marl has been extensively applied. But it is evident, that to derive this benefit fully, all the lands of the neighborhood, whether in wood, meadow or arable, should be made calcareous on the surface.

The salt marshes in this neighborhood are generally high and firm, and rarely covered by tides. The growth is of different kinds of salt grasses, which are excellent food for cattle in the spring, and beginning of summer, but which become unfit afterwards. If this is (as is alleged) from the older state of grasses, it might be prevented by mowing the marshes, and thus providing a young and tender growth in the latter part of summer. The kinds of grass which grow on these marshes affect that locality more because of its saltness than its wetness. As a proof of this, I saw the same kind of grass, of good growth, covering the lofty ramparts of the fortress at Old Point Comfort, though the marsh mud which had been used to construct them, had been for years in this very dry situation.

Marl has not been in use but a few years in Nansemond county, nor is it now used to much extent, or by many persons, compared to the great number who have it in their power to profit well by this manure. I saw, as well as heard of, striking improvements made by marling, on the farms of Mr. J. T. Kilby, Mr. Mills Riddick, sen[ior];[5] and Mr. J. Bunch, all in the neighborhood of Suffolk. But still, the example of these gentlemen has had but little effect in stimulating, or even awakening their torpid countrymen. Some persons went zealously, but not judiciously to work: they put rich marl on poor land in quantities far too heavy; and having thereby injured their land, they have taken no means to correct the error. They are content to abandon marling, and to stand by their first fruits of it, as full evidence of its worthlessness, or the danger of using it.

This county is celebrated for its crops of very fine sweet potatoes; and on some of the lower river lands, water-melons are a considerable and very profitable field crop. They are sent by vessel loads to the northern cities. For such distant markets, they are gathered before they are quite ripe, and therefore never can be of as fine flavor as when left to ripen on the vines. The greater part of the sweet potatoes are sent to the northern markets.[6]

The former owners of farms on Nansemond River have generally lived in the exercise of what has been so much lauded as "old Virginia hospitality"—and which deserves more to be reprobated, as the system or habits that have tended more than any thing else to ruin the estates and their owners, or their children, throughout lower Virginia. I speak not in condemnation of true hospitality; but of its excess and abuse. According to the long established precedents in such cases, few of the lands on the Nansemond, are now held by the children of former possessors—and still fewer by the same individuals who held them twenty years ago. A contrast to this state of things is exhibited by the people who cultivate the poor lands of the "piney woods." Deprived of all the comforts and luxuries furnished by the river, and compelled by the difficulties of their situation to be laborious and frugal, they have prospered in their estates, in spite of poor lands and a wretched system of agriculture, as much as their richer and more favored countrymen have done otherwise. They live plainly, but comfortably, have no fear of the sheriff before their eyes, nor anticipations of leaving their children in want, or dependence—and are able to exercise, and doubtless do exercise, true and kind hospitality, without ostentation, and without the waste of the entertainers' time and good habits, still more than of his victuals and drink, on idle loungers and dissipated companions and visiters.

It was stated in the early part of these remarks that the low state of agriculture in this and the counties lying to the east, was inexplicable. But the reasons given for it will serve as a partial explanation. The main one is the inducements held out to capitalists, and still more to poor men and laborers, to get timber for market, rather than to till the earth for slower-coming gains. The immense quantity of excellent timber in these counties has afforded occupation, in getting it up for market, to any amount of labor: and the northern demand for shingles, and that of the navy yard in Portsmouth, and the works at Fortress Monroe [at Hampton, across Hampton Roads from Norfolk], for every other kind of building and ship timber, have offered ready and profitable markets for all the products of labor thus directed. Even though the general and average returns of this and agricultural labor had been equal, the latter would be received only after a year's time (at shortest,) while a load of timber could be cut and sold in a few weeks. But the profits of the timber business were the best. Besides—the great public works long carrying on at the navy yard and the fortifications required a great number of ordinary hands, for which enormous hires have been always paid. This served to take away from agricultural labor much of its proper laboring force—and induced the owners of labor and capital to look to other sources of profit than cultivating or improving the soil. These advantages, have had evil, instead of good effects.

9 The Great Dismal
"Jottings Down" in the Swamp (1839)

No. 1. Blackwater and Chowan Rivers,
and the Mouths of the Roanoke

Plymouth, N.C. Nov. 19, 1839.

The singular features of the upper Blackwater river, or swamp as it is called above the limit of navigation, have long ago attracted my attention; and more than once I have attempted to induce the landholders on its borders, and the public, to appreciate its neglected *capabilities* for improvement, both for navigation, and for the drainage of the extensive swamp lands on the upper waters.[1] But it was not until yesterday that I had an opportunity of seeing its now navigable course, and of finding that as remarkable though different features belong to its lower as its upper waters. Even to this time, almost no use is made of this excellent navigable route, except by the small steamboat which runs regularly from the junction [at Franklin, Virginia] of the river with the Portsmouth and Roanoke railway, to Edenton and Plymouth; and this vessel carries little except passengers. That almost no use should be made of this river and also of the Nottoway and Meherrin, for transporting country produce, is owing to the almost closed state of the lower waters and ports of North Carolina, and the consequent low prices which can be there given by merchants.

The Blackwater, from its head spring, which is within two miles of the lowest falls of the Appomattox, is remarkable for its level and sluggish course. It has not a rock, and perhaps not a pebble, in the whole of its long route; and nothing obstructs the passage of its waters, except fallen trees, and the rubbish which has collected above such stoppages. These obstructions have long ago choked up the upper waters for many miles below the head; and have the extensive swamp covered with the impeded water until dried by evaporation; and the acts, as well as the neglect, of the proprietors have been continually operating to render the stream as useless as possible, and to render the swamps irreclaimable, (under our stupid legal policy,) and unproductive of every thing but malaria, sickness, and death. If the law of Virginia had not (effectually though indirectly) forbidden the draining of the swamps of the Blackwater, it might have been done, and 20,000 acres of the richest soil have been made dry and productive, by an expenditure of less money than has been paid in physicians' bills, for the sickness caused by these swamps, and the worthless mill-ponds at the heads of the branches. But this is digressing from my present subject.

At the steamboat landing, and for some miles lower down, the width of this river is sometimes not more than 70, and seldom is 100 feet wide, and this narrow space between the shores is still more lessened by the overhanging branches of the trees which grow thickly on the borders. Then so crooked is the course, and abrupt the turns, and so much are the changes of course shut in and concealed by the sameness of appearance on both sides, that the river presents continually the appearance, (though a continually changing appearance,) of a little narrow lake. Yet the water is deep enough throughout to float a frigate. Though as black as its name correctly purports, the river was clear of all muddiness. The still and dark waters margined and over-hung by the thick forest growth of the swamps, seemed to belong to the most silent and gloomy solitude of nature; and the contrast was even unpleasant which was presented by the introducing of a thing so noisy and so highly artificial as a steamer in rapid progress.

The entire space between the high-lands is a low swamp, elevated but a few inches above the ordinary height of the river, and of course always wet and miry. The river continually changes from one side of the highland to the other; and therefore it always has swamp land on one side, and generally high land on the other. The character of the swamp was conjectured merely by what was seen from the vessel; but from its little elevation above ordinary or low water, still more than from its being covered by every fresh[et], I consider it as both irreclaimable and worthless.

Though the leaves had mostly fallen, there were still enough of many kinds of trees and shrubs clothed in their brilliant and varied autumnal tints, to show how much more beautiful this river must be at an earlier season; in May and June especially, when the abundant woodbine and other vines are in blossom, and the dense forest has its richest verdure.

The margin of the river on the swamp side is throughout a dike formed by nature, but so regular as to appear like an ancient construction of art. It is generally above four feet high, and wide enough for a road. Indeed a part of it, of nearly a mile in length, in the lower part of Southampton county, in location happens to suit, and is used as a public road. This natural embankment is obviously the result of the deposite of sediment by the overflowing water, and which here, as on most other streams drops the heaviest and most abundant portion quickest, and of course nearest to the stream. But the wonder here is, that this operation should have been so regular—and by a stream which very seldom has suspended in its water any perceptible earthy matter, and never any of coarse or sandy kind as this is. When this ridge was deposited, the character of the stream must have been very different from what it is now.

There is but little change of appearance in the Blackwater, until it loses its name by junction with the Nottoway, at the North Carolina line. The Meherrin soon after joins also, and the united waters are known as the Chowan. The swamps, which had been more and more taking the place of the high-land margin, were now generally stretched along both sides. And the river rapidly increases in width, until, before joining Albemarle Sound, the Chowan is six miles in width.

All the former *inlets* (as they are called) of the ocean, or rather *outlets* of the rivers and of Albemarle Sound, being now completely shut up by the sand, and the pent-up floods having to seek their difficult passage as far south as Ocraco[ke] inlet, the level of the waters has been raised above their former limit. On account of the same separation from the ocean, there is no regular, and scarcely any perceptible tide in the sound, and its water which formerly was salt, is now fresh. There are no better waters in the world for the navigation of vessels of the largest size than the Albemarle Sound and the rivers which flow into it. Yet vessels of no more than five feet draught can now be used on these magnificent waters, because no greater depth can be carried, and that through a long and dangerous passage, to the ocean. The best outlet to a market is through the Dismal Swamp canal, notwithstanding its deficiency of depth and width, and the cost of tolls. One of the most magnificent, and at the same time one of the most useful of public improvements, would be to increase this canal to such size as to admit the passage of ships. Then the great rivers of North Carolina could be put to full use; and at the same time the passage of ships to the ocean would be made as cheap, and far more safe and sure, than if a ship channel could be opened through the wall of shifting sand which binds the coast.

Albemarle Sound, though crossed at the upper [western] and most contracted part, and seen on my passage only by bright moonlight, still appeared a magnificent sheet of water. Our course to Plymouth was up one of the several mouths of the Roanoke, with swamps on both sides still, and more extended and impenetrable than any passed before. These different passages of the Roanoke form several large islands, which are all of low swamp, except small knolls of firm ground which rise a little above the level of the adjoining swamp. One of these islands, Guard Island, in the interior is a juniper swamp, and has bears and rattlesnakes among its inhabitants.

No. 2. Margin of the Great Swamp. State of Agriculture

Plymouth, N.C. Nov. 20th.

Yesterday was spent in viewing some of the farms of this neighborhood, embracing some of the richest land in the world, and some of the poorest and the worst managed. For my guidance, and for the facilities and information without which the observations of a stranger would have been very illy directed, I was indebted to the kind attention of Dr. Armistead,[2] whose farm, two miles from Plymouth, was the most closely examined, though a slighter view was taken of various other properties, within a few miles distance.

Four-fifths of this county, Washington, and perhaps as much of the adjacent counties of Tyrrel[l], Hyde, and Beaufort—all the extensive peninsula, indeed, formed by Albemarle and Pamlico sounds—is one immense swamp. This whole peninsula, is about 60 miles long, by 40 wide. The part which is not swamp (estimated as one-fifth of the whole,) is composed of the narrow knolls of firm soil which are scattered throughout—islands of sand, or of clay, in a sea of black mire.

I was prepared by my former view of the Dismal Swamp in Virginia, to find this land similar in character, and fully as much of purely vegetable formation as the former. But such is not the fact. I saw one juniper swamp only; and though its under-growth was so thick that I could not penetrate it far enough for examination, no doubt its soil is of the same purely vegetable formation as the juniper lands in Virginia. But with this exception, all the swamp land seen, has been of what is called "cypress and gum swamp;" and which, judging from the small portion which has been drained and cultivated, is of very great and also very durable fertility. This soil has certainly much vegetable matter; but much of earth also; is about 2 to 2½ feet deep, and resting on a subsoil of what appears to be a tenacious blue clay, when first dug up, but which becomes pulverized by exposure, and forms fertile soil. No doubt the soluble vegetable matter has sunk into this clay, and given to it both its color and its productive power. I never saw any soil that appeared richer than this land, or that promised a greater profit upon the expense of improvement. Still, very few persons have yet begun to drain their swamp lands, and all their cultivation is on the firm knolls, which were never rich, and are now very much reduced by the unceasing cultivation of corn, year after year, which is the general practice of the country. The sole mitigation or change of this course, (if it be either,) is, that peas are always planted among the corn, and the land thus made to bring two crops in each year. But the peas are gathered as well as the corn, and then stock turned in to take all that is left; so, that nothing of the crop is given to the land as manure.[3]

The soil of these dry knolls is generally sandy; but sometimes quite stiff. The latter is much the meanest soil of the two, for product, and also the least manageable under cultivation.

The mere observation of the map of North Carolina will show this remarkable peculiarity, that all the numerous streams have their sources in the interior and central parts of the swamps, and thence flow, some in every different direction, to seek lower outlets. This is not only the case as to the great swamp region now under consideration, but also as to the other great swamps which lie south of the Pamlico. This general fact alone would clearly prove that the great body of the swamp land is generally higher than the dry and firm soil which lies without, and through which the swamp streams, which in some cases are large and navigable rivers, seek their passage to the two sounds. It is not meant that the swamps at their edges are higher than the outer firm land immediately adjacent. On the contrary they are lower; but gradually rise towards the interior, and each long and narrow ridge or knoll of firm ground that is reached seems a low step or dike to a higher elevation of swamp within. In my former slight examination of the Dismal Swamp in Virginia, I was struck with, and invited attention to, the like fact of the swamp being higher than *nearly all* the surrounding country, dry, firm and elevated as the latter was. But still there was on one side, (the western,) which I did not visit, higher land, from which streams *might* flow into the swamp, and help to supply the vast superfluity of water, which passed off in every other direction. But even if such is the case in Virginia, (which I now doubt, or suppose at least that the supply of streams from without is of but secondary importance,)[4] it certainly is not so in North Carolina. Whence then can be derived the great supply of water which not only keeps saturated this vast extent of morass, but fills, and keeps full, several lakes of considerable magnitude, and feeds, through all seasons, numerous streams and even deep rivers? Can the rain water alone be sufficient to produce such great results? Or can there be an additional supply from below, in undiscovered and unsuspected springs, and abundant enough in supply for the great effect produced? If reasoning *a priori*, [one] should have inferred that in a climate as warm as this, and having no more rain, that all the rain that fell on a before dry and level surface, and retained there for want of fall to an outlet, (and therefore to be wasted only by absorption of the earth and evaporation,) could not serve to maintain wetness through all seasons, and consequently could not form a morass. But even if this inference is erroneous, and permanent swamps would be thus formed by rain water alone, retained exactly where it fell, there is yet a much greater and more inexplicable effect produced here, where numerous and large streams are perpetually drawing away the waters of the swamp, and of which moreover the land is so elevated that the rain water may flow *off*, but cannot possibly be increased by flowing *to*, from other lands.

The channels of the streams near where they pass out of the swamps, being deep enough, may well be used to receive the canals, or main-carrier ditches, made to drain the outer parts of the swamps, which are the lowest parts; and to prevent these parts being again flooded from the interior and higher part of the great swamp, some of the numerous long and narrow ridges may be used as natural dikes or barriers. The only lands that I saw or heard of in this neighborhood, on which draining had been commenced successfully, were the farms of Dr. Armistead and Mr. Swift;[5] the former being much the more advanced and considerable operation. The proprietor, though with but short experience as a farmer and drainer, and his work therefore necessarily imperfect and insufficient still has effected a great and admirable change, and which has already been attended with great profit. His main-carrier is upwards of two miles in length, twelve feet wide, and about four feet deep; which depth is not enough by two or three feet, which addition the proprietor designs to give as soon as can be conveniently done. This canal passes through the middle of his swamp land, and serves to drain and to fit for putting to use the whole body of about 500 acres. It empties into a swamp creek. Still, the overflow of water from the interior swamp could cover the head of this canal, and render it incapable of discharging even one-tenth of the supply; but that has been guarded against by a short dike which has been made to unite the adjacent ends of two of the long ridges, and which, thus united, form a perfect defense. But it is enough, and more than it can now do, for the canal to discharge all the water from heavy rains which fall on the area it is designed to draw from. In each of the three last years there has been a storm of rain and wind in August or September, which overflowed the drained swamp, and all its cultivated portion, from two to three feet deep, notwithstanding all the discharge of the canal. This glut of water on such disastrous occasions was no doubt much increased by the wind blowing down the dead trees, and their tops falling into and partially choking the canal. The crop of corn now on the ground, though a very heavy product, offers evidence of the destructive effect of the wind and the flood of last September. The enlarging of the canal, and the entire destruction and removal of the dead trees, will greatly lessen, if not entirely guard against such evils.

The swamp in its natural state is one uniform expanse of the most gigantic and magnificent forest growth; of cypress and blackgum and other aquatic trees mostly, but with some mixture of others more usually found on dry though low land. Vines and shrubs, of various kinds, form the undergrowth. The whole surface of the earth is kept always (unless in winter) under an impenetrable shade. This dense cover of course must much obstruct evaporation, and, when removed for tillage, will greatly lessen the first supply of water to the drains.

The first operation, after digging the main canal, is to kill the trees, which is effectually done by slightly "belting" them, or cutting around through the bark and a little into the wood. Whether this is done in winter or summer, the trees die within one or two years after; and (what was different from my experience and expectation,) no new sprouts grow out from below the belt. Several persons who *intended* to drain, commenced their operations by thus killing the trees. But unfortunately, this was also the *end* of their improvement. In such cases, a new and thick growth of altogether a different kind rapidly covers the land, which if left undisturbed for a few years, would be more difficult to destroy than the original forest. I saw one piece of second growth thus produced. Nearly all the trees were of a species of magnolia, with a very deeply corrugated yet close and firm bark. These beautiful trees were from six to even twelve inches in diameter, with long straight bodies, clear of limbs to a great height; which manner of growth is common to all kinds of trees in the swamp. But if brought under tillage in proper time, the drained swamp land is easy to plough, and to manage and get in good order in all replete removal of the enormous quantity of large decaying trees.

While the trees are dying, and a year before beginning cultivation, many parallel small cross drains are cut, emptying into larger ones running at right-angles, and these again into the main canal. These small drains are from 100 to 200 yards a-part, and are covered. They ought to be dug 3 feet deep, and the wooden covering to be 2½ below the surface. For want of such depth, some have already failed, and have been re-opened for repair. There could be no better subsoil in which to make covered drains, as the bottom is always in the adhesive blue clay. In this a narrower channel is dug, to serve for the pipe, leaving shoulders of the clay, on which, and across the open pipe below, short split boards of heart-cypress, (which is almost imperishable,) are laid. Dr. Armistead thinks that, in addition to the great convenience to cultivation of *covered* compared to *open* ditches, that the former serve even *better* to take in the water by filtration, than the latter do by its flowing in; for so level, and so porous is the soil, that, even when overflowed by a heavy rain, the water scarcely runs at all into ditches, but rapidly disappears by sinking into the earth. But though a strenuous advocate for covered drains where they are suitable, and having had much experience in their construction and of their benefits, I fear that objections to them, though not now suspected, will be found hereafter; and that their present efficacy is owing to what will ultimately cause their filling up and destruction as drains—that is, ease with which the surface-water finds a passage into them. Wherever water can descend through cavities, it will certainly carry along with it more or less of fine earth; and the little descent in the course of the drains will cause this earth to remain in, and at last choke the pipe. And if such should not be the result, there may be another opposite evil nearly as great. As the vegetable part of the soil shall decay, the texture will become closer, and water cannot, as now, sink into the earth, or through the covering of the drains, and, of course, in the same proportion will the operation of the drains be lessened. Though so much more troublesome during cultivation, I should prefer a smaller number of deep open drains, into which should empty the deep and clean water-furrows of wide and high beds. However—where every thing is so new to me, it is my business to gather and receive, instead of offering to give instruction to others who are better, because more practically informed, as to the peculiarities of this soil. The main canal on Dr. Armistead's land not only needs deepening, but the smaller drains are yet to be dug for much the greater part of the land, which the canal will serve to lay dry. Of more than 500 acres which will be so drained, and for which the most expensive work is done, only 180 acres have been yet made ready for cultivation. From that, (principally, for his highland is of the general mean quality,) he counts on making 1500 barrels of corn; and certainly he has made a most splendid and profitable addition to the previous state of his property. This land is usually estimated, as I was informed, at less than a dollar the acre; and for that price any quantity may now be bought. The draining upon a proper plan and a large scale would be far from expensive; and when drained, the land would be well worth $50 the acre, if the most fertile land in the country is worth that sum. If its durability will be equal to its fertility, it would be cheap at $100 the acre. Yet almost every poor landholder in this county, who is barely feeding his family by the incessant tillage and exhaustion of the poor narrow ridges, also owns of this swamp what would make a princely estate. But it is not only the poor and the ignorant, but the rich and the better informed, who neglect this incalculable source of value. For with the exception of a very few such improvements, (the most important of which I have not yet seen,) the drainage of the swamp lands is either not thought of by proprietors, or considered as among the many visionary schemes of book-farmers.

If drainage were effected even to but slight extent, this region ought to be one of the best for grass husbandry, and for cattle. But the making of hay is not thought of, though so much of the land is so admirably suited to grass. The reeds and other grass of the swamps in their natural state would make most valuable pasturage for cattle; and still better for hogs, with what they would get by rooting. But the bears are so numerous and destructive to both hogs and cattle, that there is little gained by their ranging at large in the swamps. Bears very rarely venture into the cultivated lands; and never leave the swamp unless forced by hunger. Yet so plenty are they in their haunts in the great swamp, that one man in this neighborhood, (of course a great bear-hunter,) killed sixteen in one season, the fall of 1838.

No. 3. Journey over the Firm Land of Washington County. First Impressions of the Great Swamp, and Lake Scuppernong

Somerset Place, Nov. 21.
Having been most kindly invited by Josiah Collins, esq.[6] to accompany him to his place of residence, and promised his aid to enable me to see the swamp lands where I had heard had been made the most successful and in-

teresting improvements by draining, I readily availed his offer; and this morning we left Plymouth. Our road was necessarily on the only long and nearly continuous stretch of dry land in the county; and was very circuitous. All of the central and southern parts of the county, through which a direct route would have led; is part of the great swamp, and of course impassable. The road is on the lower yet firm land near the shore of Albemarle Sound, and for 4 or 5 miles we were within a mile of it, and generally within a quarter of a mile, and in full view of the trees on the opposite side. The water however was but seldom and barely visible, owing to the low level of the road. This firm ground is supposed to be not more than of 10 feet elevation above the water of the sound, and of course is lower than the main body of the great swamp. But whether of the sandy or clay soil, (as before described,) the land was firm, and the road excellent throughout. Though wider than the smaller ridges, the land generally was of the same kinds and alternations of soil. The swamp growth was often in sight, on one or both sides. After turning southward, from the neighborhood of the sound, we crossed the outer part of the Poplar Swamp, a projecting portion of the main body, and which, like that in general, is now covered several inches deep in water, though no rain has recently fallen. Afterwards we passed through a part of what is called the "body land," being the largest body of firm land (7 to 8000 acres,) and having uniform and peculiar quantities. This very level, and a close stiff clay soil, which, though firm, requires draining to remove surface water, and without that operation being well executed, the land is worth very little. This draining was not done, or very imperfectly, for a long time; and during that time, the people living on this land supposed that it was not capable of producing them bread. Accordingly, they either cultivated corn on some poor sandy spot in the neighborhood, or bought it with what they gained by cutting lumber, which business was, and still is, the general source of income, here and elsewhere in this county, except the little corn and less wheat cultivated. But since having been sufficiently drained, this "body land" brings good corn, as I saw standing, and very heavy crops of wheat. Some of it has produced 30 bushels of wheat to the acre.

On several of the small farms passed in the day's journey, were seen the Scuppernong grape vines, trained in the manner usual in this part of the country, and on Roanoke Island. One or more vines, (which appeared to be three or four inches in diameter,) rose perpendicularly about 7 feet, to the top of an open horizontal frame or scaffold, raised on posts, and over which the branches spread, and thickly cover. A single vine will cover, and closely shade a wide space; and as the growth spreads, the frame is extended. The Scuppernong is a native grape. The fruit is white, very large, and grows not in bunches, but singly. It is, to my taste, delicious, and is also a good grape for wine, of which some that is very palatable and much that is bad, is made hereabout.

We crossed Scuppernong river on a draw-bridge. The river is a narrow, but deep, and navigable for sea vessels for a mile above the bridge. It has the same general appearance of all the small rivers of this country, its waters being black and sluggish, spreading far over the low and wide swamp on its borders. Having passed out of this river swamp, our road again rose into high-land or firm soil, and generally under its original high-land forest growth. So far, for more than 30 miles the road had passed through lands as devoid of interest as any whatever. But soon after this, notwithstanding my having been somewhat pre-prepared by previous general description, it was with no less surprise than pleasure that the scene was seen to change, and successive works of industry as well as of nature were successively reached, none of which could have been otherwise suspected of existing in such a locality. The road, while yet on high or dry land, first brought us to the margin of a wide and deep navigable canal, down which the water flowed with a fullness and degree of rapidity which afforded evidence of an unusual rate of descent.

The canal stretched out of sight in a perfectly straight course both above and below, and apparently (as really) for miles in extent. One of its banks had been formed into a wide and excellent road, along which was our still ascending course, meeting the direction of the flow of the water. After proceeding thus, and the land obviously still increasing in elevation, our route rose from a high-land and firm soil to swamp land, having its usual and marked peculiarities. We had indeed just entered the border of great swamp region, and our course was towards its interior. The land here on both sides was a wet morass, though not inundated, and was principally covered with evergreen trees, so that the verdure of summer was generally presented, when every thing else showed the hues or nakedness of winter. The most numerous and largest of the evergreens are of the same species of magnolia before mentioned, and which is called the black laurel. Some here are more than two feet through the body. It is a beautiful tree, on account of its glossy and bright green leaves, and its long and very straight body; and much more pleasing to the eye were these trees where they formed the almost entire and thick growth.

Next we reached cleared and well drained swamp land, under a heavy crop of corn, of which the growth covered a single field of some 600 acres. The rich soil was quite dry; yet its appearance, as well as the large intersecting ditches, and the yet standing dead trees, all showed clearly enough that it had formerly been swamp of the same character as that on which still grew the native forest of gigantic cypresses, which bounded the prospect beyond the cleared land. Ahead of us, and still distant, now were in view various buildings, so numerous as to seem a village. The increased rapidity of the current, as well as the ascending course of the road, which was now manifest to the eye, showed that the land was still more and more elevated. The mills on the canal, and some other large buildings had narrowed the view ahead; but when these were passed, and the upper termination of the canal nearly reached, its great and magnificent source, Lake Scuppernong, opened gloriously to my view, and a few moments after brought us to its margin. The position of the mansion house barely permits the passage of the road between it and the canal; and the house is also so near the lake, that the spray from its billows in great storms would cover the space between, but for the few cypresses standing in the edge of the water, and other barriers against the violence of the waves.

Across the lake from the canal stretches the great savanna on the opposite shore, and which is scarcely visible, owing to the low size of its general though dense growth. The few scattering high trees thereon, seem to be standing in water, and not on land. The other sides are closed with a dense cypress forest, of the usual great height. In general appearance, at the first glance, this lake is alike to Lake Drummond in the Great Dismal Swamp of Virginia. But the first impressions of both still are very different. Over the latter, it would seem to a stranger that nature reigned in gloomy solitude that had never before been disturbed. With Lake Scuppernong, the works of man stand in strong and conspicuous contrast to the natural scenery and the general condition of the vast and impenetrable morass in which are concealed both this beautiful creation of nature and that of the rare exertion of the industry and intelligence of man; and in a region where these qualities, or their proper direction, seemed scarcely to be in existence.

Of course, I had been prepared in some measure by previous but very inaccurate report, for the remarkable scene now spread before my view. But if it were possible for a person not so informed to find his way to this spot, by a journey over the only practicable path—and still more if the last few miles were through the intervening trackless swamp, yet remaining in its natural wild state—the *unexpectedness* of the objects found would be even more striking to such an observer than even the actual romantic and peculiar beauties of the scene.

The first discovery and earliest knowledge of this lake was in 1755. Some hunters who had for years dwelt within three miles of its shore, on some of the small "islands" or spots of firm ground, had observed that the deer when pursued, usually ran off in a particular direction, from which the dogs soon returned, as if baffled in their pursuit. Thence it was inferred that some wide water lay that way, in which the deer found a safe refuge. These hunters at last determined to attempt the arduous labor of penetrating the swamp, scarcely passable as it was then even to them, and to find the great water, if it existed within the short distance that they suspected it to be. They set out for this purpose, and worked their slow and toilsome way through the day, and lodged, as well as might be, where night overtook them. The next morning, discouraged by the difficulty of further progress, and fearing to add more to the toil and danger of their return, it was agreed to abandon the adventure forthwith. But before turning their steps homeward, as a last re[s]ource, one of the party, Tarkinton, climbed one of the tallest trees, and from its top, had the first view of the broad waters of the lake, and at no great distance. His joyful exclamations, announcing the discovery, and its direction, induced one of his less zealous companions to try to seize upon the honor of first reaching the lake; and he immediately pushed on in the direction indicated, and had leaped into the water and claimed for it his own name, before Tarkinton could descend the tree and overtake him. Thus was the name *Phelps* attached to the lake, as contrary to justice, as to euphony and good taste. But in old records it was also sometimes called by the Indian name of Scuppernong, being the source of the river before and still so called; and that name, of better sound and better application, ought by all means to supersede the use of the other.

No. 4. *The Wild Swamp Lands, and the Lake*

Somerset Place, N.C., Nov. 23.
The great swamp of North Carolina, which fills nearly all the space between Albemarle sound and Pamlico river, was at first called the "Little Dismal," in contra-distinction to the "Great Dismal," in Virginia and the adjacent part of this state; and this misnomer of the former is still continued, though it has been long known that it is much the greater of the two. Four lakes are in the interior, of which Pungo and Alligator lakes are much smaller than Scuppernong, (which is 8 miles long and 5 miles wide,) and Mattamuskeet much larger, the latter being from 30 to 40 miles in circumference. The courses of all the streams, some flowing out of these lakes, and

the various head branches of the great swamp rivers Alligator and Pungo, all serve to show that the greater interior space of the swamp region is much higher than its borders, and that its surface descends on every side. Lake Mattamuskeet lies nearest the sound, and was most accessible. Much of the swamp land on its margin has long been drained and cultivated, but in a very imperfect manner. With this and a few other exceptions, nearly the whole of this great region still remains in its natural state, and very little of it has been explored, so as to be well known, even by bear hunters, and much less by any other persons.

Pungo Lake lies within two miles of this,[7] and the works for draining that lake by the government of North Carolina are now in the course of execution. I should be very glad to visit that work; but though so near to this place, it is totally inaccessible at this season; and could not be reached directly, even in the dryer time of summer, without cutting a path through much of the impervious thicket.

In the language of the residents, the lands of this great region are distinguished by different names, representing different qualities or conditions of swamp. The first is *swamp land*, which term is not commonly used, as by me, for the whole morass, but is confined to the kind covered by large forest growth, of which cypress and black gum trees are the most numerous. This is the dryer part of the general swamp, and it is sometimes so clear of under-growth that there would be no great difficulty in walking over such land, but for its wet and miry soil. *Juniper swamp*, is altogether different from what is usually meant when "swamp" simply is spoke of. But though abundant in other parts, I have met with no juniper lands in my limited view of the great swamp. This kind is merely rotted vegetable matter, and having no change or difference of subsoil, to any known depth. This is the most usual kind in the Dismal Swamp in Virginia, and was described fully in my observations thereon, published in a former volume, (Far. Reg. vol. iv, p. 513.)[8]

Savanna land is the name given to another very large proportion, and which perhaps is the larger part of the interior or highest part of the whole region.[9] Still, relatively to the nearest *swamp*, or forest-covered land, the savanna has the lower level, as it is more generally and deeply covered in water. The savanna has no trees, except a few scattering and stinted and ragged pines. The under-growth is of evergreen bushes, as laurels, bays, &c., vines of various kinds, and in some places is exclusively of reeds. None of these usually rise higher than about 10 or 12 feet. But the growth is so dense that it is almost impossible to penetrate it, without the previous slow and laborious process of cutting a passage through. The savanna which lies across the lake from Somerset Place extends to Mattamuskeet, and perhaps farther in other directions. The lower part of Mr. Collins' land, which has long been under forest growth, is remembered by Mr. Pettigrew to have been savanna many years ago. The change has probably been induced by the partial exclusion of the water, after the canal began to operate. I therefore infer that savanna differs from the other swamp merely in being more subject to water, therefore the soil, for the time, is still more of vegetable material, and therefore, as well as on account of its kind of growth, more often and thoroughly burnt over, when long droughts have permitted fires to spread, and sometimes to burn away the soil itself, as well as its growth.

For the purpose of having a closer view of the great savanna, as well as for the pleasure of the excursion, I sailed across the lake in Mr. Collins' nice little pleasure boat. We approached the edge within about 60 yards, which was as near as the boat could keep afloat over the shoal bottom. But no better view could have been gained even by landing on the miry soil, which is now scarcely as high as the present surface of the lake, and of course is full of water, and often overspread with it. But the wet and boggy state of the soil would have been the least of the obstacles to further progress. The covering thicket seemed so dense as to be impenetrable any where, and the sight could not have extended through more than a few feet of distance. If, however, the observer could be elevated above the low tops of the reeds and shrubs, there appeared nothing to obstruct the view for many miles. The few dwarfish pine trees, whether yet alive, or killed by fire, served but to deform the otherwise uniformly beautiful, though varied growth.

When this water excursion was made, the sun was unclouded, and the weather, though cold, bright, and beautiful. The color of the lake, as seen from the shore under this sun-light, was a deep dark blue. But when sailing over it, and looking down upon the water, the color appeared to be dark brown, though perfectly clear of muddiness. When seen however, in a drinking glass, it is but very slightly tinted, with the vegetable extract which gives the much darker color to all the waters of the adjacent swamp, as well as to other swamp lakes. This is therefore considered a "white water" lake, as distinguished from other swamp lakes. Contrary to what would be supposed, it is said to be the least colored when the water is lowest in the lake. It is free from all taste of vegetable or other impregnation, and is a delightful drinking water.

I also visited the savanna on the eastern side, beyond Mr. Pettigrew's farm, by walking on the margin of the lake as far as possible. The water would have then flowed a few inches deep over all the savanna, but for low dikes built along the margin to prevent the water spreading across the savanna to the cultivated back lands of the farm. And it was evident that in the present state of the lake, (which is not the highest usual in winter,) that its water, if unobstructed, would spread over all this part of the unreclaimed savanna.

Several persons though residents, and well acquainted with the ground, have lost their way and their lives in the swamp, and not far from the then boundary of cultivation. Of two who are known thus to have perished, the skeleton of one was found many years later, and identified by the remains of the iron tools which he had carried with him. Of the other no trace has ever been found. These facts, better perhaps than any general description of mine, will serve to convey some idea of the difficulties of getting through the thickly overgrown and treacherous bog, and of the lost guidance of the sun, or some other sure indication of direction.

The irregular margin of the swamp land, shows that in storms the lake has made inroads and widened its surface on every side, but mostly in the direction to which the violent winds most generally blow. Cypress trees, still living, though for ages perhaps surrounded by water, in many places stand far from the shore. And in others, where the trees have preserved the soil, little narrow promontories of high and forest-covered swamp jut out between the indentations washed out of more yielding earth. It is probably owing to this cause that the bottom of the lake next the shore is so shallow as to be left naked several hundred yards in width in the lowest state of the water. Beyond this, the depth increases suddenly, and the central and great body of water is from 8 to 14 feet in depth. The height of the surface is reduced from 1 to 3 feet by the summer drought; and the greatest reduction known, below high water, was 4 feet perpendicular.

No. 5. *The General Plan of Drainage, as Executed*

Somerset Place, Nov. 23.

About 1770, Josiah Collins, the oldest of that name, together with two other persons formed a copartnership under the name of the "Lake Company."[10] They took up nearly all the surrounding swamp land, by laying their own patents, or buying the small patents of other persons, in cases where the land had been previously thus appropriated. The only tract which is not embraced in this extensive property, (which was more than

80,000 acres,) is the farm of Mr. Ebenezer Pettigrew, lying on the lake next to Somerset Place. The interest of the two members of the copartnership was afterwards bought out by the elder Collins (the grandfather of the present occupant,) and the property has been divided among his descendants. Somerset Place, the individual and distinct property of Josiah Collins esq. is 3000 acres, extending from the lake on both sides of the great canal into the lower and still swamp land in the rear. A small part only of this has been improved and the actual drainage and cultivation of the share of another member of his family. But the whole tillage and management is as one farm, and by Mr. Collins', and therefore it is not of his separate property, but of the cultivated land that will be spoken of as his farm. This explanation is perhaps uncalled for; but it is made to avoid even the appearance of mistake or erroneous statement on this head.

The great work of draining on this and the adjoining and similarly situated estate (Mr. Pettigrew's,) and the great value thereby created, are matters of general notoriety; and long since I had heard of the work through verbal report, and, as supposed by me, from such accounts, that the object and effect of the great canal was to *lower* the level of water of the lake very considerably, and thus to leave higher the swamp land around. This is indeed the general plan proposed by the report of the former engineer of the state, Mr. Shaw, for draining the public swamp lands around other lakes, and which plan, it is said, is now in the course of execution, by digging a canal to draw off the waters of Lake Pungo. This supposition was entirely wrong. Whatever might have been the opinions and secondary objects of the first proprietors, (the Lake Company,) the great and primary purposes of their main canal, (as are its main effects,) were, first, to provide a good navigable outlet from the lands designed to be drained, without which the future products would be valueless, because they could not possibly be sent to market; secondly, to bring into extensive use, for propelling machinery, a water power of any desirable amount; and thirdly, that the canal should serve as the final and general receiver of all the water collected by the numerous smaller drains, which are made to draw from every acre of the estate. But it is obvious that this last important and indispensable operation, of being the general receiver of the drained water, could not have been effected, except by the surface of the water in the great canal being *lower* than all the swamp land to be drained on its borders, and than all the smaller drains leading to the canal. But to have the water of the canal thus lower, (as it now is,) can only be done by commanding, regulating, and limiting, the rate of supply from the lakes. For if suffered to flow unrestrained, though the highest part of the canal is but 20 feet wide and 4 feet deep, the flood which would pass through would cover permanently and deeply nearly all the area of swamp designed to be drained. The usual height of the lake water has not been lowered but very slightly by the vast and continual draught made by the canal; and the land-holders, so far from desiring to lower it very considerably, would wish by all means to avoid that result. Even if drainage were the *sole* object, and to be attained without regard to expense—and if, with that view, the canal had been made twice as deep as this and three times as wide—though such a channel might have discharged ten times as much water, it would have been insufficient always to keep down the level of the waters to the reduced and usual height; and if, by its actual operation, the canal had left free from water, and dry enough for tillage, a narrow border of some of the highest land, it would also, by throwing an enormous, variable, and therefore totally uncontrollable flood, cover ten times as much land lying on a lower level, but equally capable of being drained by proper measures. The impropriety and danger and enormous expensiveness of this mode of drainage will be evident to any unprejudiced and careful observer.

It has been already stated that the central parts of the great swamp region are the highest parts; and much higher than the lands, whether swampy or firm, that lie nearer to the outlets of the swamp waters by streams and rivers. This and other lakes are in the central parts of the great swamp, and their waters are sometimes as high as the highest parts of the adjacent swamp. Sometimes in the dryest seasons the level of the water of this lake has been lowered 4 feet; but the usual height is generally restored by the next rainy spell; and sometimes with so much excess, that the overpowering flood spreads over the outer swamp lands, several feet deep, before it is discharged into the rivers. Such are supposed to be the general facts, as merely inferred however in part. But these I now leave, to state what I have examined in regard to Mr. Collins' drained land, of which the comparative levels and other circumstances are known with certainty.

The water of the lake is now, (though not at the highest mark,) as usually, except in dry seasons, within a foot of the highest, and is overflowing the lower parts of its natural brim, or margin; and would even now flow over the highest parts of the cultivated fields, but for the low dike which has been constructed along the lake side, and which is one of the permanent roads. From this more elevated margin of nearly two miles in length, the surface of the land, (now all dry,) is naturally an inclined plane, with a very uniform rate of depression to the side of the estate most distant from the lake. The rate of declination of the surface, as shown by the canal, is much most rapid nearest the lake; and is equal to 6 feet perpendicular in the first quarter of a mile. Below, it is more gradual; but in the whole course of the canal, of something more than 6 miles, the fall is between 18 and 19 feet, and of the swamp land along side, nearly as much. It will be obvious, from these circumstances, that if, (for example) the depth of two feet of earth only were cut away, or removed horizontally from the margin of the lake, that a broad sheet of water of a foot in depth would be poured over the whole of the most elevated side of the land, and pass over the whole farm, (if no artificial obstructions existed) to escape at the opposite and lower side. To guard against any such invasion from water, the lake-side dike, though so little elevated as to be scarcely visible, and seeming as merely designed for the road, serves as a secure barrier. This very regular inclination of the surface of the land, with the supply of water at the top, would permit the easiest, and the most abundant as well as extensive irrigation of any yet effected by art, if the aid of watering should be desired hereafter. It is upon this general and regular inclination of the land that the scheme of its drainage was founded, and on which its successful operation depends.

The digging of the great canal was the first work, and this was completed about 60 years ago, by the Lake Company, at a cost of $30,000. Its course is perfectly straight, so far as visible on the farm, and is at right angles to the side of the lake where it enters, and its direction is with the greatest descent of the surface of the land. To a quarter of a mile from the lake, where the water-power is used for the mill and other machinery, the level of the water is kept up, for a head. To that point, the water is kept in by its banks, and stands at the mills 6 feet higher than the cultivated fields on both sides. Below the mills, the water of the canal is lower than the adjacent land whenever the mills are not at work, and the full flow of water restrained; but higher when that flow is used or permitted, which is usually the case, except at night, and in wet seasons, when it is necessary to afford a continual outlet to the rain from the fields. The canal receives the water collected from the land by nearly all the drains, through three several and remote channels; of which the two on the higher levels are kept shut, by close floodgates, when the mills are grinding and the canal receiving its full supply from the lake, and are opened whenever that supply is shut out by a floodgate across the canal, near its junction with the lake. Thus all the water collected by drains from more than 1200 acres of the arable land is shut in usually through every day, and only permitted to run out at night; and that the accumulation in the lowest ditch is not enough to hurt the arable land, is the strongest evidence of the dry state of the reclaimed land. The third passage into the main canal enters so low down, that it is left open at all times, the water in the canal at the junction being the lowest where this supply enters. There is also another place of discharge from some of the drains most remote from the canal, at the opposite side through a small arm of Scuppernong river, which originally was the only natural channel, in usual seasons, of the surplus water of the lake.

There are five main drains, 8 feet wide, running through the land parallel to the canal; these are crossed at right angles by "leading ditches," 6 feet wide, and at distances of a sixth to a quarter of a mile from each other. The banks of these leading drains are all thrown to the side towards the descent of the surface. Of the "lake side ditch" which is also nearly parallel to these, the earth dug out was thrown to the opposite side, towards the lake, to help to form a road there. The design of this is different from the leading ditches, it being to catch the water which oozes under the road embankment, from the much higher and adjacent water of the lake. The banks of all these main drains and of the leading ditches crossing them are formed into good and permanent farm roads; and these roads, together with the main road of the farm, amount to 24 miles in length, and of course there are as many miles of the large drains. The rectangular spaces formed by these great drains, are again intersected by three-foot ditches running *down* the slope of the land, and emptying into the leading drains below; these are crossed by small and shallow "tap ditches" (such as are elsewhere called "grips,") at every 50 yards distance. Finally, the entire surface is ploughed into ridges, for corn, of 5 and 6 feet, with good water-furrows between. Thus the water-furrows collect all surface or rain water, and discharge it into the shallow tap ditches, which empty into the deep 3 feet ditches, and these into the 6 feet leading ditches, these into the 8 feet main drains, and these into the canal, or the other and natural outlet. The tap ditches are only about four inches lower than the water-furrows, and do not obstruct the passage of the teams and ploughs. They are easy to dig, but require cleaning out after every ploughing of the field. The three-foot ditches are 2 feet deep, and the two larger kinds are 3 feet or more, and the canal, 4 feet. The water is so directed that each large drain discharges its proper share. All these kinds, from tap-ditches to main drains, amount to 130 miles in length of ditching, on the farm of 1400 acres drained and cultivated. The whole operation of the drainage seems very perfect. The soil, considering the loose texture of much of it, and its absorbent character, was surprisingly dry; as dry indeed as such soil, and of such a level surface, could have been expected to be, if free from all higher water.

No. 6. The Soil, and Its Former and Present Vegetable Products

Somerset Place, Nov. 25.
The labors executed by the present proprietor in the short time that he has had possession (11 years) have been great, even for the large force and capital employed; and the performance has been the more remarkable as being conducted by one who was very young, and totally inexperienced. He has drained, and brought into cultivation 500 acres of the now cultivated arable surface of 1400, besides other and perhaps as arduous labors, which are not required on other kinds of lands. One of these was the clearing up the dead trees of much land drained before his arrival, in addition to that labor on all the newer part. The killing of the trees of the natural swamp forest, and grubbing and removing the shrubs and small trees, is comparatively but a light job. But until the last of the gigantic cypresses so left yields to the wind, and is prostrated, there is a yearly recurring labor required to remove the fallen trunks.

The great size of the old cypress trees generally, and of other trees also, but more rarely seen, is beyond the expectation of any one who sees them for the first time. Very many are five feet through the body at 4 to 6 feet from the ground, and would carry a diameter of 3½ to 4 feet for 40 to 60 feet, the length of their trunks. Some cypress trees are much larger than these sizes. One in the field I was told was 30 feet in circumference, as high as the measurement could be conveniently made. The age to which they live, and their durability after death, are not less remarkable than their size. Mr. Collins has counted on the sawed end of a cypress log, more than 800 rings of the grain, showing as many years of growth; and as this and others that he has counted were by no means of largest size, he supposes that their ordinary and natural term of life must be 1000 years at least. But this estimate does not go back to the earliest existence of some of the dead trees, by many centuries. There has been at least one generation of cypress trees which lived and died here before the oldest of these now standing sprang up. Of this fact I was shown several sufficient proofs, of which I will state the case proving the most ancient date. A trunk of a cypress, long dead, but still standing firmly in the cultivated ground, had been found by measurement to be 33 feet in circumference at 3 feet above the present surface of the ground. The surface probably has been lowered nearly as much as from the mark of measurement stated; and this has exposed the upper parts of the roots of the tree, and also the before buried body of another prostrate and large cypress, over which the trunk of the standing tree had grown. The visible wood of the buried tree is still sound. These "ground logs," as they are called, are so numerous under the swamp soil, that it would seem as if the trunks of the more ancient forest, thus buried, were as many as the trees now standing above them.

The whole of the reclaimed land was, when taken into cultivation, of what is called "cypress and gum swamp." The soil to the depth of 2½ or 3 feet, is very much like that of the newly reclaimed tide-swamps of James river, and is as much formed of vegetable materials.[11] When new and in dry state, this will take fire, and burn to the depth of 18 inches or more. But open and "chaffy" as is this vegetable soil when first cultivated, it produces very heavy crops of corn. The subsoil is of compact blueish clay, which becomes friable and fertile soil by exposure; but for this subsoil, the soil could not, as I think, be deemed a permanent possession. But there is enough elevation to spare several feet without injury in that respect; and when the soil has rotted away, as it does in time, and allowed deep ploughing to bring up the clay subsoil to the surface, a still better, and what I suppose will be a permanent soil is thereby formed. In the oldest cultivated parts, the extent of this loss of soil and depression of level, though greater there, cannot be measured or estimated. But in the land cultivated this year for the first time, though drained long before, the mark of the former surface can be fixed by the dead trees and stumps whose roots are now so far naked as to make it evident that the surface is already two feet lower than formerly. Yet the material of this soil, though rotten, is but little decomposed; and must lose much more in bulk before it is brought to the state of fine black mould, like the land which has been brought under tillage ten years or more. The first ditches which are dug in a new piece will barely reach the subsoil; but with every year's cleaning out, a little more of it is dug up, until most of the depth of the ditch is in the subsoil. But this rotting away, which would render entirely worthless (as it does our tide-swamps and marshes) any land which cannot spare 3 feet of its upper soil, is here productive of no greater evil than the necessity of continuing to deepen the ditches, so as to keep their bottoms as much below the subsiding surface as at first. This land will sometimes bring 12 barrels of corn to the acre, and often as much as 10 barrels. This year the crop is very inferior, owing to the ravages of the chinchbug, and Mr. Collins supposes that it may not average more than 5 barrels. From the size of the stalks I should have guessed a much heavier product. The largest crop which he has made in any one year was upwards of 8000 barrels. The land nearest the lake is stiffest, (or seems to partake more of the nature of the subsoil,) and is the most productive in wheat. A particular portion has been known to produce 37 bushels of wheat to the acre. Still however, this crop has been so uncertain that Mr. Collins has recently abandoned its culture, except on a small scale, for home consumption. One important cause of failure is the great growth of partridge pea. Another plant which grows with remarkable vigor is chick-weed, which I have not seen abundant any where else. Here, after corn, it usually covers the whole surface with a thick matted though low growth, and being a vine, clings to and runs upon every thing within its reach. It is now green and in full vigor. It serves as excellent winter pasturage for sheep.

In a small part of the land, the trees had been cleared away at once, by being cut down, instead of the ordinary slow and gradual course of "deading" them, or killing them by cutting slightly around, and then letting them stand until overthrown by the progress of time and decay, aided by the force of the wind in storms. The former course has been found to be very objectionable, and is abandoned. The large cypress stumps will remain to encumber the ground for 50 if not 100 years, and more than any other tree, even if equally durable, because of their larger size. All the roots of the cypress strike downward and deeply, none running horizontally or nearly so. Hence, even after the surface of the earth around may have been lowered by decay a depth of two feet, and of course as much of the formerly underground parts of the cypress exposed, still the lowest part appears to a stranger to be not of root, but of trunk, and merely shows a still greater enlargement of the always broad pedestal to the mighty column. Thus there are but few dead cypresses, or their stumps, which do not spread across the width of a five-foot ridge, and obstruct one if not two water-furrows, and many of the largest extend across more than double that width, and of course are serious impediments to the drainage of the obstructed water-furrows. Long as the cypress resists rotting, it is a very brittle wood, and is very easily broken by storms. Scarcely can a large living tree be seen in the original forest of the swamp of which the top has not been broken off by wind, and replaced by a subsequent growth of different form. And the dead trees left in the drained land are generally soon divested of branches and tops, and reduced to a naked trunk of 40 to 70 and 80 feet high. This resists the wind much longer; but finally, by the decay and weakening of the roots, they yield, and the trunk falls. The roots are broken in such cases not far below the surface, and the blowing down of the decayed tree leaves no deep or considerable hole in the earth where it stood. I saw measured one of the recently overthrown and broken trunks, which was 72 feet long, and all of clean body, below the lowest branch. This was 5 feet through above the swell near the root, or where the regular and very gently tapering form began. Another of less length of trunk was 6 feet through the body, measured as above. These were not selected as being of uncommon size, (for very many in the forest are said to be much larger,) but merely to serve as standards of comparison. For without using some such mode of comparison, a stranger would be deceived as to the general size. If one only of these great trees were found elsewhere, its size would strike every beholder with astonishment, and it would be supposed greater than it really was. But here, they are so numerous, and the forest growth and every thing else is on so gigantic a scale, that the eye has no accustomed and known objects by which to measure dimensions so large, and, by a slight and hasty observer, all objects would be supposed smaller than they are.

Rice was cultivated here by the Lake Company, to considerable extent, and with good success. But the culture was found to cause so much sickness among the slaves, that it was abandoned. The successive parallel slopes, ditches and embankments, formed by the "leading ditches" which run across the ground, afforded great facilities for flooding the land, and drawing off the water when desired, for rice culture. Subsequently, corn was the main crop, and cultivated for a long time successively on the same land. The present proprietor has found the very long continuation of this crop objectionable, and has commenced, as a regular course, to cultivate corn for three years in succession—and then to let the land lie out of tillage three years, and be grazed in the middle one of these three years. He is very careful to cover up in the soil for manure all the rank growth of grass underneath, when corn is again to follow the next year. To effect the object, and enable the plough to cover so heavy a growth, it is first weeded off the beds (5 and 6 feet wide) by broad hoes, and drawn into the water-furrow, and then two furrows of a plough are run to throw the slices so as to meet on the weeds. This is very roughly and imperfectly effected; and after standing a while, two more furrows are cut in the same places, and the earth thrown again over the weeds, and then the balance of the old ridges reversed and the new ones finished. The corn-stalks are cut off and drawn with the grass into the water furrows, in the same manner.

No. 7. Uses of Water Power. Mr. Pettigrew's Farming and Improvements.

Somerset Place, Nov. 25.

The extensive use made of water to save labor on this estate, is one of the most interesting subjects for observation. It has been already stated that the descent of the canal gives a head of 6 feet of water at the mills, at the distance of a quarter of a mile from the lake. Part of this power works a saw mill, and a corn and wheat mill of two pair of stones, with the bolting, and other machinery, &c., proper for the making of flour. Also the corn is shelled and fanned, and, though not now, formerly the wheat was thrashed, and cleaned by the water-power conducted to the barn and one of the great corn houses. Besides these more important operations, and for some of which there is daily use made of the water-power, it is also directed to crushing and grinding corn in the ear for horses and other stock, the working a circular saw, turning grindstones, and may be substituted for hand labor in various other ways. When it is desired to prepare a cargo of corn for the Charleston market, there is no need for commencing until notice has been received of the vessel having arrived in the river below. The shelling of the corn is then commenced, by a shelling machine of immense power, then fanned, next lifted up by elevating machinery, from the first to the fourth story of the house, there measured, and then emptied through a spout into a large flat boat lying in the canal, which, as soon as loaded in bulk, is conveyed along the canal to the vessel. Thus the risk of keeping a large quantity of shelled corn in bulk is avoided, and, by the aid of water, all the operations necessary to load a vessel may be completed in a very short time.

It is not only the main canal that is used for navigation. The "leading ditch," nearly two miles long, which passes through the barn-yard, is made 12 feet wide for that purpose also. When crops are made on that part of the farm, that ditch is flooded by letting in water from the lake, (by a ditch communicating with the lake, and commanding the water by means of a guard gate,) the wide ditch is kept full by closing another gate at its outlet, and the crops of corn or wheat are brought to the barns in flat boats, with comparatively little labor.

There are three barns, or rather houses for holding grain only. Two of them, for corn, are of great size, and constructed with all the care and strength of materials necessary to resist the pressure of the weight of the contents. One of these is 100 by 60, and three stories high. The other is 80 feet square, and has 4 floors or stories above ground, two in the body and two in the roof.

It is to me a matter of regret that I cannot see Mr. E[benezer] Pettigrew, and acquire something from his store of experience and varied information, and especially as it regards this place, where he has passed his life, from boyhood to old age. He is absent from home. Mr. Pettigrew, the elder, commenced his labors, some 40 years ago, under all the disadvantages of his neighboring proprietor, and with the great additional ones of very limited capital and a small and weak laboring force. Under such circumstances, the extent and value of his drainage, clearing and cultivation and other improvements, are wonderful, and his labors have been as profitable as they are admirable. The general plan of drainage is the same on both estates. Mr. Pettigrew formerly drained [his croplands] into Mr. Collins' canal; but has since constructed his own canal, of 15 feet width, which he uses in like manner both for navigation and for propelling mills and other machinery. He cultivates wheat successfully and on a large scale. As I am informed, he is an excellent manager and cultivator, and, besides his swamp improvement here, he has recently drained and brought under good and productive tillage a large tract of the "body land," and thereby created a fine farm, and an entirely new and rich source of agricultural production. If he who merely makes "two blades of grass grow where one only grew before"[12] is a benefactor to his country and to mankind, how much greater is his service who drains a swamp, or converts a worthless waste into a fertile farm!

Mr. Charles Pettigrew, the son,[13] was also from home during the first days of my sojourn at Somerset Place. His return however enabled me to give a slight and rapid glance at his father's lake farm, but, which was too much hurried by want of time, and also by bad weather, for me to attempt doing justice to its description, and therefore the attempt will not be made. I will merely offer a few notes on some of the most striking objects.

The farm of Mr. E. Pettigrew lies alongside of his neighbor's, and their mansions are not half a mile apart. In both, there is the like position of the buildings near the border of the lake, and the same general plan of drainage. Mr. Pettigrew's main canals being smaller, (15 feet wide) he requires more of them. Accordingly, he has already two, parallel to each other, and together about nine miles in length; and he is about to dig another which will be nearly half as much more. Part of his drained and cleared land was of savanna, as is still the wild land immediately adjacent. The soil of this field is as black, as rich, and seemingly as valuable, as the best swamp. It is now suffering from the access of water from the unreclaimed savanna, which will be remedied by the designed canal. But this land has produced 10 barrels of corn to the acre, and when the remedy is applied, will do so again. The rotation here is the three-shift, without grazing; or 1st year, corn, 2nd, wheat, and 3rd, the natural weeds, which grow so rankly, (6 to 8 feet high, and very strong,) that the three-horse ploughs, which are used to break up the field for corn, cannot possibly turn them under without a previous operation. This is effected by the "weeding down" process mentioned in his neighbor's practice. The land is kept always in beds of 6 feet width with deep and clean water-furrows between. The weeds are cut off by broad hoes, and drawn into the water-furrows, and then well covered by the meeting of the first two furrow-slices, made in reversing the beds.

The lake side of both the adjoining farms is alike protected by a low dike. This dike, with the bridges across the canals, forms a continuous road for two miles along Mr. Collins' farm, and apparently as much more along Mr. Pettigrew's. This road keeps near the lake shore, and forms, both by its situation and its decorations, one of the most beautiful and extensive promenades, either for walking or driving, that I have ever known. The road is perfectly level in its course, firm and dry. It is planted throughout on the side from the lake with rows of trees, which are of different sizes according to the date of the clearing of the fields along which they are planted. One of these rows, on Somerset Place, a mile in length, is of tall and noble sycamores, all of the same age and of very equal size. Another very long row of as large sycamores is on Mr. Pettigrew's part of the road. On another part of his, young cypresses have been left where growing naturally, and where deficient set out, on both sides, so as to form an avenue. Between the road and lake is a narrow and irregular margin, under its natural and the usual swamp growth, among which are some trees of very large size, and others as remarkable for their grotesque form.

Neither the road dike nor the margin outside, fringed as it is with trees, could serve as a protection from the violence of the waves, were it not for the shallowness of the lake for several hundred yards from the shore. On this wide shoal the billows are broken, and their violence expended, before they reach the land. Still, were it not for artificial safeguards, there would be yet left enough of power in the dashing of the broken and scattered spray, to produce great changes and do much damage to the land. This is exhibited along the lake shore where the swamp and its forest have not been touched by man. There, the deep indentations of water in some parts of the shore, and the ragged points of high swamp, wood-bound and defended, stretching out at other places into the lake—with the monuments of more ancient and extensive devastation and change presented in the position and forms of the old cypresses which stand alone and still living, though far out in the lake, or serve to protect by their roots and spreading base some yet remaining points of swamp—all show the great encroachments which the water has made upon the land. After suffering much damage from much slighter operations of this kind on the drained land, and the most strongly constructed defences having been found insufficient to resist the waves, Mr. Pettigrew discovered a mode both effectual and cheap, and which is used on both estates at every exposed point, or wherever the bank is so low as to need earth to raise it. The means used are "brush-bars," which are formed in the water close to the shore, by merely driving down perpendicularly a double row of small stakes, the rows two or three feet apart, and the stakes as far apart as will serve to hold the limbs of trees, or any rough and small brush laid between the rows and packed down closely, and rising a little above ordinary high water. The whole is made in the roughest and apparently slightest manner. Still, this feeble barrier serves not only to prevent the invasion of the water, but to repel it by forming more land. Perfectly clear as the water usually is, its violent action on the bottom in storms renders it muddy; and its sediment is left landward of the brush-bars, in such quantity as to form a considerable though slow accretion to the land by the deposite. It may well be conceived that the obvious and continual operation of the winds and waves for thousands of years may have greatly enlarged the surface of this and other such lakes, while it also increased the elevation of the swamp land on the margins by throwing over them and leaving the earth which had been suspended in the water. The sediment being derived from the bottom of the lake, is doubtless composed principally of such earth as forms the stiff subsoil of the swamp lands. The earth most recently deposited thus by the water has a singular appearance, and shows that the earth must have very peculiar texture and qualities. The fine and fluid mud left in the hollows of the ground behind the brush-bars, after sufficient exposure to dry weather, hardens and cracks, and separates by contraction into numerous pieces of a few inches across. When these pieces are not more than a quarter of an inch thick, they will become curved or curled by the further drying, and in color, shape, and almost in hardness, seem much more like a piece of old shoe-sole, or some other piece of long exposed leather, than earth or soil. After being again wetted by exposure to steady rain for 24 hours, I found these pieces to be softened, and made flexible and somewhat elastic; but they had not lost anything by the washing of the rain, and hardened as before, by drying. This same earth is found in small pieces, like angular gravel, on the surface and throughout the upper soil of the lands recently brought under cultivation.

No. 8. The Negroes' Chapel and Religious Services. A Short Lay Sermon to the Clergy.

Somerset Place, Nov. 26

If my notes were other than agricultural, or on such matters only as are in some degree connected with agriculture, and if it were ever allowable to speak thus publicly of the private scenes and social habits to which confiding and frank hospitality had afforded free access, I might mention many more subjects presented in the mansion of my host, and the circle of his family and visitors, in strong contrast to the savage wildness of the natural locality, or to the general absence of intelligence and refinement in the surrounding population. Far be it from me to treat of such matters, even though nothing should be uttered but in commendation. But there was one subject of observation which may be made a proper exception to the foregoing rule of conduct, and which offers a lesson of worldly policy, as well as on duties of immeasurably higher importance, to the agriculturists and slave-holders of the south.

On Sunday, the family and visitors designed to attend the performance of the customary public worship at the old neighborhood chapel a few miles without the swamp. But just before the time to set out, rain began to fall, and going abroad was thus prevented. It was therefore that I was invited to attend with the family the service in the negroes' chapel, which Mr. Collins has erected for the use of his slaves.[14] A clergyman of the episcopal church resides here, principally for the purpose of imparting religious instruction to the slaves, but who also officiates at the church in the neighborhood.

The negroes' chapel is a rude and rustic, but neat building, white-washed without and within, and provided comfortably, but in the plainest style, with the accommodations necessary for the congregation, and for its sacred purpose. There is room for 200 persons, and I was told that sometimes it is filled. But at this forenoon service, the house was not half filled, though the number still was as large as country congregations usually are. There was too about the customary proportion of females to the males present, or about thrice as many of the former as of the latter. This alone would have told, if I had not been so informed otherwise, that the attendance of the slaves is altogether voluntary. Every thing is done to invite and persuade, but nothing to compel their attendance. One means, and I should think the most efficient, is that the master and his family regularly make part of the congregation, when not at the more public chapel, and they participate fully in the services, and acts of worship.

The clergyman being engaged abroad, as always at the forenoon service, his place was supplied by another member of the family, who is licensed as a lay reader, and is also a candidate for the ministry. The service was according to the episcopal form; and never on any other occasion has it appeared to me more impressive.

In the evening, the congregation again assembled for a still more solemn service. The black audience was much larger than before, but still composed only of slaves belonging to this and the adjoining farm. The bishop of North Carolina, was a visiter of the family, and his presence, I presume, had induced this appointment for administering the sacrament. He was present in his episcopal robes, and after the ordinary service, performed by the resident clergyman, the bishop delivered an address on the service about to be performed, equally and admirably suited to the dignity of the subject, and the humble powers of comprehension of the greater number of his hearers. The officiating priest then administered the sacrament to all the communicants present, including the bishop, the master and several of his family, and about thirty of the slaves.

It was the first time that I had ever been an attentive witness to the performance of this solemn ceremony; of course I am altogether unworthy to appreciate it, and unfitted even to describe it. But never have I elsewhere been so strongly impressed by any religious service, or by any pulpit eloquence, as by the unostentatious ceremonial of this humble house of worship. Nor have I ever before had so forcibly urged upon my understanding and sense of duty, what we owe for affording the means of religious instruction to our slaves, and the high importance, and the excellent effects as here proved, of such instruction.

Why is it that, by the many who have undertaken the sacred duties of preaching the gospel, and laboring for the salvation of souls, so few and such feeble efforts have been made to give religious instruction to our slaves? The excuse which I have heard made, in answer to this question, is, that the owners of slaves might probably view any such efforts with suspicion, or perhaps meet them with hostility. As a slave-holder, and one professing to know something of my fellow slave-holders, I pronounce that there is no ground for such an opinion, or excuse. The assertion of the existence of any such general or considerable opposition, shows either gross ignorance, or otherwise is a foul slander, for which there is not a semblance of foundation. And even if there were some few narrow-minded slaveholders who would object to such advantages being offered to their slaves, is that any reason or excuse, for one who assumes to be a minister and messenger of God, to withhold his instructions from those slaves whose masters would receive them with welcome and thankfulness?

It is a fact universally admitted, and deplored by all considerate persons, whether christians or not, that throughout lower Virginia, the ministers of the gospel, of all denominations, are very few, and their usual congregations, or number of hearers, very small. Still there are many such persons who are passing their lives and complying, as they think, with their holy calling and vows, in regularly preaching once a week to a few dozens of hearers. Yet if these preachers deemed it their duty to labor as much for the salvation of blacks as of whites, there is scarcely one of them so situated that he could not easily find hundreds to instruct, who, unless so sought and invited, will perhaps never hear his voice, and certainly not profit by his ordinary and public ministerial services. Nay, there is scarcely one of these country ministers who at his very place of residence, or temporary sojourn, could not assemble an audience of slaves belonging to the farm, of greater number than he usually can collect of whites from all the neighborhood. And if any true zeal were felt and manifested for so excellent an object, ministers would find very many slave-holders, and even irreligious slave-holders, who would not only *permit* the performance of these truly apostolic labors, but would rejoice to have them performed, and (what I fear is more to the purpose,) would also be willing to *pay* for the performance.

If the owner of a hundred slaves, or a hundred thousand dollars, shows any disposition to receive religious instruction, what assiduous care, what devoted zeal, will not be shown for his eternal welfare! But while the greater part of the time and all the ability and power of persuasion of a minister would be willingly devoted to clear away the darkness, and strengthen the feeble striving, of such a penitent, and to aid his novitiate in a better course of life, his numerous and ignorant slaves, on the same ground and perfectly accessible to the preacher, would be scarcely thought of by him, and would derive no benefit whatever from his presence!

If an inference were to be drawn from the general course of procedure only, or of omission and neglect in this respect, by the greater number of those who have undertaken to perform *all* such duties, and who claim the sacred character of God's ministers, and messengers of salvation to all sinners—(and if in no degree corrected by the knowledge of any such meritorious and admirable exception as was described above—) the inference would be, either that it was a general article of belief among christian preachers that negroes had no souls to be lost or saved, or, if otherwise, that the souls of a hundred poor slaves were of less value, and had less claim upon their spiritual care, than that of a single individual of elevated grade in society.

No. 9. The Belgrade Farm. A Corrected View of the "Body Land."

Nov. 28, 1839.

Having postponed my departure to the latest day that the claims of my business at home would permit, on the 27th I left Somerset Place to take passage in the steamboat at Plymouth at night, and to proceed homeward as rapidly as I could be conveyed by the several connected lines of steam navigation and railways. Mr. Pettigrew the elder had arrived at his Belgrade farm (7 miles from the lake,) the evening before, and had sent me so kind and urgent an invitation to call on him the next morning, that I could not do otherwise, although there would be less than three hours to stop there. The short time so given was spent in conversation which was necessarily very hurried, and at the same time walking rapidly over the farm and taking a glance at its condition and improvements; and thus at the same time giving sufficient employment to our eyes, feet, and tongues. But however gratifying it was to me, even in so hurried a manner, to make the personal acquaintance and listen to the interesting remarks of Mr. Pettigrew, the opportunity was altogether insufficient for me to gather either, by observation or other mode of information, any thing in detail, to which I could do justice in reporting. I have requested, and hope the request may not be slighted, that Mr. Pettigrew will hereafter furnish me in writing something of what I had not time to obtain by personal inquiry and observation.

The weather was very cold for the season and latitude. The earth everywhere in the shade was hard frozen, and in the sun-shine the surface had just thawed, and of course was wet and adhesive. Both these conditions were as unfavorable as possible to the exhibition of the characters of the soil. Still, it required but a few minutes' observation to show me that I had been totally mistaken in my earlier opinion as to the nature and constitution of the "body land," formed upon merely traveling over another part which is in very inferior condition. Merely from the appearance of the surface on which I walked, and still more from the deeper soil and subsoil exposed in Mr. Pettigrew's ditches, I would have been convinced that this "body land" is not a clay soil proper, nor naturally a firm or high-land soil of any kind, but that this also is of swamp formation, though greatly and for a long time altered by the operation of natural causes. The whole surface is even more level than the existing swamps in general; so much so, that from one part of the tract which I saw, the surface water formerly and naturally was discharged towards every side, to seek outlets in different streams. The now artificial discharge by deep ditches is directed just as convenience requires, and might be made without much additional difficulty towards any point of direction.

The general plan of drainage is the same as that used at the lake, by drains of different sizes crossing each other at right-angles, and the smaller emptying into the next larger. The water-furrows between the narrow beds may be considered as the smallest drains, which first gather and discharge the surplus rain or surface water, though merely made by the plough in the ordinary course of tillage; and the ten-foot ditches are the largest and ultimate receivers which pass through the fields, and the fifteen-foot canal without, into which all the water is discharged, and thence into Scuppernong river. The main canal also serves for navigation for delivering the crops. Indeed, as every such main canal must connect with a navigable river, there must be the means of intercourse by navigation between every two properly drained farms in this part of the country.

The soil and subsoil, though, as stated above, apparently of swamp formation, have no resemblance to any of the drained swamp near the lake. The "body land" in its forest state, has no cypress or other aquatic trees, but has a considerable proportion of large oaks, intermixed with poplars, sweet-gum, beech, and other trees usually found on rich and low or alluvial, but yet firm and usually dry lands. Nature and time seem to have produced here such a change as may be supposed would take place in the great swamp, by its being relieved from superabundant water for the duration of some thousands of years. The upper soil being principally of vegetable material, would, in the course of time, rot down to a state of finely decomposed parts, and close texture; while the former forest of cypresses would die and give place to trees preferring dryer soil. Such seems to be just the natural state of the Belgrade land, and I presume it was that of the great "body" of which it forms a part. And for just such a change of natural circumstances to be made, it is only necessary to suppose that Scuppernong river, the present great outlet, was formerly so obstructed that the level of the back water was kept several feet higher than its present elevation.

The ditching of Mr. Pettigrew has been executed in the best manner, and his labors of cultivation seem to be scarcely inferior. The soil, though so stiff as to be best suited for wheat, yet will bring as heavy average crops of corn as any known in this country. I saw a field of 40 acres, from which had been measured a crop of 10 barrels[15] average to the acre, and part of which probably brought 11 or 12 barrels, as another part fell much below the general average.

The supposed ancient change of this land, from its still more ancient condition of a swamp saturated or overflowed by water, is my own hasty speculation, and which I should not leave it to be inferred that Mr. Pettigrew is responsible for. But some interesting facts, of which he informed me, go far to sustain the supposition. In his now cultivated fields there are several shallow ponds, varying from half an acre to two acres in extent. After these places had been laid dry by ditches, and brought under cultivation, by digging into the soil, charcoal and ashes were found, showing conclusively that these depressions had been made by the burning of the earth to that depth, such as occurs to some extent in the great swamp and especially the savannas, in every very dry season.

Next, as to the extent of operation of such action as might have lowered or consumed the surface and consolidated the soil of a swamp when relieved of its former overflowing water. Mr. Pettigrew told me that in the years 1805 and 6, the longest and greatest season of drought which he has had experience or report of, the great swamp was dry for 18 months together, and the water of the lake sunk 1 ft below its highest mark of a full head. During this long time, fires were continually raging in some parts of the swamp, and even passed over the same parts more than once in the time, as the state of dryness increased, and the combustible material was thereby deepened. In a particular part of the land adjoining Mr. Collins' canal, which was before described as being now covered with laurel trees, Mr. Pettigrew knew that the soil continued slowly burning for 12 months together. When at last the rains again saturated the swamp, and the superabundant water flowed thence into the lake, the water of the lake was at first so deeply colored as to stain cloth that was washed in it, and it was feared that its previous quality was lost forever. But this condition was merely transient; and it was not long before the water was restored to the state of purity which it had before, and has since retained.

And now I have to bid farewell to the swamps. And if I may venture to imitate the manner used by old Fuller in his different "farewells" to the counties of England, after discoursing of their "worthies" and their peculiarities,[16] my farewell would be of this fashion:

May the proprietors of the great swamp region, who are poor, copy the untiring industry and "indomitable perseverance" (to use his neighbor's words,) of the older of the two lake-side proprietors; and may the rich copy the liberal and judicious expenditures, and the enterprise of the younger; and may all profit by the intelligence and foresight, and enjoy the success, of both of them. May the swamp waters be diverted from their present operation of drowning land, and destroying fertility, health, and life, to the conveying the richest and most abundant of new products to good markets; may the soil be changed from being the poisoner of thousands, to be the feeder of millions. May the people of the good commonwealth of North Carolina acquire and maintain the energy requisite for these and other great works to make useful the now obstructed and dormant but mighty resources of the state; and may their government possess the wisdom and sound discretion which may safely direct the good work.

10 Carolina Swamps and Heroic Reclamation
Notes of a Steam Journey (1840)

Wilmington and Its Railway

At 1 o'clock, A.M. April 14th, I left Petersburg, Va., in the southern steam carriage, and at 7 P.M. reached Wilmington, N.C., which was later than usual, owing to several causes of delay. Distance, 181 miles.

The newly finished Wilmington railway (of 163 miles) is the most level and straight route, of any of considerable length, in the world; and being well planned and constructed, as well as on so remarkable and admirable a location, it necessarily is an excellent road. It seems also to be well managed—a matter which is as important to success as all the other requisites put together.

The construction of a railway of such great length is a rare instance of bold enterprise on the part of a community so small, and necessarily so deficient in wealth, as the people of the little town of Wilmington; for to them is entirely due the credit of the enterprise, and principally the successful consummation of the work. This road makes, together with the Raleigh and Gaston, and parts of the Petersburg and Roanoke, Portsmouth and Roanoke, and Greensville [Greenville] and Roanoke rail-roads, more than 300 miles of railway, finished and in regular use, within the state of North Carolina. Making every allowance for the aid of investments made by the citizens of Virginia in these great works, the results, brought about too in the last few years, speak loudly in favor of the enterprise of the "old north state," which has been jeered as being the Rip Van Winkle of the American confederacy, or as *asleep* in regard to public improvements. Perhaps Rip was aroused from his long sleep rather too early, after all. At any rate it would have been far better for his immediate neighbors, both north and south, if they had remained asleep full as long.

There being not sufficient accommodation at the public houses for the entertainment of so many visitors, we were met, upon our arrival at the depot, and divided among the principal inhabitants, like so many captives, except for the intention not being hostile, but hospitable, and entertained in the kindest manner during our stay.

The next day was the festival, to celebrate the recent completion of the rail-road, to attend which, as an invited guest, I had so far diverged from the point to which my previous engagements called me. An enormous length of tables was filled by upwards of 600 guests, and the dinner passed in the usual manner of all such entertainments.

Of Wilmington it may be said, almost literally, that it stands on a mere sand bank. This I had heard before of its site, but did not realize the truth of the description. The ground is not level, as I had supposed from the very level surface of the country in general. On the contrary it is quite hilly. The elevations are hills of almost pure sand; and having been formed by the wind on the ancient shore of the ocean. Judging from the absence of all vegetation in the open parts of the town, I at first thought that I had, for the first time, seen land either too poor or too sandy to produce (naturally) a blade of grass. Such, however, was not exactly the state of the case, as I found afterwards by walking out in the adjacent country.

The enterprise of Wilmington is conspicuous in other things besides its great rail-road. The thickest settled and business part of the town was all burnt but a few months ago; but there are already indications that the ruins will soon be replaced by new and good buildings. A new church, of the Gothic order of architecture, is, to my eye, the most beautiful structure, and the most appropriate in design to its purpose, of any modern building known.

Steam Mills of Wilmington

There are five steam saw-mills in operation in Wilmington, which together saw 100,000 feet of plank a day; and another nearly reconstructed, which had been destroyed by the late fire.

There is also a steam mill for hulling rice, and one for planing, and tonguing and grooving flooring plank. This last operation is as wonderful, to one who sees it for the first time, as inexplicable before seeing it, on hearing it described. The planks are first cut to even width by a circular saw, which passes through with great rapidity. The plank is then passed through the machine for the main operation; and in a few moments it comes out finished. The plank passes at a regulated course through two rollers. The planing is effected by four cutting edges, which turn around an axle, and which cut like broad adzes, but with an inclined edge and stroke. These whirl around with such rapidity as to appear to the eye a solid iron cylinder, and to leave the spectator doubtful as to how the planing is done. On each side are other circular cutters, striking horizontally, at the same time, and with cutting edges differently shaped, so that one cuts out the groove, in one edge of the plank, while the other, working opposite, cuts so as to leave the tongue which is to fit into the groove of another plank. This mill, as well as one of the saw mills, we understood to be the property of Mr. Lazarus, our hospitable host.[1]

Rice Fields and Their Culture

The tide of the Cape Fear river, at Wilmington, rises about four feet perpendicular. The opposite shore (in Brunswick county) was an extensive marsh or tide swamp, which has been ditched, and is cultivated every year in rice. This is indeed the principal culture in this part of the country, for market, as not enough corn is made for the consumption of the inhabitants, and very little other grain raised. The ditches and banks and flood-gates are so arranged that the tide-water can be brought over the ground, or let off and shut out, at discretion. This command of fresh water is essential to productive culture. The water is then let on, by some planters, and by others it is kept off until after the plants have come up, and have been once weeded. In the other case, the shallow water is drawn off for the weeding. In both cases, the water is then thrown on again; and while its cover helps the aquatic plant, rice, it effectually keeps down all the weeds that would otherwise injure the crop. The water is kept on nearly through the whole time of growth, but not quite, as that would render the growth too "spindling" and feeble. The product is very great, and also the profit. Still I cannot but consider the culture of rice as a curse to any region, as it must cause the production of malaria and its consequent diseases. This did not seem to be admitted by the inhabitants of Wilmington, and the rice planters, or, if admitted, but little importance is attached to it. An intelligent young planter, of whom I made some inquiries on this head, told me that he expected, as a matter of course, to have an attack of bilious fever every year; but he did not appear to think that of much account.

I crossed the river to see a rice field. The plants had but just come up, and there was nothing to be seen that was interesting. About 15 miles lower down the river, at Dr. F. Hill's[2] plantation, when passing in the steam vessel, I saw the rice ground flooded. The river water is there mixed with the salt of the ocean, and would kill the rice; and therefore fresh water is brought by a canal from a large pond, distant several miles in the country.

As I supposed, from my experience of embanked tide lands on the James river, these rice grounds, wet as they are kept, and therefore the soil less liable to decomposition, are annually rotting away and sinking; and some have been thereby made worthless, and thrown out of cultivation. But, as to these grounds at and above Wilmington, there is a counteracting operation of the river freshes, which bring down and deposit on the land mud enough to compensate for the waste of the original soil by decomposition. The rice grounds lower down the river are therefore most subject to this lessening of height; and these are *black* soils, while those higher up the river are brown, or "mulatto" land, that color being given by the deposite of sediment left by the river during freshes. The present and ordinary color of the river is the clear coffee color caused by vegetable dye, and which belongs to all the rivers of lower Carolina, except the great Roanoke, which flows from limestone and other clear sources in the mountains of Virginia.

Steaming to and from Charleston

On the 16th, at 2 P.M., our party proceeded, by the steam sea line of the Wilmington rail-road company, to Charleston, S.C. 180 miles, which full distance was passed over in less than 15 hours. At 5 P.M. of the 17th, we set out on our return, and the next morning, before 8, we were again at the Wilmington wharf. The steamer is one of four belonging to the company, the best for speed, safety, and accommodation; and the weather was as delightful and as favorable as could have been desired. There was nothing for me to desire, except relief from the necessity of leaving Charleston in so short a time, and upon the first visit. But so far as could be in the space of the few hours spent there, I was highly gratified with all that was seen and met with. Even that short time did not prevent the offer of hospitable attentions from entire strangers, which though but the common characteristics of warm-hearted southrons, were not the less appreciated.

As a hearty well-wisher, though personally a stranger, to the commercial capital of South Carolina, I was as much gratified as surprised to see how far and how well it had already recovered from the desolating effects of the great fire which but a few years ago swept over so large a portion of the city. But for the knowing of that great calamity, and the looking for its destructive effects, I should scarcely have guessed that so much of the city had been so recently destroyed.

The inertness and indolence and improvidence of my southern country men are unfortunately so conspicuous in my own state, as well as elsewhere, that I had taken up the opinion that these defects were to be seen the more strongly displayed as we proceed southward; and hence I had unconsciously taken up a very improper idea, and made much too low an estimate, of Charleston. Without touching on other matters, the buildings and general appearance of this city are more pleasing to me than of the more populous and splendid cities of the north. In the usual style of this one, there appears to be an unpretending dignity and beauty, and in some cases even grandeur, combined with simplicity. In this respect, Charleston may be likened to a gentleman born and bred, simply but perfectly well dressed, compared to a mustachioed dandy and exquisite. Indeed, it seemed to me that the population contained a larger proportion of those who appeared to be gentlemen than I had ever seen in any other city; and, for some time, I believed that not a pair of *moustaches* disgraced the city; but I was compelled to learn otherwise, by seeing before my departure several specimens of these ape-imitating "lords of creation."

The Battery is a newly constructed and delightful promenade along the border of the Cooper river, extending to the point of junction with the Ashley. The pavement of the wall is of very broad flat stones, with a low railing next the water, and open to the prospect of the harbor and islands below. The beauty of this view was enhanced still more when seen from an upper apartment of a splendid old mansion to which we were invited by its owner, and in which I was told had been entertained my distinguished countryman who was sent as mediator and ambassador from Virginia to South Carolina, both acting in their capacities of sovereign states. I had always felt proud of that occasion, as to both commonwealths, and was gratified to have now re-awakened, by the narrative and the scene, emotions which I had felt so vividly.

In the rail-road and steamboat route from the Roanoke to Charleston, there is nothing to be desired by the traveller, except the removal of the existing cause of detention of travellers and the mail, at Wilmington for 17 hours. This remarkable delay in a journey so rapid in all its other parts, is now made necessary by the want of lights to direct at night the course of the difficult navigation of the Cape Fear river. If this were done, it would only be necessary for the southern train from Charleston to wait two hours later than at present, and it would receive the mail and passengers a day earlier than now, and the whole journey would be expedited 15 hours to Charleston and all southward, without any additional speed while moving. If the Cape Fear river was north of Washington, it would have been lighted by government, and properly so, even if only for private commercial benefit. As it is, it has been long asked for in vain, even though there would actually be a great national benefit secured by the government, in the more speedy transportation of the great southern mail.

A Sandy Desert. Predaceous and Carnivorous Plants

The country adjoining Wilmington had to me an entirely novel and very remarkable appearance. The original large growth of pine trees had been cut down for fuel or timber, and had been succeeded by an almost unmixed growth of thinly set dwarfish "shrub" oaks[3] which rarely rose higher than six feet. The surface of the earth was but half concealed by the fallen leaves, and the scattered tufts of coarse grass; and pure and perfectly white sand was visible in so many places on the surface, amidst the green vegetation, that the general appearance was as if a snow had recently covered all the ground, which had melted in some places, while it still remained on others. I never saw any land so nearly approaching a sandy desert. This worst appearance here does not extend far back from the town.

But this barren and unsightly soil, and the adjacent country which is not greatly better, is the paradise of botanists. I was informed that there are found more species of plants growing naturally within ten miles of Wilmington, than in all Massachusetts; and one-third as many as are given in Elliott's catalogue of all the plants of South Carolina and Georgia.[4] I had new cause to lament my ignorance of botany, when in a region so interesting to the better informed. Still, I did not lose all the gratification to be derived from some of the most remarkable of these beautiful works of nature.

In the grounds around Wilmington is found in abundance the wonderful *predatory* (if not actually carnivorous) plant, called "Venus' fly-trap," (*Dionoea muscipula,*) of which a description was published in a former volume of the Farmers' Register. This plant is very small, and was not yet grown, and the largest of the *traps* were scarcely more than half the size they will be hereafter. Still, though not yet possessed of their full degree of *sensitiveness* which I hope to see, (as I had dug up and sent home a box of growing specimens,) they closed with a quickness, and operated with a degree of effect, far beyond my previous conceptions. The catching apparatus is an extension of the leaf. It is in shape much like a very diminutive steel trap, set open for catching, except that the valves are close in the plant, and open frame work in the artificial trap. The teeth around the circumference (forming a circle when lying open,) are long, and when they meet, interlock regularly and perfectly. All the interior is not sensitive. It is only three small and short filaments, on each side, which are scarcely perceptible, that serve as triggers. I saw every other part of the valves touched with a blade of grass, and even with some force, without its affecting the plant. But as soon as one of these filaments was touched by a small bug, the valves instantly and quickly closed together, their surrounding teeth interlocked, and enclosed completely the unlucky intruder, and will remain so closed until its struggles cease with its life. Then, or after the purpose of the death of the insect has been effected, the trap opens, and is ready to make another capture. I cannot believe that this wonderful and admirable apparatus has been contrived by nature, and kept at work, without some object. I cannot but believe that the death of the insect furnishes some benefit to the plant, and that, so far, it may be said to feed upon its prey. These plants grow on the borders of the wet places among the sands. Formerly it was supposed that they were found only in the neighborhood of Wilmington; but they have been since found in Florida, and elsewhere.

There is another bug-catching plant which grows abundantly hereabout, which though not possessing the power of animal motion, and not seizing its prey by an act of mental volition and design, as almost seems to be the case with the other, yet this one is scarcely less curious in its mechanical structure, and its adaptation of form to its object. This is the plant which bears the beautiful yellow flower vulgarly called "side-saddle" or "trumpet-flower," (*saracenia flava.*) It is a large pendent flower, on an upright stem of a foot to eighteen inches high. The very singular and beautiful form of the flower could not be described by me so as to be understood, and therefore will not be attempted. The trap is the leaf of the flower, and of a later growth, for the flowers now are generally fully blown, and some even on the decline, and none of the trap appendages are more than half grown, through the structure is fully shown in them, and also in the old and dry, but still well preserved leaves of last year's growth, which remain, and nearly all of which contain the scaly wings of small beetles, and other undecomposed remains of their insect prey. This leaf is in the shape of a slender trumpet; of very thick tough texture, which is closed at the ground and gradually enlarges to an inch in diameter at about fifteen inches high. On one side of this mouth of the trumpet, a part of the leaf extends and spreads over the mouth, so as effectually to keep out any rain water; and for greater precaution, in the middle of the umbrella is a deep groove, which, as a gutter, conveys the rain water off. The lower and closed end of the trumpet is always filled some inches deep with a limpid and tasteless water, secreted by the plant, which seems to attract insects. They descend into the trumpet easily; but can never return, as the sides are beset with numerous minute hairy filaments, which point downward, and effectually prevent any progress upward of the insects enticed to enter the receptacle.

This magnificent and curious plant is said not to be found much farther north. There is however another of very similar general structure, the purple *saracenia*, which grows as far north as New England. However, I never saw it before, and it would have been considered admirable for its beauty, and its curious mechanical structure of the flower, and its water-holding and insect-catching apparatus of the leaves, but for the inferiority in all these respects to the more common *saracenia flava*. As little knowledge of and taste for botany as I have, a longer residence in this neighborhood and sufficient leisure would tempt me to begin the study. It is surprising that this region is not more visited by the lovers of curious plants and beautiful flowers, and that the ready means now offered by the railroads are not more used to convey these rare treasures northward. I sent on by the train, in boxes, growing specimens of all the above plants, with their native soil, just dug up, and which must have reached Petersburg in perfect condition; and might so be carried as far as Philadelphia, even should they fail to be kept alive afterwards. In this manner, by proper arrangements, hundreds of rare plants, which even in green-houses cannot be produced in perfection at the north, might be sent, in all their native splendor and vigor, to floricultural exhibitions a thousand miles distant.

General Appearance of the Lands of New Hanover County

One of my first objects after reaching Wilmington was to seek out, and make personal acquaintance with Dr. James F. McRee,[5] a gentleman whom I knew merely as the only subscriber to the Farmers' Register in or near Wilmington. I drew the inference, in which I have rarely been mistaken, that a man who had thus appreciated my labors for seven years, would probably be one both able and ready to aid me in acquiring information. I was not mistaken. In addition to the pleasure derived from Dr. McRee's general conversation, and to his kind and serviceable attentions in other respects, I was indebted to him for much of the information in regard to the peculiarities of this region, and its products, which I was enabled to acquire. My first inquiries of him were in regard to the extent of the marl formation of this region, (having previously seen it in Wilmington,) and I heard from Dr. McRee such interesting facts on this subject, that I was very willing to accept his invitation to accompany him to his farm, Ashemoore, near Rocky Point, the most favorable portion for observation. For this place we set out on the 17th.

In a mile or two from Wilmington, along the ordinary carriage road, the excessive sandiness of the soil changes to rather less of that character, and to such as is general through the pine lands for many miles. The general growth is long leaf pine, thinly set, with very little undergrowth of trees or shrubs. The surface of the earth is set with tufts of what is here called "wire grass," but which (as might be inferred from its locality,) is altogether different from what is called by that name in Virginia. This grass, as Dr. McRee informed me, (and to whom, by the way, I am indebted for all the botanical names here used,) is the *aristida stricta*. It grows in small tussocks. Each spire is a single straight upright stem, cylindrical and as regular in shape as a wire. This is the most general grass in the poor pine woods, and open grounds of same quality. When burnt off in the spring, as is most usually done, the young grass springs out more rapidly, and furnishes good grazing all over the country.

The surface of the land, like that of the last 50 or 60 miles seen along the rail-road, is almost a level. The slight depressions are all of swampy character, and are called "bays" or bay land, because the loblolly bay tree (*gordonia lasianthus*) is always found in such places. But whether dry or wet, all these pine lands, and the shallow "bays" intersecting them, are very poor, and, without being furnished with the calcareous ingredient which they want, are, and will continue, worthless for tillage. It is from such pine woods, which cover so large a part of eastern Carolina, that the great suppliers of turpentine and tar are obtained, the making of which is almost the sole business of the residents of the pine lands.

The Calcareous Lands of Rocky Point

Upon reaching what are termed the Rocky Point lands, the marks of soil changed from the indications of the basest to those of the most valuable calcareous lands. Rocky Point is a low bluff of the limestone peculiar to this remarkable region, jutting out on the north-east branch of the Cape Fear river. The stone or the softer marl (or that which may as correctly be called *chalk*, as the harder is *limestone*,) forms the continuous and nearly horizontal substratum of all this neighborhood, and comes so near to the surface of a body of some 6 or 7,000 acres, as to give to it a peculiar and well established character for great fertility, and power of long endurance of continued cultivation of the severest kind. A small tide water creek, which passes through the Ashemoore farm, marks the middle of this body, and on each side of it are the best of these celebrated Rocky Point lands. They are there best, no doubt, because the surface is lowest, and the calcareous substratum is nearest to the surface. It is sometimes so near that the plough turns up a white slice upon the top of the black soil; and in these, and many other places, where the cause is the same, though not so manifest to the eye, the productiveness of the land is greatly impaired. This injurious effect, however, is greatly increased by the too great wetness of the soil, owing to its level surface, its absorbent and sometimes close and adhesive texture, and especially to the solid calcareous sub-stratum, which prevents the surplus water escaping downward by filtration. The land too has not been cultivated in beds (or ridge and furrow,) which, if properly executed, and with the necessary ditches, I am sure would make the lowest acre abundantly dry. There are no springs rising to the surface that cause wetness, but only the surface water from rains to guard against; which is a very simple and sure business. The soil, in quality and in level, and in its calcareous substratum, is more like the famous low grounds of Gloucester and Back River in Virginia, which have been described at length in former volumes of this journal, than any other known lands. I never saw soil that seemed of better natural constitution and quality than some of the fields of Ashemoore, or which promise better rewards for their cultivation.

The limestone is but an accumulation of pure shelly matter, solidified into stony hardness. The marl is the same in chemical composition, but about as hard as chalk, and has very much the texture of an impure chalk, and is soft enough to be used as manure without pounding, burning, or other mode of reducing. I analyzed, on the place, three specimens which were selected as fair samples; and Dr. McRee, before, with the aid of my portable apparatus, had examined several others. The general proportion of carbonate of lime in both kinds was fully 95 per cent. One specimen only, of marl just dug and hastily dried, yielded as little as 88 per cent.; and as this was the only one so poor, by 6 per cent., I suspected that it had not been sufficiently freed from water, in drying. However, at any rate it is the richest and most valuable marl I have ever known, and the easiest to be used. The stone, of course, would require to be burnt; and it will yield excellent lime, for cement or for manure.

The lowest parts of this body of land, not yet cleared, are swamp, called "white oaks;" not because white oak is their general growth, and indeed it is very rare there—but because such are the only places on which a white oak tree can be found in this part of the country. These swamps are covered with the trees that are most favored elsewhere by the richest, stiff, alluvial, and wet bottom lands. The calcareous bed lies near the surface of all these swamps or "white oak" lands.

Dr. McRee has taken great pains to introduce grass husbandry on his land, and his clover is wonderfully productive. That sown on the first of March, 1839, was fit to mow, and was mown for hay the same year, in July, and the hay sold and delivered in market within eight months of the sowing of the seed. This is a remarkable proof of the admirable fitness of the soil for clover; and it was particularly valued by me, as the strongest known proof of what I have so often maintained, that if the soil be but made calcareous, the warmth of climate in North Carolina, or even farther south, is no bar to profitable clover culture. Before my practice proved otherwise (after marling) it was as firmly believed that lower Virginia was too hot to produce clover to profit, as it is now generally (and as erroneously) believed of lands 200 miles more southward.

But though the calcareous deposite beneath the Rocky Point lands is richer and more easily accessible, than any known elsewhere, it is but the most remarkable case of a formation that is spread through a vast region of the state, accessible throughout a great part of its extent, and which would be highly profitable to be used wherever it can be obtained. I knew before that marl had been found along the Neuse and some of the upper waters of the Chowan, and that it had been used to some small extent by a few individuals; and I inferred, that if sought for, it might be found at some greater or less depth, almost every where between the granite range and the sea coast. But I had never heard of a single actual discovery farther south than the borders of the Neuse and Trent. In addition to what I saw in Wilmington, (though the stratum of marl there is thin, and the limestone poor,) and at Ashemoore and the surrounding lands, I learned from Major Gwynne,[6] the able engineer who directed the construction of the Wilmington railway, that marl was found in the wells dug at the water stations of that road, through the distance of 60 miles from Wilmington. And this marl lies either under, or near enough for transportation to the wretched pine lands, which, wretched as they now are, need but to be marled to become valuable and productive under tillage. I may now, as heretofore, urge this improvement, for this region, in vain; but a time will come when the value of this neglected means of improvement will be properly appreciated in North Carolina, and when the putting it in use will add millions of dollars to the productive wealth of this region, which, of all within my knowledge, is most favored by nature, and the favors so offered are most slighted by man. And though I have not yet seen the continuation of this region through South Carolina and Georgia, I entertain no doubt but that my remarks would there be applicable.

It would seem as if the Rocky Point land, so deservedly noted hereabout for its fertility, owes its value to its being so thinly spread over the calcareous deposite, that the two earths have necessarily become mixed, by various natural causes. When the roots of trees, and even small plants, can strike through the upper poor soil and into the marl below, the parts of the latter which are taken up into the plants, at their death and decay are finally left on the surface. Thus, in the lapse of ages, the surface, no matter how destitute of lime, and how poor, must thus be made calcareous and rich. But not so if the surface soil be but six or even four feet above the marl, and cut off by a barren intervening subsoil, which the roots of plants are not able to pass through. Then the soil will remain poor; and so it would, even if within a foot of the marl below, but for the operation of plants or animals in bringing up the marl to the surface.

In accordance with these views, where the land is higher, it is very inferior to the best kind; and at a few miles from the river, a still higher elevation of surface becomes either the ordinary poor pine forest land of New Hanover county, or savanna, of which I shall speak presently.

The texture of the calcareous substratum of the Rocky Point lands is altogether different from any of the numerous marl beds I have seen in Virginia. In chemical constitution, and in hardness, much of the former may be properly called by its common name of "limestone;" and by the same tests the balance might be called *chalk*, slightly adulterated, and tinged with a very foreign matter. But geologists, I believe, do not admit any true chalk to be in this country; and the concretion of shells to a stony hardness, cannot make the *limestone* so called in mineralogy. However, in agricultural sense and use, they are truly what these names would imply.

If the people of Carolina who have the means of marling would apply the lesson afforded here, it might be considered that here God had marled, and thereby enriched the land, and had thus revealed to man the mode of improvement. The enriching of these lands was effected simply by natural marling, with the additional aid of freedom from exhaustion, and thereby the accumulation, and fixing in the soil, of as much vegetable matter as the calcareous ingredient could combine with.

The limestone is not disposed with any regularity as to the softer marl. Isolated masses of the former, of various sizes, are seen scattered over the best fields; and sometimes the stone, and sometimes the chalk is nearest to the surface, or the one over the other. The ditches generally reach the calcareous substratum. When in the chalk or marl, the excavation is easy enough. But when the stone is opposed, blasting by gunpowder is necessary to open the ditch. I saw in two ditches where this last had been done, for stretches of 200 to 300 yards each. Still, Dr. McRee told me that this ditching by blasting was not very difficult, and, as he thought, was not more expensive of labor, than to ditch through newly cleared forest land. There are some other unexpected peculiarities of a limestone region in this neighborhood, and even at some miles from the calcareous soils. There are subterraneous caverns and subterraneous streams. The former are like the limestone caves of the mountain region, except for their very narrow dimensions, which forbid any passage or examination. A small dog has been known to make his way through a subterranean passage for several hundred yards. I saw, at Ashemoore, one of the "natural wells" which show subterraneous streams. This was in a field not of calcareous soil, and was about the size across of an ordinary artificial well. The water stood at about two feet from the surface of the earth, and is supplied by a perpetual stream passing through below. The water, however, is not higher than usual, owing to additional supply from rain.

I heard that a little marl had been used in some of the gardens of Wilmington; but not the least use of it has as yet been made on fields, by any south of the Neuse and Trent, of whom I have heard.

The Savannas

There are, in many places in New Hanover county, and other of the south-eastern parts of North Carolina, interspersed among, and surrounded by forests, tracts of open grass land, called savannas, as clear of trees as if made and kept so for cultivation. These meadows, however, are altogether formed without the designed aid of man; and if cultivation has ever been attempted on them, it was found profitless, from the unproductiveness of the soil, and the attempt was abandoned, after the first trial. The railway passes immediately through one of these savannas, about 27 miles from Wilmington, which is said to contain more than 4000 acres in one body. My passage through this piece was so rapid, that nothing could be distinguished as to its particular growth of grass, or the nature of the soil. It only presented to the eye one unbroken and unvaried expanse of level surface, as much elevated, apparently, compared to the road-way, as the forest lands, and covered with young grass, the deep verdure of which was made the more conspicuous by the old grass having been burned off early in the spring. To examine this kind of land more closely, was one of the objects of my visit to Ashemoore, under the guidance of Dr. McRee. Part of his estate is one of these savannas, which is partly seen from the rail-road just north of the Rocky Point depot, and 15 miles from Wilmington. This savanna is of small extent, being only 200 to 300 acres, but in every other respect is similar to all the others, and would serve, as I was told, as well for observation as any number of them.

I found that but a very small part of this savanna had been burnt over this spring, which gave a better opportunity of judging of its usual growth from the still abundant remains of the dry grass of the last summer. This was of many different kinds, all of which seemed new to me, or if not new, so altered by difference of soil as not to be recognized. However, my examination in this respect was necessarily very slight, owing to my ignorance of the scientific arrangement and descriptions of plants. The principal growth seen consisted of the *aristida stricta*, (before mentioned as being general through the pine-covered land,) and a dwarfish kind of broom sedge, which is smaller than, and different from, those so well known in lower Virginia, but which was not the less manifestly one of the members of that family of grasses, (*andropogon*,) all of which, I infer from my experience of the habits of some of them, must indicate by their growth an *acid* soil. The old growth, notwithstanding all the eating and trampling by cattle, seemed but little diminished, and had stood about two feet tall.

I dug into the soil of this savanna, and at the place examined, a spot rather higher and therefore then drier than the general surface, found the depth to be about a foot. This, however, was more than usual. Along the side of the ditch made to form the rail-road, where it passes by the out-skirt of this savanna, the soil was about 6 to 9 inches deep, and less generally than 6 inches through the adjacent pine land. The soil of the savanna land is as black as earth can be, is moderately stiff, showing the presence of enough clay to form a productive soil for grasses and wheat, and would be supposed, if judged merely by the eye, to be as rich, and valuable for tillage, as in fact it is poor and worthless, as now constituted. The subsoil is a pale yellowish sand, with barely enough clay intermixed to be adhesive, and not to prevent water passing through and saturating it. Still lower, as I was told, lies a bed of clay, which, no doubt, forbids the descent of the rain water, and causes the general wetness of these lands, though they lie high, and are free both from springs, and from floods from any higher lands.

There does not seem to be any great difference between the savanna land and the great body which is covered with pine forests, except in the difference of covering plants. Indeed, in many parts of the pine lands, where the trees stand but thinly, the savanna grasses cover the ground, and where burnt off, present the appearance of a true savanna, with the addition of a sprinkling of trees. The same general causes operate with more or less force on all this immense region of pine forest land; and, according to the degree of force, certain places either become open and perfect savannas, or remain as forest land.

I was the more interested in examining the savanna land because of my former investigations of the causes of the formation of the western prairies. I was not before aware of the existence of any lands of that character in all the Atlantic slope of the United States. But the Carolina savannas are certainly true prairies, though differing so much both in nature of soil and in degree of fertility from those of the west; and the first formation and continuance of these savannas may be satisfactorily traced from the operating causes which were set forth in the essay on the formation of prairies, commenced at page 321, vol. iii., of Farmers' Register.[7]

The whole of the great region of pine forest in lower North Carolina, with the exceptions of the intervention of rivers and swamps, is high land, of as level a surface as can be conceived. Without any springs rising to the surface, which would be the sources of marshes, and with a climate in which dryness and warmth predominate greatly over moisture and cold, it might be inferred that such lands, and sandy as they are withal, would be very dry.

But though no springs rise to bring water from subterranean sources to the surface soil, and though the vast extent of surface alone is security against any considerable effect, even on the western borders, of inundation from higher lands, still there are sufficient causes why these lands should be very wet through the greater part of the year. The almost perfect level of the surface prevents any considerable quantity of the fallen rain-water from flowing away to the rivers and swamps; the sandiness of the soil, and still more of the subsoil, serves readily to drink in the rain water; and the close nature of the clay below forbids the excess of the water thus taken in to sink lower, and therefore it must fill the entire soil and subsoil, for several feet in depth, until drawn up and carried off slowly by evaporation, during the driest part of the summer and autumn; and which evaporation is much retarded by the general shade of pine forest over nearly the whole of this great region.

These circumstances are sufficient to show why wetness of the earth should generally prevail, even in the drier times of much the greater part the year; and why every heavy rain should so fill and overflow the surface of the land, whether forest or savanna, and serve to make dry walking impossible soon afterwards, in any direction. During two or three months of the driest season, it is true that all these lands, savanna as well as forest, become perfectly dry. But that does not prevent the growth and general circumstances of the land being governed and directed by its opposite conditions of wetness, which continues for three-fourths of the year.

This general state of moisture of the land is particularly favorable to the growth of grass; and though all the grasses both of calcareous and rich soils are necessarily excluded by the chemical constitution and poverty of this soil, there are enough of other grasses which are best nourished by the acid quality of soil which must exist here, and which is poisonous to all the most valuable plants of cultivation. The poverty and acidity of the soil also forbid the growth of the kinds of trees which flourish best on rich and well-constituted wet lands, as the poplar, ash, black-gum, maple, &c. Pines, then, the most appropriate and favored growth of dry, acid, and sandy soils, form necessarily the almost sole growth of this great region, though the general state of moisture is unfavorable to their greatest vigor. Thus it may be supposed that there must be a general disposition of the growth of trees to give way to the growth of grass, whenever there are additional and strongly operating auxiliary causes of support given to the latter. These causes are presented in the several following cases: 1st, whenever the soil is more stiff, or the lowest clay sub-stratum rises nearer to the surface, and either, or both, serving to retain more of the rain water at and near the surface; or, 2d, when the vegetable matter in the soil is deeper and more abundant, and being no[t] fully decomposed, is of a peaty nature, (a consequence of the preceding named condition, and which is always the case in the savannas,) and which therefore is still more unfriendly to the growth of trees; and 3d, the burning over of all the forest land in the spring, which is almost universally done, and which must damage a weakly growth of trees more and more by every repetition, and thereby allow a greater growth of grass, and of course more of the fuel furnished by the grass for succeeding burnings. It is where all these circumstances are combined, that, by the gradual destruction of the trees by fire, the savannas have been formed, and will be maintained as long as the same causes continue to operate.

The savanna I examined most closely was within three miles of already discovered marl, of the rich quality above stated; and it would be a cheap improvement to bring and apply it to the savanna; and I am confident that marling, and draining, (which also could be easily effected,) would make this savanna excellent land for wheat and clover, as well as for corn or oats. But the improvement might probably be made still more cheaply, as the bed of marl doubtless lies under it, and most of the other savannas, and might be obtained either immediately from below, or elsewhere, more cheaply than by carriage for even three miles. The railway offers sufficiently cheap transportation for lime, or the richest marl, to considerable distances.

The drainage of savanna[s], might be effected by the plough, and by the same operation that would best prepare it for cultivation. If attempted, I would advise the ploughing of the land during summer, when dry enough, by a three or four-horse plough, into very wide beds, (say of 26 feet, if for 5 corn rows, or 19 feet if for 3 rows each,) with deep and clean water-furrows between the beds. As soon as the surface had become somewhat closed, another ploughing in the same order and manner, would raise the bed still higher, and nearly double the depth of the water-furrow. It would only then need that all the water-furrows should discharge into one or more cross-ditches, cut in the lowest ground, and having sufficient depth and good outlets, and the draining would be complete. The ditches designed to receive the water discharged from the water-furrows, and also any cross ditches required at depressions, might likewise be made principally by the plough, if done before the bedding. Full direction for both these operations have been published in the Farmers' Register; that for ditching by the plough, in an article on draining, in numbers 7 and 12 of vol. i., and for the wide bedding in the observations on the Gloucester lands and culture, in vol. vi.[8] I have never seen lands to which both these modes of drainage could be better applied; and if followed by marling or liming, (and without that, all improvement of this acid land would be profitless,) I know no lands which promise better recompense for the expense of the improvement.

But even as they are, the savannas are excellent grazing grounds while the grass is young and tender. The burning was early on the large savanna, (as is usual,) and it already is a beautiful expanse of green, regular and unbroken to the eye of the traveller as a grazed lawn. As yet but few kinds of the numerous wild flowers of the savannas have opened; but in June they are seen throughout, and of various kinds, and successions of kinds continue to adorn every part of the ground until frost.

State of Agriculture, and Products of the Country

Except rice, made on the embanked marshes of the Cape Fear river, there is but little if any surplus grain made in the counties around Wilmington. The corn required for the support of that town is partly brought from abroad, and none is exported. There is no where any wheat sown, except lately by Dr. McRee and one of his neighbors on Rocky Point, and the oats are all consumed on the farms where raised. There is but little cotton made; and scarcely any thing for market made from the fine grazing afforded by the savannas and forest lands. As evidence of the abundance of pasturage, I understood that very few persons give any food to their cattle during winter, as they can obtain grazing throughout in the woods and swamps. But the mortality, in spring, from this starvation system, is very great.

There is not much land cleared or cultivated except on the borders of the rivers, and of that I saw but little. In the pine lands there are but a few small patches cleared, and those very poor in product; the almost sole business of the residents being to make turpentine and tar. Even where better lands and better farming are found, the same crop, corn, is generally cultivated without the intervention of any other, either every year, or every second year, as long as the land can bear it. The nearest usual approach to a rotation of crops is, 1st year, corn, and 2d, oats on the best part, and no crop on the worst; and all grazed after the oat harvest, until near planting time next spring, usually, though sometimes the grazing is avoided. Of course no turf forms on land so treated; and it is not ploughed flush, before planting corn, but only "listed,"[9] and checked across, and the "balks," broken up after the corn is well up. Yet there is good demand for all grain and grass products, and extraordinary prices paid for articles of small culture, because of their scarcity. For example, the first person who made a business of sending turnips to Wilmington market, sold them readily throughout the whole season for $1 the bushel; and though another producer has come in since to share the furnishing of the market, the price still has kept to 50 cents the bushel. Both Irish and sweet potatoes grow well; the latter as well as could be desired, and they are found difficult to keep later than mid-winter, and therefore are not raised very largely for stock food.

Field peas are planted through every corn-field, by planting 10 or 15 peas on one side, and sometimes on both sides of every stalk or place of corn. This is done before the last ploughing of the corn, and that is the only benefit of tillage which peas receive. As many peas are gathered as are necessary for use, and on the balance, as they stand, the hogs designed for slaughter are turned in, (the corn having first been gathered,) and are allowed to feed at will. Peas alone, when so abundantly consumed, are supposed to cause disease to hogs, (a weakness in the loins,) which is prevented by having a patch of sweet potatoes in the same field, for change of food. The red and cow peas are the kinds most generally planted. It is the prevailing opinion that the making of this *secondary* crop of peas does not materially lessen the product of corn from the same ground; but no fair experiment has been made to test this important question, and I therefore greatly distrust the correctness of this opinion.

The sweet potato is most favored by a southern climate and a sandy soil, and therefore I should infer must surpass in value, for stock food, and for greatest product of nutriment from a certain space, all the roots of northern climes, such as Irish potatoes, turnips, beets, carrots, &c. But though the culture of this most valuable southern root is universal on every farm and clearing, as is its consumption on every table, still it is held as a product of secondary and minor importance, and there is nothing clearly established (that I have heard) as to its amount of product, or the preservation of the crop. The risk of rotting in the cellars or other receptacles, and the great cost of planting roots, are great obstacles to extensive culture. Some persons, however, and by various modes, succeed in keeping potatoes sound, until late in winter. But it seems owing to peculiar, and not well ascertained causes, or otherwise it could be done always. Small quantities in cellars made and kept both dry and warm, and also well ventilated, are kept much later.

Pine Forests

From the few fine specimens I had seen of the long leaf pine, on the southern border of lower Virginia, I had expected to see magnificent trees forming the greater part of the vast pine forests of North Carolina. But such is far from the fact. Some few indeed present the noble elevation of straight trunk, and great size, which I had looked for; but most generally, the trees are much smaller, and far from being pleasing to the eye. The long leaf pine is the original and almost sole growth of the drier lands of New Hanover, and even much farther north, as seen along the railway. The short leaf pine, as called here, and which is the "old-field" pine[10] of Virginia, is here, as with us, the universal second growth of the drier lands wherever the first growth has been cleared off. As to the short leaf or common "woods-pine"[11] of Virginia, I have seen none, though I hear that some few may be found.

It is a curious fact, and one much to be lamented, that there is scarcely any reproduction of the long leaf pine throughout the vast region of which it now forms the almost exclusive cover; and that when the trees now living shall have died, there will be almost a total extinction of this beautiful and valuable tree. It is a tree of very slow growth; and probably most of those now living are older than the general settlement of the region, or the existence of the circumstances which now oppose reproduction. The large cones or "burs" of this tree contain numerous seeds, which drop out as the cone opens, and which are eagerly sought and eaten by hogs. This "pine mast" forms the great resource for feeding the hogs while ranging at large, as is the usage. In this way, very few seeds escape to sprout. And of the few that do sprout, scarcely any of the young trees survive the after attacks of the hogs, which root up the young trees, to eat the roots, even when the trees are several years old. Hogs ranging in the woods are quite fond of the tender roots, and the bark of the roots of older trees, and live on this food principally in the winter and spring, after the pine seeds are consumed.[12]

From these causes I presume it is that very few young long leaf pines are to be seen. I noticed them no where except a few in the neighborhood of Wilmington, in riding from that place to Rocky Point. And the appearance of the young long leaf pine is too remarkable not to have been observed if seen. It is indeed as remarkable for its singular and beautiful form when very young as for its noble size and form when old. One of three seasons' growth (for example) was only a little more than two feet high, and retained at once all the leaves of the three seasons, though in very distinct states. The leaves of the first year, dead and dry, but still firmly set and strong, covered the plant thickly for about a foot above the ground; then the green and vigorous leaves of the last year's growth, for about the same extent; and lastly, the new upright and compact shoots of this spring, one or two and sometimes three, from which the leaves, though formed, had not yet unfolded, and which shoots are each about the size of, and not very unlike in general appearance, a large shoot of asparagus, as it first rises out of the ground. The older leaves are about a foot long, and as thickly set on the single and upright trunk as can be imagined; and the whole plant, of this or of less size, is much like an artificial military plume of enormous dimensions.

The Turpentine and Tar Business

The making of turpentine and tar is the almost sole business of the thinly settled population of the pine lands. They are generally poor and indolent; yet this business affords good profits even at the present low prices, and enormous profits were made when naval stores were more than double their present prices. Turpentine now sells at $1.80 the barrel at Wilmington, and it has sold for upwards of $4. Mr. Lazarus told me that he had paid to a poor white man, who worked singly and unassisted in making turpentine, $1000 for the fruits of his labor of one year. It is understood that a good hand can attend to 9000 trees, and can secure 200 barrels of turpentine in a year.

In commencing the operation on trees before untouched, a receptacle (or "box",) is cut by the axe on one side of the tree, and about six inches above the ground, which is large enough to hold a quart of the fluid turpentine which exudes from the cut sap-wood, and which flows into this hollow from the upper part and sides. The flowing of the sap begins of course in the spring. At the end of a few days, (according to the time and state of the season,) the laborer visits all his trees, dips out the collected turpentine and puts it in barrels. He then cuts from each side of the tree a shallow groove, inclining downward to the box, through the bark and a little into the wood. Into these new cuts the turpentine exudes, and flows down them into the box. The tool by which this operation is performed is called a "shave." It is a circular piece of iron like the eye of a weeding hoe, with the lower edge sharp, and which is attached to a shaft or handle, so as to cut its groove like a gouge, but by being pulled to, instead of being pushed from the operator.

Every time the box is emptied of its turpentine, the "shaving" is extended upward, and thus gradually making the tree bare of bark and of the outer surface of the sap-wood as high as can be conveniently reached, or 15 feet and upwards. This shaving rises about two feet in a year, and thus it takes about seven years to finish one side of a tree. The side edges of the bared surface are carefully kept perpendicular and straight, and not quite to embrace the half of the trunk of the tree. Next, the opposite side is "boxed," and treated in the same way, taking care to leave a strip of an inch or two of bark on each side between the old and the newer work. Without other cause of decay or destruction, the trees will live and yield well until the sides can be shaved no higher. But the spreading of accidental fires seldom fails to kill the tree earlier. For the entire face of the cutting being encrusted with turpentine, and the wood below being converted to solid lightwood, no trees can be more inflammable; and the fire burns so deeply in, as to kill the strips of living bark by heat, or to weaken the trunk so much that it yields to, and is prostrated by, the next storm. The trees, or parts that escape being burnt, are finally cut up into billets, and the tar extracted from them, by burning them slowly in a close kiln, made by covering the lightwood with earth in the mode well known in every pine country.

It is only the turpentine that retains its fluidity, and is collected in the box, that is considered first-rate. The part that sticks to and hardens above has lost its most valuable part, (the oil or spirits of turpentine,) by evaporation, and when scraped off, which is the last part of the process, is sold at half the price of the fluid turpentine. Of course the expense of land-carriage is a sufficient bar to the production of so heavy and low-priced products, where the distance is considerable.

The turpentine getters are careful every spring to rake away the leaves from the foot of every tree, and to burn the collected trash when it can be done slowly and safely. But they cannot always command the progress of the fires; and from that, or other less carefully made fires, great havock is often made among the boxed trees.

Where vicinity to market, or cheapness of carriage, permits this business to be in full operation, it cannot last long, as the long leaf pines will be destroyed and will not be renewed. The other kinds of pines are not worth working for this purpose.

On the morning of the 20th, I joined the railway train at Rocky Point depot, and before 2 P.M. reached Goldsborough, (distance 65 miles,) near Waynesborough, in Wayne county, and the necessary place of departure for Newbern [New Bern]. Here I had to wait until 1 A.M., the next day, when, by a stage coach drawn by two horses, but loaded for four, I reached Newbern after dark. Distance 70 miles.

This day's journey was along the general course of the Neuse river, though no where in sight of it, except at Kinston, in Lenoir county. By another mode of travelling, this might have been a very interesting route; for the calcareous formations, to examine which had been the first object of my intended visit to Newbern, are scattered all along this river. But I could hire no mode of conveyance, other than that of the public stage coach; and even if it had been otherwise, I could not have profited by the change, having no well informed companion to guide and instruct me as to localities. I could therefore do no better than go on to Newbern, trusting to the chapter of accidents for my subsequent procedure.

Judge Gaston's Reclaimed Pocoson

On the morning of April 22d, the day after my arrival at Newbern, I accompanied Judge Gaston[13] to his farm, about ten miles distant, for the purpose of seeing the large *pocoson* [pocosin] of which he has reclaimed and has under cultivation between 200 and 300 acres. I had not before had an opportunity of observing, or being correctly informed as to the nature of this class of wet lands, and I was glad to avail myself of the request and invitation which afforded such an opportunity, with the additional gratification and advantage of the guidance and conversation of a companion so interesting and instructive.

The appearance of the still unreclaimed part of the pocoson is very similar to that of most of the great swamps in the drier time of the year. However, none of the uncleared parts adjacent to this improvement are as wet as the whole was at first, as the ditches have had a partial and considerable effect in drying the forest part. The effect of this partial drying is seen in the dying of the large laurels and other trees which require very wet soil for their most perfect condition.

The surface of the land is so level, that it is drained to outlets in two very different directions. The water carried off by both these main ditches is but little, to be collected from so large a body of land which had been absorbing rain water all the winter. The discharge was not greater than I should have expected to find from the existing circumstances, even if there were not a real spring in the whole ground.

The surface of the pocoson is higher than most of the surrounding firm pine lands. The opinion of Judge Gaston is, that the pocoson was originally the ordinary level pine land of the country; that the passage of the small streams of water from its flat surface was afterwards impeded by fallen trees or other obstructions: consequently a new character gradually was assumed, different from its former state of dry land, and approaching first to that of swamp, and next to that of peat, in so much as the progress of decomposition of the vegetable products did not keep pace with that of their production, and therefore that the accumulation was yearly increased and the surface of the land kept growing higher by the increase of the undecomposed vegetable deposite from the trees, and other plants growing thereon. In this state, the waters of rains, as well as those of impeded small streams, would be necessarily sucked up and retained by the great mass, as by a sponge, and the surplus served to overflow the surface during the greater portion of the year. The proofs to sustain this opinion were presented to him in the two facts, 1st, that the surface of this vegetable soil is actually higher than the surrounding firm land; and, 2dly, that light-wood logs and knots, the evident remains of a former growth of large long-leaf pine trees, were dug up often in the ditches, though not a tree of that kind remained standing, or could possibly live on such soil as the pocoson now has. The subsoil is clay, as is the surface soil of the adjacent firm pine lands. Independent of the ditching, the labor of clearing the dense forest of large swamp trees, which covered all this land, must have been enormous. The whole clearing and improvement has been gradually made, and some of the land cleared long ago has been so much exhausted as to be turned out. The balance, now in a single field of corn of more than 200 acres, had the appearance to me of being very rich soil; and some of the best, I was told is as productive as almost any land in the country. This reclaimed body of land, extensive as it is, is but a small part of the whole great pocoson. Much of the nearest adjacent land of the same kind, as before stated, is partially dried by ditching, and the large trees have been killed by girdling, for the purpose of extending the clearing.

There is an immense amount of this kind of land in lower North Carolina, very little of which has been reclaimed, or in any way improved. This is not surprising, notwithstanding its great fertility, when the great and expensive labor required is considered, and that the far cheaper improvements which might be made by marling, are almost totally neglected in this part of the country.

I had seen, in passing, the day before, through Jones county, some pine lands approaching in their appearance, as to their grassy turf, the savannas of New Hanover. On inquiry in regard to them, I learned from Judge Gaston that such lands and more perfect savannas were also common in this part of the country, and that one of the latter was on another part of his land, close by where we then were. We rode over it, and I was still more delighted than with those seen before, with the beautiful meadow-like turf which this exhibits throughout. This had been burnt early, and the young grass is more forward, and there are more wild flowers already open or about to open, than on the more southern savannas lately seen. With no labor, save fencing and mowing, I should think that this land would make very good though rather coarse hay, and profitable for sale, as northern hay is a regular article of import and sale in Newbern; and if merely well limed on the surface, and slightly ditched, would be excellent meadow, and produce the finer grasses. I before expressed my opinion as to the advantage and manner of carrying the improvement of such land still farther, for tillage.

I was not aware until recently that alligators were found so far north as this. Formerly they were often seen in the Neuse, and though very rare of late, they are supposed still to inhabit this river. A small one was caught in one of the ditches of Judge Gaston's pocoson. It was also on this farm, not very long after the first clearings of the wild pocoson land, that occurred a memorable conflict with a bear, which was so severely contested against great odds, and with such destructive power or resistance, that the narrative of the circumstances was published at the time throughout the United States, and even in the papers of Europe. Though at the time surrounded and attacked by the overseer and the negroes and their dogs, and assailed with every weapon that the men could suddenly command, the bear seized on the overseer, then the foremost of his assailants, and tore his flesh so as to endanger his life, and would have killed him outright, but for his being drawn away by the negroes by main force. At last, however, the bear was killed, after a longer and more desperate defence than could have been supposed possible for him to make. And yet it was only in defence that he fought; for h[e] continued slowly to retreat, even after being fired at twice, and slightly wounded at each fire, and to the last, always singled out for his attacks the particular enemy that had last struck or wounded him.

Rattlesnakes are also found hereabout, though but rarely. They are found more frequently near Wilmington, as I heard when there. These also are most forbearing enemies to man, though among the most brave, and dangerous in their defence, when attacked. They not only give warning to the intruder or assailant by their rattles, but will avoid a conflict, or [attempt] to use their deadly means of defence, as long as endurance is possible.

Most of the lands seen on our ride from Newbern had much more of clay in their composition than I had supposed belonged to this country. On the south border of the Trent, where crossed by our route, the lands have an undulating surface, are very light, and were once among the best high lands in the country, but now are very much exhausted by scourging cultivation, and neglect of all means of resuscitation.

Marl and Limestone of the Neuse and Trent

My visit to this neighborhood could not have been worse timed, for the accomplishment of its designed object. The court had but just commenced its session and would not rise in a week, engaging in some manner the attention of most of the intelligent farmers, and of course the continual attendance of every member of the bar, of which class was James W. Burton, esq., the gentleman whose information and invitation had first induced me to promise to visit this region; and which promise has so long remained unperformed, as to admonish me never to make such another engagement[.] [T]wo other gentlemen who had most kindly and urgently requested my visiting their farms and calcareous deposites, in Jones county, were both absent. Under such circumstances, I would not permit the sacrifice of the time and services which were notwithstanding offered to forward my objects; and after remaining between two and three days at Newbern, I turned my course homeward. A letter from Mr. Bryan[14] on this subject, previously received, and which will be published in this number, will serve to give better information than I could have obtained personally under the existing circumstances.

Very few persons have as yet even commenced to use the marls of the Neuse and Trent. I heard mentioned Messrs. Isaac Taylor, J. W. Bryan, S. Biddle, and J. C. Burgwyn, and the first beginner, the late L. Benners.[15] Mr. Taylor has marled about thirty acres, and must have derived much benefits from it, as I heard of his having good clover on that land. This old gentleman's good farming and improvements were in other respects well worth seeing, and I was sorry that a visit to him, which I had promised him and myself to make, was prevented by the necessity of leaving Newbern a day sooner than I had then intended, so as to be able to avail of the use of the public conveyances.

I have no doubt, however, that the visible and early effects of marl have been small hereabout, compared to the best that might be derived from the use of this manure. Such would be the result that I would anticipate and

254 Travels and Expositions

predict, merely from knowing the general system of tillage in usage, which is more or less exhausting in almost every case, and which therefore does not permit the accumulation of vegetable matter in the soil by rest, and by the prohibition of close and general grazing. Neither the increased general products nor the net profit of marling on land under continual or very exhausting tillage, can compare with those of the like application to lands under meliorating culture. This I would as earnestly wish to impress on the few persons who are beginning to marl in this region, as to induce them to extend their applications of the manure in proper manner.

Odds and Ends

The journey from Newbern to Plymouth (69 miles) is through a low, level country, much of which is more or less wet, though firm land, and some of it is swamp. The western part of the great Dismal Swamp, is touched by the road.

Within a few miles after passing the town of Washington, the road passes among a number of basin-shaped depressions, which appear precisely like the "sink holes" of the mountain limestone region of Virginia, and which I should attribute to the same cause, that is, to lime-stone caverns having formerly existed below, which by breaking in, under the weight of the super-incumbent earth, have made these depressions. If searched for deeply enough in these sinks, I think that limestone rock will be found.

At about 14 miles from Plymouth, the pines of the forest are generally of the great size which I had been disappointed in not seeing elsewhere. All are deformed by being skinned for extracting turpentine, and in a manner much more wasteful of the product, and promising to be more speedily destructive of these noble trees, than is the case in the great turpentine region seen farther west [south?].

Here I again reached the Scuppernong grape neighborhood. The settlements are much more frequent, and almost every dwelling house has its adjacent arbor covered by one or more Scuppernong vines, forming an object even at this early season very pleasing to the eye, and which in summer must be as beautiful as profitable. It is very strange, that when the expressing the juice of this grape is done so generally, and to such great extent, in this part of the country, that no one attempts to make true wine from it, by proper fermentation. For though what is called Scuppernong wine, if of the best kind, is a delicious drink, it is not a fermented liquor, nor does it indeed deserve to be called a wine. I wish that some of those who prepare the best of this liquor would try to make wine of it, by the process recommended and practised by Mr. Herbemont, in his treatise which was republished in vol. ii. of Farmers' Register.[16]

This snug little steamer Fox, (which may claim naval supremacy on the Blackwater by the same kind of right that Robinson Crusoe had to the sovereignty of his island,) arrived at Plymouth, as usual about 11 o'clock, the night of my arrival there, and in a few minutes was again on her return voyage. The next morning at 10 o'clock, I reached the landing at the Portsmouth and Roanoke railway [at Franklin, Virginia], expecting, as formerly, to be able to reach Petersburg by way of North Carolina, the evening of the same day. But this convenient and speedy, though circuitous course of travel, had been interrupted by a new arrangement of the mails, (and derangement of travelling,) and I had to choose between waiting for the next train some 18 hours, where there was neither accommodation nor any other gratification, or to consume the interval by keeping with the steam train in its course to Portsmouth, to return again with the next; which was done. Thus to travel 80 miles out of one's course, without losing time by it, is one of the strange things attending the present rapid mode of travelling, and its interruption on cross routes. This delay gave me an opportunity of spending from 2 to 11 P.M., in Portsmouth and Norfolk, and of viewing some of the (to me) interesting subjects of the following notes.

Silk Culture in Operation and Preparation

Who could have anticipated, even as late as two years ago, that this heading would be suitable for any notes of observations made in Virginia? Yet it *is* so now; and I feel confident that, before this present season expires, so many and such successful operations will have been completed, as to leave no room for the experimenters to doubt, and which will serve to establish silk-culture among the regular and profitable directions of labor and capital in the southern states.

The authorities of the Alm-house of Norfolk have done honor to themselves, and furnished a profit, and rendered a distinguished service to the institution they govern, and to their country, by introducing silk culture, upon an extensive scale. Mr. Duesberry,[17] the intelligent and zealous superintendent, has either ready fitted up, or in progress, the upper floor of the Alms-house. He has now feeding, in various stages of growth, half a million of silk-worms, and intends to feed as many more, during the season, as will make four millions in all. Mr. D. is less a novice than most other experimenters in Virginia, he having fed silk-worms on a small scale for the three preceding seasons. The oldest of his worms (the eggs of which had been prematurely hatched by March 12th) had begun to spin on April 24th, and by the evening of the next day, when I saw them, the cocoons seemed completed in outward appearance and in firmness, so that I could bring away a few as specimens of such unusually early product. A fire had been kept in the fire-place of the room generally, until within a few days previous, when the weather became warm. The oldest worms, doubtless, were somewhat injured by such early hatching, and the consequent exposure to cold, and even want of food; but had on the whole done well, and were forming fine cocoons. The greater number, hatched later, were in the best possible condition.

I was surprised as well as highly gratified to learn from Mr. Duesberry that the preparations already made, or in progress towards sure completion, for feeding silk-worms this season, were so extensive. He named the sundry individuals who are about commencing the business in the environs of Norfolk and Portsmouth, and the number of worms that each counted on feeding. There will be at least a dozen proprietors so engaged, and who expected to feed collectively more than twenty millions of worms this season. It is true that this is *yet to be done*; and therefore I shall not make statements at greater length as to mere intentions. But there is no doubt that a business to this amount will be undertaken, and I hope to be enabled to state results, and successful and profitable results, of what now is but preparation and promise. Norfolk county, though having natural agricultural advantages which few other regions can compare with, has heretofore been the most sluggish in the march of agricultural improvement. But this county is now decidedly ahead of all the south in the silk-business; and I trust that energy and success in that will produce the energy which will certainly lead to success in the exercise of all its great and neglected means for improvement and agricultural wealth. In North Carolina, though I met with but few cases of regular silk-culture either in operation or preparation, I heard of sundry, and some on a very large scale, at places not in my route; and I witnessed several striking proofs of the admirable fitness of the climate to enable the worms to endure privations and rough treatment. At Wilmington, in Dr. McRee's garden, I saw silk-worms feeding on morus multicaulis bushes[18] in the open air, and apparently as healthy and vigorous as if sheltered and well cared for. The eggs, in a hatching state, had been merely placed on the bushes, and nothing more done to save them. A few of the oldest were in their fifth age, and had endured one snow (on March 25th) and all the rains, one of which occurred on April 15, the evening before I last saw them, and was very heavy. The hatching having been very unequal, most of the worms were much less advanced than the most forward. Doubtless most of them had perished, and perhaps all then remaining will perish, before attaining full growth, by the depredations of birds and ants. But still, the good condition of those which remained when I saw them, after such early and long exposure, was a very satisfactory evidence of the climate being altogether as kindly as if *natural* to silk-worms.

In Newbern, several persons were feeding very successfully, but mostly on a small scale. The largest operator was Mr. W. R. Street, the keeper of the hotel where I lodged, whose establishment I visited. He had feeding, as he supposed, 200,000 worms, in an old dilapidated frame building, which had no glass or close windows, but plank shutters on hinges instead, and was more open to outer air, and to cold and dampness, than any house I ever saw used for the purpose. Two negro women were the feeders, with very slight superintendence from Mr. Street, in short visits two or three times a day. There had been fires kindled on three of the coldest days only. The progress of the worms had been slow, owing to so much cold weather; but they appeared to be doing well, and to have suffered from nothing except too much care, or illy directed care, from their feeders. I learned from my inspection, and a free conversation with the two feeders, to be much more confident than I had been as to trusting the details of the business to such ignorant as well as inexperienced operators as we must now find in our negroes.

At 11 P.M. (25th) I left Portsmouth by the return train, and reached the junction of the rail-roads (78 miles) to early breakfast next morning. There I had again to wait through a tedious day for the next train to Petersburg, which started at 8 P.M. and reached Petersburg after midnight.[19]

11 Pines as Monuments and Commodities
Notes on the Pine Trees of Lower Virginia and North Carolina (1859)

Pines made a large proportion of the trees of the primitive forests of the eastern and lower lands of Virginia and North Carolina. And when any of these lands had been cleared and cultivated, exhausted and abandoned, then a new growth of pines formed the universal unmixed cover. As nearly all the lands of lower Virginia had been thus treated, and in succession had reached this second growth, which thus covered all the then poorest and most worthless lands, a general cover of pines, and the term "pine old-fields," came to be generally understood as indicative of the poorest meanest of lands. For this reason, and also because of the growth of pines being so common and pervading, these trees were not only undervalued, but despised. If a natural forest of various trees was thinned out to make an ornamental grove near a mansion, every noble pine would be certainly cut out, as if a deformity, and a worthless cumberer of the ground. In planting trees for the embellishment of homesteads, if any proprietor had in part selected any of our native pines for that purpose, his taste would have been deemed as ridiculous as it was novel and strange. For the most magnificent pines, or the unmixed evergreen of a pine forest in winter, to be admired, it was requisite that the observer should be a stranger, from some distant region, in which pine trees and pine forests were not known. Then, indeed, and in all such cases, their remarkable beauty and grandeur would be fully felt and acknowledged.

All of the many species of pines have the properties of being resinous, and bearing their seeds in cones; which, however varying in size and form have a close general resemblance. And there is a like general similarity of shape, differing from all other trees, of their peculiar evergreen leaves. These spring from s[h]eaths, or are held in clusters of two, three or more leaves to each sheath, according to the species of the tree. The leaves, differing from all others, except of the kindred family of the larch, are long and slender, almost as thick as their width, and of equal diameter throughout their length, except immediately at the extremity, which is a sharp point. The new leaves as on other trees, grow only on the new twigs (or "water-sprouts") which shoot out in the spring, from the last year's buds. But the leaves of the preceding year's growth remain attached to the older branches through a second summer, if not the autumn also. In some species the leaves sometimes in part remain into the third year before dropping off entirely.

Some of our species of pines are of such distinct and marked appearance, that the most careless observer would not fail to distinguish them. Such are the Southern long-leaf pine, (*pinus australis,*) the Jersey pine (*p. inops,*) and the white pine, (*p. strobus.*) But many farmers who have long lived on cultivated lands, among pines, have not learned always to distinguish other still more common species. And even when this knowledge is not wanting, still there is such confusion and misapplication of the vulgar names of all the kinds, that it is difficult for any one to speak of or to inquire concerning any one pine, by the vulgar name of his own neighborhood, without the name being misapplied by an auditor from another locality. Thus, the name "yellow pine," in different places is used for three different species, of all of which the heartwood is more or less yellowish. The name "spruce pine" is used in Virginia for one species of pine, and farther south for another. And the several designations of "long-leaf pine," "short-leaf," "old-field pine," &c., are merely terms relative, or used in contrast with other different growths, and are each applied to different kinds in different places. Even the botanical names, though serving generally for exact designation, in most cases have either no special application, or are entirely erroneous as to their meanings. Such are the designations *"mitis,"* *"inops,"* and especially *"palustris,"* as descriptive terms of species. Further, the qualities and value for timber, and even appearance of pines of the same species, are so much varied by different conditions of situation and growth, that some of the most experienced and intelligent "timber-getters" (or "lumberers") consider as two distinct species, trees which belong to the same. I have myself, until recently, been under some of these mistakes as to the species with which I had longest been familiar. Under such circumstances I cannot even now be confident of avoiding errors. But even my mistakes, (if corrected by others better informed) as well as my correct descriptions and designations, may serve to clear away much of the obscurity and error in which this subject has been involved.

One of the most remarkable and valuable qualities of some of the pines is, that their winged seeds are distributed by winds to great distances, and in great numbers, so that every abandoned field is speedily and thickly seeded and the kind of pine which is most favored by the soil and situation, in a few years covers the ground with its young plants. The growth, especially of the most common second-growth pine, (*p. taeda*, [loblolly]) is astonishingly rapid, and even on the poorest land. And while other land might still be bare of trees, that which favors this growth would be again under a new and heavy, though young, growth of pines. This offers, (especially in connection with the use of calcareous manures,) the most cheap, rapid and effectual means for great improvement of poor soils. And besides this greatest end the cover of the more mature wood, if marketable for fuel, will offer the quickest and greatest return of crop that could have been obtained from such poor and exhausted land.

I will now proceed to remark on each of the several species of pines found anywhere in the region in view, and will commence with such as are most easily and certainly to be distinguished, before treating of those less distinguishable, or in regard to which there may yet remain any doubt or uncertainty.

The Long-Leaf or Southern Pine. (Pinus Australis of Michaux,[1] *Palustris*, of Linnaeus.)—The name *palustris*, notwithstanding its high authority, is altogether inappropriate, as this pine prefers dry soil and is rarely seen, and never in perfection, on wet or even slightly moist ground. *Australis* is peculiarly appropriate, as this tree is limited to a Southern climate.

This species barely extends a few miles north of the southern boundary of Virginia, in the south-eastern counties of Southampton and Nansemond.[2] Few, if any, stand in the lower and wetter lands of the more eastern counties in the same southern range. The long-leaf pine prefers dry and sandy soils, and is found, almost without interruption, says Michaux, "in the lower Carolinas, Georgia and Florida, over a tract of more than six hundred miles, from N.E. to S.W. and more than one hundred miles broad;" but not, (as that author also says), from the sea to the mountains, or near to either, in North Carolina. In that state it extends westward not much higher than the falls of the rivers, and towards the sea, no farther than the edge of the broad border of low, flat and moist land. Its general and best growth also equally indicates a sterile soil. The mean size, sixty to seventy feet high, with a nearly uniform diameter of fifteen to eig[h]teen inches for two-thirds of the height. Leaves ten to twelve inches long, (fourteen and more on some young trees,)[3] growing in threes, (to each sheath,) and about 1-16th to 1-13th of an inch in breadth. The cones from 7 to 8 inches long, and 2 to 2½ broad before the opening of the scales or seed-covers, or four inches when spread open. The seed-covers of the cones are armed with short, strong and not very sharp spurs. The seeds, when stripped of their shells, are white and larger than a common grain of wheat, and are of agreeable taste, though having resinous flavor. They are so eagerly sought for by hogs, that scarcely any are left on the ground to germinate. For this cause, as well as the great destruction of the trees in tapping them for turpentine, these pines are rapidly diminishing in number, and, if not protected, this noble species will almost disappear from the great region which it has heretofore almost exclusively covered and adorned. This tree is especially resinous, and is the only pine that is tapped for turpentine. Scarcely a good tree in North Carolina has escaped this operation, unless in some few tracts of land where that business has not yet been begun. This tree also has furnished the best of pine lumber; but its durability is said to be much lessened by the tree, when living, having been made to yield turpentine. The heart is large and the grain of this timber is close, and only inferior in that respect to the short leaf yellow pine, (*p. mitis* or *variabilis*.)[4] For naval architecture, timber of this tree, when large enough for the purposes required, is preferred to that of all other pines.

The broad belt of land stretching through North Carolina, which has been covered by the long-leaf pine, except for the borders of rivers, is generally level, sandy and naturally poor. Even if it had been much richer and better for agricultural profits, the labors of agriculture would still have been neglected in the generally preferred pursuit of the turpentine harvest. But so poor were the lands and so great the profits of labor, and even of the land, in the turpentine business, compared to other available products, that capital thus invested has generally yielded more profit than agriculture on the richest lands. Therefore, it is neither strange nor censurable, but altogether judicious, while these great profits were to be obtained, that nearly all the labor of this region was devoted to making turpentine, instead of enriching and cultivating the soil. But the effect of the course pursued has been not only to limit agricultural labors to the narrowest bounds, (as was proper,) but also to prevent almost every effort for improving the soil and the productions of the small extent of land under tillage. However, the juncture is now reached when this formerly most profitable turpentine business must be gradually lost; and then agriculture and improvement of fertility will not only be attended to, but will be especially rewarded in many portions of this now poor region, which yet possesses great resources for being fertilized. The rapid destruction of the forests of long-leaf pine is not only the necessary result of the two causes before stated, but the work has been still, more rapidly forwarded in some places, by another cause. At one time, in years past there was a sudden and wide-spread disease of this kind of pine, caused by the attack of some insect unknown before or since. Fortunately the operation, though far extended, was not general. But whatever it was, the destruction of the living trees was nearly or quite complete. For thousands of acres of pine forest together, and in a single summer, every tree was killed. The evidences of such destruction in the still standing dead trunks, are now seen in many places, and most extensively, as I lately saw, along the route of the Wilmington and Manchester Railway, not many miles south of the Cape Fear river. Similar extensive, and as transient destructive visitations, had occurred long before. One of these I remember to have read of forty years ago, in a communication to the Memoirs of the Philadelphia Agricultural Society: Partial as these depredations have been, as to species, any one proprietor, or many adjacent proprietors, in the route of these ravages, might have the whole value of their pine forests utterly destroyed in a few weeks.

The great beauty and striking appearance (to a stranger) of a southern pine tree, of great size and fine form, are owing to the long and straight and slender trunk, and to the very long leaves and large cones. In the close growth of forests, the branches, like other old and good timber pines of other species, are crooked, irregular, rigid and unsightly. But these and all defects are overlooked in their forest growth, when all the numerous trees make but one great and magnificent object, their tops meeting to make one great and thick canopy of green, over the open space below, by innumerable tall columns of the long and straight and naked bodies of the pines.

The Cedar Pine. (Pinus inops.)[5]—This pine, like some others, has sundry names and some of which are also applied elsewhere to other species. In

Virginia it is known in different places as the "spruce" or "river" or "cedar pine." The last vulgar designation, which will be here used, has been applied because of a slight general resemblance of the growth and appearance of the tree to the cedar; at least more so than or any other pine; and so far the name is descriptive and appropriate. The most general vulgar name farther north is "Jersey pine," which is adopted by Michaux.

This pine is generally seen only of young growth and small sizes. Where long established, and of largest sizes, in Virginia, it is rarely found exceeding fifteen inches in diameter. The trunk is not often straight enough for sawing into timber. The bark is very thin, and also smooth compared to all other pines of this region, and the sap-wood also is very thin. Of the older trees, nearly all the trunk is of heart-wood. Though the tree is but moderately supplied with resin, it makes good fuel, and much better than other pines of Virginia, of the new growth and but moderate sizes, such as are mostly used for fuel, for market, and especially for the furnaces of steam engines. The leaves of this pine grow in twos, (from each sheath,) are generally shorter than any other kind, usually from one and a-half to two inches, and about one-twentieth to one-sixteenth broad. The cones usually are from one and three-fourths to two and one-fourth inches long, and three-fourths to one inch thick, when closed. The separate seed-covers on the cones have each a small and sharp prickle, curved backward. The cones are set drooping backward on the branches; and they remain so long before falling, that the old and the new together sometimes stand on a tree as thick as the fruit on an apple tree. The branches are much more slender, tapering, and flexible than of other pines, and the general figures and outlines of the well-grown trees are more graceful and beautiful. When making the entire growth of a thick wood, and on the slope of a hill-side, where the tops of the higher trees are seen above the trees next below, and all thus best exposed to view, the foliage and the whole growth, so disposed, are singularly beautiful.

I have not observed this tree anywhere in North Carolina. It is but sparsely set and mostly of young growth in the south-eastern parts of Virginia. But the growth is there increasing and spreading. In Prince George [County], on and near James River, the young trees are far more numerous, and more widely scattered now than was the case forty years ago, when I knew them there only on some spots near the river banks. On the lower Appomattox, in that county, this is now the principal pine growth, and of its large sizes. In Westmoreland, and the other parts of the peninsula, between the lower Potomac and Rappahannock, this is now the main growth, and the great supply for market fuel, which is so great a product and labor of that region. Yet I have heard, from Mr. Willoughby Newton,[6] that it is remembered when not a tree of this species was to be seen in all the extent of that peninsula. It is now there the regular second-growth pine, which first springs on and occupies all abandoned fields, as do the other "old-field" pines, of different species, in other parts of Virginia and North Carolina.

The White Pine. (Pinus strobus.)—This tree, of beautiful foliage and general appearance, and which grows to a magnificent height, is not known in eastern North Carolina, and is so rarely seen anywhere in Virginia east of the mountains, that it scarcely comes within the limits of my designed subject for remark. However, it is named for the contrast it presents, and thereby setting off more strongly the opposite qualities of other species. But its description need not occupy more than a small space. This is the great timber pine of the northern States. In travelling westward from the sea coast through the middle of Virginia, this tree is first seen in the narrow valleys of the North Mountains in Augusta county. It is there called the silver pine. The small trees are beautiful and the large ones magnificent. The bark of the young trees is very smooth, (in this differing from all other pines,) and the branches spring from and surround the young stems in regular succession, and three or four from the same height, on opposite sides, as do the young side-shoots of dogwood. The leaves grow in fives, (from each sheath,) about four inches long, and very slender and delicate, and of a bluish green color, and silken gloss.

This pine, different from all of the other species growing in our region, prefers such fine soils as are found on the alluvial but dry margins of rivers and in mountain glens.—(*Darlington's Agricultural Botany.*)[7]

Short Leaf or Yellow Pine. (Pinus variabilis. P. mitis of Michaux.) Cones, length 1¾ to 2 inches. Breadth, (as closed,) ¾ to ⅞. Nearly smooth, the prickles being very short, slender, and weak. Leaves, length, on different trees, 1¾ to 3 inches; breadth, 1-24 to 1-20. The leaves grow mostly in twos (from each sheath,) and many trees, if but slightly examined, might seem to show that this was the universal law of this pine. But on most trees there are also leaves, in much smaller numbers, growing in threes, intermixed with the others. This variation is especially apt to occur, partially, on very young trees, of rapid growth. On one tree, of eight inches diameter, cut down to furnish specimens of cones, I found so many of the leaves in threes, that those in twos did not amount to one in twenty. The leaves in threes being in greater number, I have not observed elsewhere. Generally, the leaves in twos on any one tree, are very far the most numerous. All the specimens, from which the measurements were made, I gathered in the old forest-land of Marlbourn[e] farm, Hanover [County], Va. The lengths of leaves on different trees vary much, and in some cases, even on the same tree and twig,—and also the sizes of cones on different trees,—as well as the proportions of leaves in twos and threes. From these marked variations, I am disposed to believe that some trees are of hybrid generation, or crosses between the pure short-leaved tree of the species, and the *p. taeda*. But whether this surmise is correct or not, and however great and many may be the variations, this species, notwithstanding its variations, is easily distinguished by its short leaves in twos, from any of the tree-leaved species—and it cannot be mistaken for the cedar pine, (*p. inops*,) the only other short and two-leaved species, because of the great difference of general appearance. The short-leaf yellow pine, (*p. variabilis*,) in middle and most of lower Virginia, is the great and valuable timber pine of that region, and makes the best timber of all, because of its more resinous heart-wood and very close grain. The most beautiful and highly valued floors of lower Virginia, and which are no where equalled, are made of plank of this tree. Old trees, in original forests, are from two to three feet in diameter, and usually are mostly of heart-wood. This is very durable. But the sap-wood, if exposed to changes of moisture, soon rots, as with all other pines. Formerly, nearly all the pines of the original forests in lower Virginia, and in dry and medium or stiff soils were of this kind. But as these and other trees have been cut out, and the forests thinned, other kinds, (mostly *p. taeda*, and in fewer cases, *p. inops*,) have made most of the later growth. And still more, and almost entirely, is this the case on abandoned old fields, whereon, though speedily covered by pines, very few of this species are to be seen. Yet in the upper country, at some distance above the falls, (as in Cumberland, Amelia, &c.,) though the abandoned fields are there also occupied by a second growth exclusively of pines, yet all these are of this kind, and scarcely a tree is seen of the *p. taeda*, or the "oldfield" pine of the lower country generally. The same thing I have seen in Orange, N.C., on abandoned high land fields, near the head affluents of the Neuse river.

When of recent and rapid growth, and especially when of second growth on land formerly cleared, this pine is mostly of sap-wood, in that respect like the *p. taeda;* but still the former has more heart, and is of more durability, when exposed to the weather than the latter.

The yellow pine grows, (or formerly grew,) in great perfection, but in detached and scattered and limited localities, in sundry of the upper counties east of the mountains in Virginia. But, generally, in the Piedmont region, at fifty miles and farther above the falls, neither this nor any other pine grew in the original forests. In the range of counties next below the falls, it was formerly almost the only pine, and also the most common of all trees, of the original forest growth. It lessens in quantity, or in proportion to other species, as we descend towards the sea coast, and also as we go southward. After reaching the low, flat lands near the sea coast, and the southern region where the long leaf pine first appears, the yellow pine is seen but rarely. But as far south and east as Pitt County, N.C., at one place, and in Beaufort County, near Washington, I saw that nearly all the forest pines, on some spaces, were of this species, and of large size and fine form. The spots on which they thus show, are of dry soil, and, probably, also more clayey than in general, so as to favor more the growth of this than of the long-leaf pine. Also, between Plymouth and the great swamp in Washington County, N.C., this pine, of large size, and very perfect form, and with long and straight trunks, is the main original forest growth, on level, stiff soil, which, though firm land, and called dry, is so low and moist that I was surprised to find thereon this kind of pine. These facts, and especially the last case, go to show that a close or clayey soil, or sub-soil, has more power to promote the growth of this pine, than it is opposed by the increased approach to southern climate, and low and damp soil, both of which are unfavorable to this pine, and very favorable respectively to other species. This pine is also seen, in few cases and of bad growth, in the always wet and miry, and often over-flowed, swamps bordering on Blackwater River in Virginia, south of the Seaboard Railway.

Loblolly Pine. (Pinus taeda.) This is called "long-leaf" in the Piedmont counties of Virginia, where the "short-leaf" is common and this is rare—and "oldfield" pine in most of the lower counties, where that designation is correctly descriptive. But as both these provincial names are elsewhere applied to other pines, I prefer the vulgar name used in South Carolina, of "loblolly," which, though unmeaning,[8] will not mislead by having more than this one application.

The loblolly pine (*p. taeda*) is rarely seen north of Washington, D.C. I saw a few on exhausted land near Bladensburg, Md., within a few miles of Washington. Proceeding southward they become more and more abundant, but do not extend westward many miles above the line of the fall of the rivers. I shall again refer to this supposed western limit of its growth, and the supposed cause of this boundary. On all the exhausted and abandoned naturally poor soils, both dry and moist, certainly, and much, also, of the naturally, good, but exhausted, south and east of this upper limit, the loblolly pine springs soon and speedily, and thickly covers the surface. With some exceptions already named, where the cedar pine is the common second growth, the loblolly pines make the almost entire, and also abundant, second growth, on these abandoned lands. In the original forests, probably, it was formerly rather a scarce tree, as it is still, where there has been not much cutting out and thinning of the natural forest. It is only as a second growth that this pine has become abundant, and only on all the poorest and worst natural soils that it has taken almost entire possession of the ground, and seems to exclude other trees, and to thrive in proportion to the base quality of the soil—and more especially in proportion to the deficiency of lime in the soil. But, also, sandy soil and warm climate are further promotive of this growth; and, therefore, as proceeding southward, through eastern North Carolina, the loblolly pine, as a second growth, thrives more and more in general. I have even seen some few large and flourishing pines of this species, on the Rocky Point land, which seemed to be certainly calcareous.

As it is a disputed question, which will be considered hereafter, whether the great Swamp or Slash Pine, a valuable tree for lumber, is of the same species, or different from this, for the present I will speak only of all such trees as are undoubtedly of the kind known as "loblolly," pines.[9]

These make the general, and in many places the exclusive, second growth from some ten or twenty miles above the lower granite falls, to the sea coast. Within these extreme limits, almost every exhausted and abandoned space is soon covered by this growth, whether naturally poor or rich, of medium texture, or sandy, wet or dry. The only known exceptions are spots of old, cleared lands, which, from some cause, were highly calcareous, on which the loblolly pine refuses to grow; or if growing, shows plainly an unhealthy and unthrifty growth.

The cones on different trees are from 3 to 5 inches long, and from 1 to 1⅝ inches thick, (as closed.) The prickles on the seed covers, stout and strong, and not pointed very sharp. The leaves from 5 to 7½ inches long, and from 1-16th to 1-13th broad. They grow in threes, and, as I believe, universally so on trees of considerable size. But on trees of but a few years' age, of rapid and luxuriant growth, some few of the sheaths will be found to contain four leaves. But this is the exception, and a rare one. The general rule is that the leaves grow in threes. By this rule, though these threes may vary from each other in the lengths of leaves, and sizes and shapes of cones, still, all are readily distinguishable from any specimen of the short leaf or yellow pine, (*p. variabilis.*) however near such specimen may approach to other usual characteristics of the loblolly pine.

The grain of this wood is very open, the wide intervals soft, and the wood, as timber, of the most worthless description. There is very little heart-wood in large trees—none, or almost none, in the small—and the heart-wood is but little resinous, solid, or durable, as timber. The sap-wood, (when growing) seems much more resinous than the heart. Trees of two feet in diameter usually have but two to three inches of this poor heart-wood. It is only when of small growth, and but rarely then, that the trunks can be riven by wedges, without more labor than profit. When split before growing too large, and after being seasoned or well dried, this wood makes quick burning fuel, of which immense quantities are sold to the north, as well as at home, for the furnaces of steam engines and other uses.

Worthless and despised as is this tree for timber, and for most other uses, it is one of the greatest blessings to our country. It rapidly covers, and with a thick and heavy forest growth, the most barren lands, which otherwise would remain for many years naked and unimproved by rest. By the fallen leaves, which from this tree are very abundant, the impoverished soil is again supplied with the deficient vegetable matter, and, with other aid, may be restored soon to fertility. And the crop of wood, where near enough to market, may be worth three-fold of what would be the value of the land, if without this product.

It is not only on dry or arable land that this tree grows vigorously and to a large size. Such may be seen on land much too wet for tillage, and t[o]o low for drainage—as on some of the abandoned lands near Lake Matt[a]muskeet, where the surface of the ground is not more than 18 inches above that of the adjacent waters of Pamlico Sound—and where, also, the salt water is raised by violent winds and strong tides still higher, and sometimes so as to cover the land on which the pines stand. The power of these trees to resist such unnatural visitation and changes of condition, and without apparent injury, is remarkable.

The Great Swamp Pine; or, the Naval Timber Pine. The Slash Pine.—During my first visit to the low lands of North Carolina, bordering on Albemarle Sound, in 1856 [1836? 1839?], I first heard of and saw pines of unusual large sizes and peculiar character, and which were understood by all of the most experienced and intelligent lumber-cutters to be of a different kind from any of the species I have described, or any other known in North Carolina or Virginia. My principal source of information and instruction, in regard to this pine, was Edward H. Herbert,[10] of Princess Anne [County, Virginia], a gentleman of much intelligence, and who has for twenty years been principally and very extensively engaged in contracts to supply to the navy yards of the government, timber suitable for the construction of ships of war. In this business he has examined the whole country and has bought, cut and supplied to the government naval stations, much of the largest and best timber, (such only being fit for the masts and other spars of the largest ships of war,) that could be procured in lower Virginia and North Carolina. He has found no pines of any kind except of that now under consideration, large enough and having enough of heart-wood, to make the masts, spars and other timbers, of the largest required size. It should be observed that the proposals advertised for, to supply, by contracts, timber for the United States navy yards, mention and recognize but two kinds of pine timber, "white" and "yellow pine." The former is of the northern white pine, (*p. strobus*,) and the latter designates especially the long-leaf southern pine—but which in usage includes also the short leaf yellow pine, (*p. variabilis*,)[11] and the great pine now to be described. This tree grows only on low and moist land, and is the better for timber, and grows larger, in proportion to the greater richness of the land. It is the principal and largest timber pine in the original forests of all the low, flat and firm, but moist lands, bordering on Albemarle Sound, and also farther South—and I have seen it growing as well, but much more sparsely, on the rich swampy borders of the Roanoke, and in the best gum lands bordering on the Dismal Swamp, and some on the low bottom lands of the Tar River. Among the other gigantic forest trees on the rich and wet Roanoke Swamps, (on the land of Henry Burgwyn, Esq.,) mostly of oak, gum, poplar, &c., the few of these pines which yet remain, tower far above all others, (twenty feet or more,) so as to be seen and distinguished at some miles distance. I have visited several standing trees and the stumps of others that had been cut down, which measured either nearly or quite five feet in diameter, and were supposed to have been from one hundred and fifty to one hundred and seventy feet in height. But the sizes and heights of the trees may best be inferred from the list below of hewn (or squared) stocks, which was furnished to me from Mr. Herbert's timber accounts. These stocks were cut in Bertie [County], N.C., made the whole of one raft which was then (May, 1856,) on its passage through the Dismal Swamp Canal to New York. The stocks were thence to be shipped to Amsterdam for naval construction, under a contract with the Dutch Government.

	Length	Inches Square	Number Cubic feet
1	47	25	204
2	66	19	165
3	86	30	537
4	79	31	527
5	88	23	337
6	65	20	181
7	74	26	347
8	80	26	376
9	68	24	272
10	58	22	195
11	86	30	537
12	58	30	363
13	74	26	347
14	74	26	347
15	70	28	381
16	70	27	368

But even the longest of these stocks do not approach the magnitude of one which was cut at a previous time in Bertie and sold in New York by Mr. Herbert. This was eighty feet in length and thirty-six inches square at the lower end. He sold it to a dealer for five hundred dollars, and the buyer re-sold it for six hundred dollars. This stock did not retain its stated diameter (at the butt) to its upper extremity, but was there from twenty-eight to thirty inches square. All these stocks were nearly all of heart-wood. It is required that two-thirds of the surface of each side of every stock shall be of heart-wood. Of course this condition permits but little sap-wood, and that only in the angles of the squared stocks. Thence, also, it follows that the proportion of heart-wood in these trees must be very large. The timber must be resinous or it would not be good, and it must be durable, or it would not serve for the masts and other great spars of ships of war, exposed to alternations of wetting and drying, and for which the best materials only are permitted to be used. The grain of this heart wood is not generally very coarse, but more so than the short leaf yellow pine. Mr. Herbert, the better to aid my investigations, procured from the navy yard of Gosport [adjacent to Portsmouth], a thin cross section of the stock used for a mast of the U.S. war steamer Roanoke, which also he had cut in Bertie. The section is of the stock hewed to twenty-seven inches square, and of which but a very little sap-wood was in the two corners of one side only. As the tree was not entirely straight, the centre of the heart is thrown considerably to one side of the centre of the end of the stock, where the section was cut off. The heart wood was 34½ inches diameter, and contained 186 rings, (as measured and counted on the wider side, or radius, which, from the centre of the heart, measured 17¼ inches.

The remaining sap-wood, 3¼ inches, contained 116 rings, or 32⅓ average to the inch.

Whole number of rings left visible in the stock 302.

A radius of three inches from centre, of heart-wood, took in 19 ring marks.

A radius of six inches from centre of heart-wood, took in 34 rings, or 5⅔ average to the inch.

The outer inch of sap-wood, (not outside of the tree,) 49 rings.

The outer rings in the sap-wood, visible in the corners, were so very close as to be indistinct; and, perhaps, some of them were omitted in the counting, though the examination was aided by a magnifying glass. In addition, and which makes a much larger omission, neither corner extended to the outer part of the sap-wood of the tree; and, therefore, if only an inch was cut off, it made the loss of at least fifty rings and years' growth. It is probable that this tree had considerably more than 300 rings, indicating as many years of life and growth. How much older must have been the tree which made the largest stock named, or other trees of five feet or more in diameter!

With such size and value of this tree, and such marked differences from every other pine known in the same region, it is not strange that nearly all opinions of the residents, and of those of most practical acquaintances with pines and their timber, should have agreed, and without exception or doubt, that this was a peculiar species. So I learned from every source of instruction, and so I believed until recently, when the comparison of all my information and personal observations made me not only doubt the fact of this being a distinct species, but induced me fully to believe that this tree, of the most magnificent and superior size and valuable and remarkable qualities for timber, is identical in species with the universally despised loblolly pine, which is almost without heart-wood, and is the most worthless and perishable material for timber; and that great age and slower growth, and in some measure a better and a moister soil, are all that have caused the different qualities and the great superiority of the old swamp pines. I know that this opinion would be deemed absurd by persons the most acquainted with these different trees and their timber. I will proceed to state the grounds for my change of opinion.

When, at first, fully believing (as instructed by others) that this swamp pine was a different kind, it was necessary thence for me to infer that Michaux, who personally and carefully examined so many of our forests and trees, and also all other botanists, were ignorant of the existence of this noble tree, which exhibits its superior magnitude over so much extent of our country. It is probable, indeed, that even the laborious and careful Michaux did not, in his travels, pass through, even if he entered, the lowland region on and near the Albemarle Sound—a region which is still almost a *terra-incognita* to all other persons than the residents and near neighbors. For if these trees had been seen on the natural soil, in their most perfect conditions of size and value, whatever might have been their species, they could scarcely have passed, as they have done, without being mentioned by any botanical writer. If not the *p. taeda*, these trees cannot belong to any other of the species of this country; and, therefore they would the more attract a botanist's attention, and induce particular notice and description, as presenting a new and before undescribed species—or at least new in this locality. And if they had been observed, and recognized as the *pinus taeda*, a scientific observer, like Michaux, could scarcely have omitted all notice of the remarkable differences between these large and valuable timber-trees and the ordinary and understood general character of that well known species. If the usually accurate Michaux had known this tree, its great size and value for timber, and its preferred moist and rich soil—and if he had also known that it was the *pinus taeda*, or loblolly pine—he could not have used the following expressions, in describing the latter species, as he has done, without limitation or exception. He says of the loblolly pine: "In the lower part of Virginia, and of North Carolina northeast of Cape Fear River, over an extent of nearly two hundred miles, it grows wherever the soil is dry and sandy." And again: "It exceeds eighty feet in height, with a diameter of two to three feet," &c. "In trunks three feet in diameter, I have constantly found thirty inches of the sap-wood, and in those of a foot in diameter, not more than an inch of heart."[12] "The concentrical circles of the long-leaf pine (*p. australis*) are twelve times as numerous in the same space" (as of the loblolly pine).[13] "This species is applied only to secondary uses [for inferior purposes]; it decays rapidly when exposed to the air, and is regarded as one of the least valuable of pines. Though little esteemed in America, it would be an important acquisition to the south of Europe," on account of its rapid growth and fine appearance, and use of the timber for "secondary" purposes.[14]

The only pines of the higher range of country which resemble, or even approach, the lowland swamp-pine, in character, is what is there called the "slash pine," common in the higher tide-water counties, and growing on high land, but only either in the narrow, oozy bottoms, or in the forest "slashes," or shallow depressions of the table or nearly level ridge-lands. These depressions have a close and stiff, though still sandy, soil and subsoil, serving to hold the rain-water and to convert the depressions to shallow ponds in wet weather, in winter and spring, until the collected rain-water evaporates in summer. In these very limited spaces, only, grow the few slash pines—of large size, and of coarse-grained, but durable and large, heart-timber. This, and also the swamp-pine of the low country, have their leaves in threes, and both the leaves and cones of the like sizes and general appearance with those of the common loblolly pines. For want of botanical knowledge, or any aid of instruction from others better informed in these respects, I could not compare these trees by their marks of botanical description and distinction of species. Experienced lumber-cutters can readily distinguish these trees by their general appearance, in respect to their value and fitness for timber; but I have found no one who could certainly distinguish them by any differences of their growth, and the sizes or shapes of their leaves or cones, from the *p. taeda*. Further, no one can certainly designate either a young swamp or slash pine. They are only known as such when old enough to have large heart-wood.

If the loblolly pine will become by sufficient age on rich soil, a "swamp pine," it may seem very strange that even the largest of the former (known to be the loblolly) never show large heart-wood. But nearly all these largest trees are of second growth, on abandoned fields, and few have ever reached sixty years old before the land is again cleared. And even if left to stand much longer, which I have never known, no second-growth pine can date farther back than the exhaustion and abandonment of the earliest cleared lands, or about two hundred years. In the case of the pine for the mast of the Roanoke, the latest found ring of heart-wood is certainly of growth one hundred and sixteen years old, at least. Of the few loblolly trees (admitted to be such) standing in original forests, the growth was slower, and, for their size, their heart-wood is of larger size than those of second growth, on land

	Number	Description of Soils	Diameter of Trunk (Exclusive of Bark) at Height of Stump	Diameter of Heart-wood	Total Number of Rings in Tree, at Stump	Number of Rings in Heart-wood
Second Growth	1	Dry, sandy slope	20	2	48	7
	2	Dry, sandy slope	21	4	44	8
	3	Dry, sandy level	10	0	..	..
	4	Dry, sandy level	11	1¼	..	..
Land Never Cleared	5	Dry, sandy slope	17½	2	40	3
	6	Dry, sandy slope	22½	6	48	7
	7	Dry, sandy slope	19	4½	49	5
	8	Level, rather moist	18	5¼	75	7
	9		21½	9	74	18
	10	Sandy and oozy	21½	6	58	13
	11	Sandy and oozy	32	8	95	32
	12	Sandy and oozy	21	6	96	43
	13	Stiff, sandy bottom	26½	9¼	97	28
Forest	14	Oozy slash	39	32	141	63
	15	Oozy slash	37½	27½	204	85
	16	Oozy slash	37½	31	269	187
	17	Low but firm, sandy	42	35½	283	207
	18	Firm, low, and moist	60	47	280	170
	19	Low and rich	41	31½	302	156
	20	Firm, low, and moist	46	39	..	181

formerly under tillage. Some of these trees will be offered as examples; and, in some cases, it would be difficult even for a timber-cutter to pronounce whether particular trees, which will be named, should be classed as old loblolly pines, or swamp or slash pines, (according to localities) too young, or of too rapid growth, to have large hearts, or to be good for timber. Even where the best of these swamp pines are cut, there are some trees of so much smaller-sized heart-wood that the cutters have found it necessary to designate them by such terms as "yearling (*i.e.* young) swamp pine," and "bastard swamp pine." All these things go to confirm my position, that there is no specific difference between the loblolly and the swamp and slash pines.

The dimensions, &c., of sundry trees of this species, which appear in the following statement, with but one exception, were observed and noted by myself. The list includes trees of second growth, which all persons would

Maximum Width of Rings (in Heart) to the Inch	Minimum Width of Rings (in Sap) to the Inch	Number of Rings in Outside Inch of Sap-wood	Remarks
1–2	1–6	6½	Former cultivated
..	..	8	and worn out;
..	..	..	still poor
..	..	..	
..	..	..	
1–2	1–14	12	Less than medium fertility
..	..	15	
1–2	1–39	..	Not oozy but would
3–5	1–39	12	require draining if tilled
1–4	1–13	6½	On flat at foot of and
1–3	1–16	9½	near to oozy hill-side;
3–20	1–25	12	all the above in Hanover
..	..	16	Prince George County
3–5	1–28	17	Tree 130 feet high; Hanover
1–1	1–18	15	Tree 110 feet high; Hanover
1–5	..	66	Hanover
..	..	..	Tree 148 feet high — Washington Co., N.C.
..	..	..	Tree 170 feet high
1–3	1–60	49	Mast of the Roanoke steamship-of-war, from Bertie, N.C.
..	..	..	Near Tarborough, N.C.; these dimensions at 39 feet high—the lower part having been removed for timber, and stump damage

pronounce to be loblolly pine; others, of original growth, which are undoubtedly such as are deemed swamp or slash pines, and good timber-trees; and others, which it would be difficult for those persons who maintain there are two kinds to say to which they belong.

The trees numbered 14, 15 and 16, may unquestionably be put with the "swamp pines" of the low country. Those numbered from 7 to 12, of much less age, only approach, in sizes of heart-wood, to good timber, which they might have attained to, if left to grow two more centuries.

It is not only the loblolly pine that is extremely deficient in heart-wood until of advanced age. Though in less degree, this defect is often found also in the short-leaf pine, (*p. variabilis*) which, generally, is the best yellow pine timber-tree of the higher country. Some trees of this kind, of original forest growth, of twenty or more inches in diameter, have less than four inches thickness of heart. If of second growth, these trees would have had still less of heart generally.

It is not always plain where to fix upon the dividing line in a tree, between the heart and sapwood; nor is the line of junction always regular or parallel with the rings of grain near the earth. Also, in trees like No. 16, which are nearly all of heart-wood, the little sap is so resinous that it can scarcely be distinguished, except as being living wood, when the tree is first cut down.[15]

Pond Pine. Pinus Serotina.—Michaux says that this pine is "rare and fit for no use"—and states the "ordinary size, thirty-five to forty feet in height, and fifteen to eighteen inches in diameter."[16] By these and other indications, I sought in vain for this pine, by such slight and distant observation as is afforded to a traveller, through wet lands,—and in some cases failed to distinguish it, even when my later and more close inspection showed that it formed the principal, if not sole forest growth for miles together. This great oversight was caused to me by the inaccuracy of Michaux's description of the height, and also by the actual general resemblance of the trees to the *pinus taeda*. And between these two, as species, the residents best acquainted with both have not observed any difference. It is not true that, differences of general appearance, and of growth, are recognized by all—and even a different name, the "savanna pine," is commonly applied to the species now under consideration, where the trees make the general growth, on the wettest savanna or boggy swamps. But the usual smaller sizes, and apparently more imperfect or stunted growth, and ugly shapes of the "savanna pines" are ascribed to the exposed unfavourable and unnatural situation in which they stand, in mire and water, and not to any fixed difference of kind, between these and the *pinus taeda* on dry or dryer soils. Indeed, the cones furnish the only certain indication of the pond pine. They remain on the tree, and unopened, for six months (or perhaps a year) after ripening—are very compact, and some of them (but not always, as we would infer from the description and figure given by Michaux,) are perfectly egg-shaped. But more generally, while they approach this shape, they are rather broader near the base, and more pointed at the top, so as to be about midway in shape between conical and oval. The cones, three or four together, often grow out from and surround a twig. Their close surface, and their remaining closed so long, and also their peculiar forms make these cones more beautiful than any others. The cones, and especially those in clusters, would be valued as mantel ornaments. The cones are about two and a half inches long, and one and seven-eighths broad. The leaves grow in *threes;* and are from five to seven inches long; and very like those of the loblolly pine. I have never met with these pines in Virginia, though, from description, I infer that they are found, in numbers, in parts of the Dismal Swamp.[17] I first was enabled to recognize and identify the tree, and the *pinus serotina,* in the low swamp lands north of Lake Mattamuskeet, along the canal to Alligator River. There it grows in considerable numbers, mostly from eight to twelve inches in diameter, and rarely eighteen. They form the sparse but unmixed forest growth on large surfaces of wet savanna land [on pocosins] on both sides of Pungo river. These were peat lands, which had been burnt over, and are so low and wet as to be deemed worthless. But, also, on the rich swamp land near Lake Scuppernong, (the farm of Charles Pettigrew, Esq., in Tyrrel[l] County,) which had not yet been brought under culture, and which had been burnt over and left naked, many years ago, the next succeeding forest growth was wholly of the pond pine, and of which many of the largest appeared to be eighteen inches in diameter, and eighty feet high. Also, on the thinner swamp soil near the canal of Mr. McRee [Dr. Armistead?], in Washington County, (near Plymouth, N.C.,) the general forest growth, for a mile or more, and generally of large size, is of this particular pine. Yet neither Mr. McRee, nor any of the neighbouring residents, had suspected that these trees were of different species from the ordinary loblolly or "old field" pine; and under this mistaken impression, this body of swamp land is generally supposed to be of little fertility, because covered (as supposed) by a growth, which indicates poor land. I do not pretend to pronounce, on my very cursory view, that this land is not of inferior fertility—nor that the pond pine may not grow on poor land, provided it is peaty and very wet. But, this pine growing and thriving, and either generally or exclusively making the forest cover, is certainly no indication of poor soil, because it grows thus on the richest, of which the case cited above of the Scuppernong swamp land is full proof.

    This tree has more heart, and more resin in its sap-wood, than the loblolly; and very different from the latter, the pond pine furnishes good and durable timber, for such purposes as the small trunks will suit. Masts for small vessels are made of those growing on the low and wet swamp of Mattamuskeet. As a wet (and perhaps, also, a peaty,) soil is most favourable, if not essential, to the growth of this pine, it is probable that on the wettest land it may have the most heart-wood, and serve for the best timber. Where it grows on dryer (though still wet) land, near Lake Scuppernong, it had been understood that this pine had more heart-wood, and was of more value, than the *pinus taeda* of the neighbouring dry and poor lands—but the superiority was not so marked, or appreciated so highly, as I heard of in other places, where the pond pines grew on much wetter lands.

Pitch Pine. Pinus Rigida.—I have seen and recognized this tree (as supposed) in but very few cases in Prince George's Co., Md., and in Culpeper, Va. But all that were observed were trees of young growth, and therefore the only indications of the kind were in the leaves and cones. The trees which I saw and supposed to be of this kind, had leaves thicker and more rigid than usual of other common kinds, three to four inches long, and growing in *threes*. The cones (in Maryland) about two inches long, nearly spherical in general outline. In our Alleghany region, this tree supplies much of the pine timber used in buildings, and in planks exposed to view, would attract notice by the great number of knots. But except in small trees, which only were accessible to me, and which do not offer good and reliable specimens of growth, &c., I had no opportunity for fully examining the growing trees, and comparing them with others. I have never (with certainty) seen and known this tree in lower Virginia or North Carolina.[18] But as it would seem from some of Michaux's words that it is in this region, and as, possibly, I may even have seen trees of this species without distinguishing them from some other kind, I will abridge the description given in the American edition of Michaux's work. Some passages of this description seem to contradict others, to which contradictions I will invite notice by marking them in italics. Michaux says of the *Pinus rigida* that it is "known in all the United States by the name of 'Pitch pine,' and sometimes in Virginia as 'Black pine.' *Except the maritime parts of the Atlantic States*, and the fertile regions West of the Alleghany mountains, it is found throughout the United States, but most abundantly *upon the Atlantic Coast*, where the soil is diversified, but generally meagre."[19] "In Pennsylvania and Virginia the ridges of the Alleghanies are sometimes covered with it. Near Bedford in Pennsylvania, where the soil is more generous, the pitch pine is thirty-five to forty feet high, and twelve to fifteen inches in diameter."[20] "Its most Northern localities are Maine and Vermont, where it does not exceed twelve to fifteen feet high."[21] "*In lower parts of New Jersey, Pennsylvania and Maryland*, it is frequently seen in the large swamps filled with red [white?][22] cedar, which are constantly miry, or covered with water in such situations it is seventy or eighty feet high, and twenty to twenty-eight inches in diameter."—"It supports a long time the presence of sea-water, which, in spring-tides, overflows the *salt meadows*, where sometimes this tree is found alone, of all its genus." The buds are always resinous, and its triple leaves vary in length from 1½ to 7 inches, according to the degree of moisture of the soil."— "Size of cones depend on nature of the soil, and varies from less than one to more than three inches in length. They are pyramidal in shape, and each scale is pointed with an acute spire about two inches [lines?] long."[23] A note to this text of Michaux, by J. J. Smith, says that the *p. rigida* sometimes attains the height of 100 feet, and four or five in diameter.[24] J. J. Smith also adds a characteristic of this pine, which I have not known in any other. "It differs from other trees of this family in its stump throwing up sprouts the spring after the tree has been felled; but these do not attain any considerable height. The fallen trunk also throws out sprouts the succeeding summer."[25]

Michaux further says that the *p. rigida* is remarkable for the number of branches which occupy two-thirds of the trunk and render the wood extremely knotty. The concentric circles widely distant; three-fourths of the larger stocks consist of sap. On mountains and gravelly land the wood is compact and surcharged with resin; in swamps it is light, soft, and composed almost wholly of sap. From the most resinous stocks is procured the lamp-black of commerce. Tar is made of this pine in the Northern States and Canada, as it is of the *p. variabilis* in lower Virginia.

Perhaps the foregoing description may enable some observer to be more successful than myself in finding and distinguishing this pine in the low country of Virginia or North Carolina.[26] Also it may prevent from being confounded with this pine either the *p. serotina* (which Michaux says "strikingly resembles" the *p. rigida*,) or the *p. taeda*, when in low and wet ground, or exposed to wet, or sometimes reached by salt water.[27]

Having now described separately each species of this region, and some others for better distinction, I will return to more general remarks, or the consideration and comparison of different species in connection.

The short leaf yellow pine, (*p. variabilis*,) is the principal tree of the original forests of the upper range of the tide-water region of Virginia, and also above the falls as far up the country as the usual growth of any pines extend continuously. For, at some distance above, as supposed from change of soil, the entire growth of pines ceases and gives place to a general growth mostly of different kinds of oak. Proceeding South-eastward to the low and wet country, this pine becomes more scarce, and is more and more substituted by the swamp or loblolly pine as original growth; and more Southward and on higher lands, and throughout Eastern North Carolina, the long leaf pine generally is the principal pine of the original forests. When any of these several forest growths were cleared off for tillage, and the lands were afterwards worn out and then thrown out of cultivation, several different pines in different places, as second growth, entirely occupy these second lands, and in most cases the second growth is entirely different in species from the pine of the first growth. Thus, in nearly all of the tide-water region of North Carolina and on most of that of Virginia, the almost universal second growth pine is the loblolly, or "old field" pine, as thence called, which succeeds to the original long leaf pine in North Carolina, and occupies, exclusively, in the abandoned former places of both, the ground which this pine had originally, but partially shared with the short leaf and other trees. In the Northern Neck of Virginia, on some other lands near to rivers, and also in the more Northern counties above the falls, (as Fairfax,) the cedar pine (*p. inops*) is the principal second growth, or is the "old field" pine of those lands. Further, the Southern and lower Piedmont lands of Virginia, but not so low as the line of the falls, when abandoned, also are covered and exclusively with *their* "old field" pine, and which is so termed in Amelia, Cumberland and that range of counties, and in Orange, in North Carolina. But the second growth pines of this higher range of country is not like that of the lower range, but is no other than the short leaf yellow pine, (*p. variabilis.*) Thus it is, the loblolly, which is the almost entire second growth of nearly all the tide-water region, refuses to grow at a short distance (generally varying from five to twenty miles) and at an irregular line of termination, above the falls, while the short leaf pine continues thence and covers all the abandoned fields for some distance farther up the country, after which that particular pine growth also ceases. Yet, because of the same name of "old field" being used in both places, many farmers and residents suppose both pines to be of the same species. And very many farmers of the lower country where the first and second growth pines are of different species, (*variabilis* and *taeda*, respectively,) suppose them to be the same kind, but altered in appearance and manner of growth by the difference of the lands and other circumstances. Of these facts, in regard to remote localities, I have to rely more on information than on my own limited personal observation. But in Prince George and Hanover counties, in which I have resided, and in more of the upper and middle range of the tide-water country, I have seen much, and have noted such general facts as these: In the original forests of the ordinary poor soils, or of medium fertility and dry land, not one pine tree in fifty is a loblolly, and all the others are short leaf pines. And the few loblolly pines there found, they are of smaller and younger growths, if scattered among the short leaf pines. Or if (as rarely) a number of loblolly pines are seen near together and occupying the ground either partially or exclusively, it is either when the short leaf pines had been formerly cut out or otherwise destroyed, or where the moisture of the soil forbade their healthy growth, or where the ground, (in soil, sub-soil and all below for sundry feet,) was so sandy as to be unfavourable to the short leaf pine though not to the loblolly.

As particular observations, made with a view to certain objects, are always more accurate and reliable than far more extended and general observations made without any particular object, I have recently made for this purpose a particular examination on parts of the forest and waste lands of Marlbourn[e] farm [on the tidewater-piedmont border]. First, in a body of original forest land; high dry, of sandy soil, but having clay below, and of but moderate productive power, (or below medium fertility,) short leaf pines made the principal growth, and all of the largest pine growth. The loblolly pines were not one to fifty of the former, and nearly all of these few were of small size. On one side of this body of old forest land is a very poor old field of similar soil, abandoned from eight to ten years past, and now covered thinly with young pines of five years old or less. (The earlier of this second growth had been cut down.) Of these young trees, perhaps one in ten to twenty is a short leaf pine, and these are always of smaller size than the much more numerous loblolly pines. On the other side of the forest land there is another small body of "old field" pine growth, the largest trees being about ten inches through, and mostly of different smaller sizes. Of these not one in three hundred was a short leaf, or any other than a loblolly pine, and the few others, of short leaf, were so small that if all are let alone to stand, these last will certainly perish, because being so over-topped and shaded by the others of much larger sizes and greater vigour of growth.

From these and other more general observations, it would seem that in this region the loblolly pine was more lately introduced (or the winged seeds transported here from abroad by the winds,) than the short leaf, and could not obtain a proper seed-bed and maintain a healthy growth in lands already and completely occupied by other established pines and other trees. But when worn out vacant lands were offered, the opposite result followed. The seeds of both these kinds of pines were everywhere numerous enough, and were so readily transported to great distances by the winds, that there was no deficiency of either kind on any land. But, in such vacant fields, or when these two kinds of pine were equally in possession, the loblolly pine is much the fastest grower, and in a few years over-tops the smaller short leaf pines, which, therefore, are unthrifty, and in time are over-powered and die under the shade and crowding of the large and more vigorous loblolly pines. Hence, in a thick and long standing second growth, however numerous the slower growing short leaf pines may have been at first, not one might live when the eldest of the others had reached to forty years. On the particular abandoned lands where pines of second growth thrive best and grow fastest, they usually stand so thick, when young, that many of the smaller and weaker necessarily must die, and thus make room for the more vigorous. In such cases, of course the short leaf trees, of slower growth and smaller size, would certainly be among the first to perish. It is only when the growth is thin, owing to some unfavourable conditions of the soil, that in this region the short leaf pine can live in numbers, intermixed with the loblolly, as second growth; there being, in that case, enough space for both to live.

But in the higher range of country other causes operate. The land there is naturally much richer than the dry land in the lower country, the soil red, more clayey, and having not enough acid, (or having too much lime,) to permit the growth of the loblolly pine, which is especially favoured by the most acid soil, and also by sandy soil. But the short leaf pine can grow and thrive on soils stiffer, richer and better constituted for fertility, and therefore can occupy such land to the entire exclusion of the loblolly pine. But still, even the short leaf species does not thrive as well on a good agricultural soil not very deficient in lime. Therefore, according as the soil is better constituted for tillage crops, these pines are more sparse and slow in growth, and on the best natural soils they will not grow at all, as on the South West Mountain lands and the limestone soils of the more Western mountain country, and rich alluvial bottoms everywhere.

I will here present an opinion on this subject which will not be maintained by argument, to do which would require too much space, and would be here out of place. This opinion is, that the soils and upper layers of all the tide-water region of Virginia and North Carolina, and also an adjacent strip, of irregular breadth and outline, above the falls, are of drift formation, the materials of the drift having been washed by an enormous flood from the lands lying above, and which were denuded in supplying that material. That the whole region so formed by drift is extremely deficient in lime, (and much more so than the denuded region above,) and therefore naturally acid, consequently especially favourable to the growth of loblolly pines. If this opinion is correct, it will be much more important than merely for assigning the necessary localities and actual limits for the healthy growth of loblolly pines. For the ascertaining of the drift formation and the places where it is present or absent, will serve to indicate where lime, as manure, will either be highly beneficial, as in all the low country, or where it will probably be of little benefit, or none, as is said to be generally the case on the red Piedmont lands. This subject of drift formation and the drift-formed region and its localities, I have treated at length elsewhere, and therefore will pursue it no farther here.

From the various facts and opinions stated in the foregoing pages, it will have appeared incidentally that some (if not all) of the species of pines, are especially good and reliable indications of the character and constitution of the soils on which they grow, and in some cases of climate also. Thus all the pines common in this region, prefer to grow on soils, if dry, of but moderate or a low degree of natural fertility. The white pine (*p. strobus,*) which, however, is not of either the lowland or the Piedmont region, is the only species known to prefer well constituted, rich, and also dry agricultural soils. The long leaf pine, (*p. australis,*) requires a Southern locality or climate, and with that, a dry, sandy, and poor soil, and also sandy sub-soil, and its healthy and general growth is an indication of the presence of all these different requisites. The short leaf pine, (*p. variabilis,*) prefers stiffer soil or underlying earth, both to be dry. This will bear more of lime in the soil than either the preceding, (except *p. strobus,*) or than the loblolly. The cedar pine, (*p. inops,*) is more rare, and its habits less known to me. But this would seem, (as a second growth,) to prefer and indicate still better original soils, however exhausted subsequently, than either of the preceding pines of this region, and also of more clayey constitution. The loblolly grows well both on dry, sandy and poor soils, and on moist, deep and rich soils. But in both of these very different positions it must have acid soil. And this last condition is caused and provided by the great deficiency of all forms of lime in the poorest natural soils, and also by the great excess of vegetable matter and swampy or peaty lands.

12 The Geology of Low Country Progress
Agricultural Features of Virginia and North Carolina (1857)

I. General Remarks. The Public but Slightly Informed of the Region in Question, and with Lower North Carolina in General

The eastern portion of North Carolina presents a large region, of remarkable features, topographical, geological, and agricultural. The enclosed broad sounds, and other waters, are not less interesting, for their recent and great changes; and, besides, they have been the scenes of some of the minor but romantic and interesting incidents of history. Into Roanoke Sound, by the then broad open passage from the ocean, which is now dyked across by dry land, Sir Walter Raleigh's ships entered, and on Roanoke Island they planted the first, though but ineffectual, settlement of British colonists in America. In another portion of these now almost land-locked waters, there occurred many of the acts of Teache, or Blackbeard, the celebrated pirate, and finally, the naval engagement in which he was defeated and killed. If the lands of this region were even worthless, for agricultural and economical uses, they would deserve and reward the investigations of the exploring and laborious geologist; and if destitute of all scientific interest, they would deserve far more attention than ever has been bestowed on them, for their peculiarities of agricultural character, and capabilities for high improvement and profit. Yet, there is no equal space of territory in all the

States of the American Union that has been so little visited or seen by other than its residents, and of which the character and values have been so little noticed or known. It is rare that any stranger enters this *terra incognita*. And even of the residents of other parts of North Carolina, of the class inclined and accustomed to travel for business or pleasure, where one such has seen this portion of their own country, one hundred have visited the remote States of the North, and South, and West.

The region here referred to, except as to the line of seashore, has no exact geographical limits—or at least there is no present information upon which to designate the extreme southern and the whole western boundary. I would include all of the low-lying and very level land, which is the universal character of all the coast lands of North Carolina, and for a breadth of two to five or more counties westward. As soon as the surface begins to lose its apparent almost perfect level, and to swell perceptibly into rising slopes, there should be placed the western or upper boundary of the low and flat region which is here referred to generally. The same character of country extends northward to the Chesapeake bay and its lowest western affluent rivers; and how far southward I am not sufficiently informed to say. In addition to the one universal feature of low and level surface of the highest and firmest lands, it is much intersected by narrow strips of lower and swampy but also firm ground; and, also, immense spaces are occupied by large and boggy swamps, which were impassable, and almost impenetrable by man, until his improvements and labors had produced practical passage-ways.

This great region affords sundry somewhat connected, but yet substantive subjects, for separate treatment. Such are the now cultivated land and its agricultural condition, and the improvements most needed—description of the great swamps, and such agricultural improvements as have been there made—the geological origin and structure of the different great classes of lands—notices of the ocean sand-beach, and the enclosed sounds, and other navigable waters, and the changes that have occurred in both, &c. Some others, or perhaps all, of these several divisions of the whole great subject may be hereafter discussed. For the present, I will confine myself to sketch the agricultural features, condition, wants, (and errors of culture,) and capabilities of the particular and peculiar agricultural region which lies between the Chesapeake bay and Hampton roads and Nansemond river, on the north, the ocean on the east, and Albemarle sound on the south. On the west, the outline would include all the Dismal Swamp. But all the great space, and the circumstances of that swamp proper, will be passed over now, to be resumed and considered in another and substantive article. The further extension of the western boundary would include the lower Chowan, and the basin of the lower Roanoke. The area designated includes some of the oldest agricultural settlements and oldest towns, and (on the Roanoke especially) some of the richest lands on our Atlantic border. It is also intersected by sundry lines of public travel, and some of which (the land and water steam-lines to Norfolk) have long been used by numerous passengers. Still, all these circumstances do not make this particular agricultural district an exception to the general rule or condition of all the great low-land region, of being unseen, unknown, of travellers who visit, or pass through, Norfolk or Portsmouth, on the great routes, scarcely one ever treads the soil, except in the towns—or ever sees any of the lands of the country, except in the rapidly changing glimpses afforded from a steam-car, or the more distant and uncertain views from a steam-vessel. Princess Anne county, which reaches within three miles of Norfolk, and Norfolk county, lie wholly in the designated section; and these counties, out of the towns, are as little known to the residents of all other parts of Virginia, as any counties west of the Alleghany mountains. Yet, within the heart of one of these counties, and within a few miles of the other, are the important towns of Norfolk and Portsmouth, and the noblest harbor, and one of the most important Government dock-yards and naval stations of the United States. And the country has been as little appreciated as it was little known; and even by its residents, until recently, and by those who knew it best, as well as by strangers, who had only heard it spoken of and described in the most contemptuous epithets. And, though recent improvements of prices of lands, and in fewer and more remarkable cases, of products and profits, and still more, and longer, in some of the North Carolina counties, indicate much actual improvement and higher appreciation, still very few, even of the most intelligent proprietors, are yet fully aware of the true and great wants of their lands, and their great capability for improvement. Proper drainage alone would double the productive value and the profits of the whole great area of what is usually considered the *now dry* land, and of the firm and partially drained swamps. In addition to the peculiar grounds for agricultural improvement and profit in the land itself, no known region possesses such great facilities for navigation, and for choice of markets. And, in every respect, no where is there a region where agricultural improvement is more needed, and is more available, and offers more prospective profit; and no where have the great advantages offered by Nature been more neglected, or seem to be less known.

For the present, my remarks on this region will be applied especially and particularly to the portion lying east of Perquimons [Perquimans] river. My personal observations have not yet been extended further west; and most of whatever may be here said of the country extending beyond Perquimons, and including the lower Roanoke valley, will be on report deemed entirely reliable.[1]

II. Peculiar Characters of the Low-Lands, in Surface and Qualities of Soil

The most striking feature of this firm low-land region, is its very low and level surface. Large bodies, say of 1,000 acres or more together, are more uniformly level than any as large space of alluvial, or other bottom land, on any of the great rivers of Virginia. Such bottom-land as borders the Pomonkey [Pamunkey] river, for example, might be called undulating, compared to the general greater flatness of the whole great region under consideration. The numerous smaller swamps, interspersed, (which receive and conduct off the overflowing surface water,) are, usually, not much lower than the adjacent highest ground. So far as the eye would indicate, changes of level of even so much as a foot of difference, can rarely be perceived, except in the swamps and depressions which convey the rivers and smaller streams, or temporary rain-floods. But changes of level which are barely perceptible to the eye, are usually made abundantly distinct by the gathering of water on the slightly depressed surfaces, which serve to make the numerous swamps of *firm soil*. A stranger, if travelling through the country in any and different directions, might suppose that the surface of the land was nowhere higher than ten feet above ordinary high tide, or the usual height of the navigable and level waters; but the real heights are greater than would thus appear to the eye. In the interior of Princess Anne county, at Level Green, (the farm of Edward H. Herbert, Esq.,) where the surface seems to the eye as low as any—the elevation, as determined by levelling instruments, is about twenty-one feet above tide. Still, the variations of surface-level are so gradual, (except as to the beds of water-courses,) that it is often difficult, if not impossible, to reach any outlet for drainage of a few feet of fall, without conveying the water by a ditch of some miles in length, and through as high, or higher ground. This feature of the surface presents the greatest impediment to the drainage of the interior lands, and especially upon the ordinary method of mere surface drainage, by open and shallow ditches.

But with all the slight undulations of surface levels, there is nothing to obstruct the view, except the standing crops and fences on the farms, and the trees on swamp or other forest lands. Except for these obstructions, any object of the height of a man, or horse, could be seen over miles of intervening space and distance. In all the great area now under consideration, there is not (native to the locality) a stone, or even a small pebble; and, in few cases, but a little of small gravel.[2] The soils vary, in different places, between open and light sandy loam, and very close compact gray clay (so-called); or, perhaps, more correctly, extremely close and compact soil and subsoil, composed mostly of the minutest particles of sand, and which, therefore, are stiffer, closer, and more intractable under cultivation than the finest or true clay elsewhere. Of such red and yellow clays as make many of the best soils and subsoils of the upper country, (above the falls, or among the mountains,) none are seen here.

III. Peculiar Characters of the Rivers, and the Many Fit for Navigation

The water-courses are numerous, and many of them are deep enough to be navigated by sea-vessels. In some of the smaller rivers, in parts too narrow and crooked for the ordinary small vessels to turn about or to pass each other when meeting, there is enough depth of water to float a ship. A glance at this section on a large map of North Carolina will show the great number and close neighborhood of these rivers which flow, nearly parallel to each other, into the northern side of the Albemarle Sound. The lower parts of these rivers, where of widths, severally, from one to five miles, are more properly estuaries or large creeks, (in the proper sense of that word, and not as usually misapplied,) kept full by the refluent water of Albemarle Sound—just as they would be, and to equal height, if there was no other supply of water from head-springs or rain-floods. But even as ascending these rivers, and after they are contracted to very narrow widths, and, as appearing on the map, the upper channels might be inferred to be merely shallow and insignificant streams, they are, in fact, deep, though narrow rivers, of level and slow-moving water, and continuing deep almost to their visible head-sources, and offer good facilities for navigation to such extent, in number and in length of rivers and their sundry branches, that one-half of them are superfluous, and, therefore, are not put to use. If any obstructions exist they are made merely by trees fallen across, and are easily removed. The whole country, and especially from Perquimons county to Currituck Sound, is pervaded by broad and deep estuaries near to the sound; and their headwaters, extending near or into the Dismal Swamp, make, with their many branches, a net-work of natural still-water canals, narrow and crooked, indeed, but as deep, as smooth, and as sluggish as artificial canals, and free from the changes of levels and the obstruction of lock-gates, which accompany the benefits of canal navigation. Most of these rivers receive their head waters from the Dismal Swamp or other swamps. The water of all is black as seen in the rivers, and the color of brandy or Madeira wine as seen in a glass, being thus deeply colored, as are all the swamp waters, by the vegetable extractive matters in and on the boggy swamp soils. This discoloration is not entirely lost in the salt tide water of Elizabeth river, at Norfolk, nor in Currituck Sound, where nine miles wide, below the former (and now closed) Currituck inlet, which, not many years ago, admitted deep seavessels.

In travelling along the public road from Elizabeth City, North Carolina, to Currituck Court-House, within the distance of seven miles, we passed four navigable water-courses, including the Pasquotank and two of its branches. Three of these had draw-bridges for the passage of masted seavessels. The fourth stream had no draw-bridge, because it was not needed in such close vicinity to others; and, also, because, though this branch had abundant depth and an open channel for sea-vessels, it was so narrow and crooked that the banks and trees standing on the borders would entirely obstruct the masts and yards. Such great and numerous natural facilities for navigation, as in the many rivers of this region, are unequalled; and they are exceeded by the aid of art, only in the canal navigation of the Dutch Netherlands.

IV. General Want of Drainage and of Proper Views on the Subject

Level as is the general surface, and slight the variations of height, in adjacent spaces of all the peninsula between the waters of the Chesapeake and Albemarle, still there are frequent slight changes, and these, more than great changes elsewhere, are marked by consequent differences of character. Every farm of a few hundred acres has some of its surface of swamp, and usually undrained. What is called high or dry land is, indeed, the highest and dryest, but mostly still and always suffering more or less for want of sufficient drainage. The parts which may be only from two to three feet lower than the neighboring highest surfaces are, because of the depression only, swamps of wet though firm ground. These swamps are very generally of firm soil, and the boggy swamps are of entirely different materials and formation. In all this flat country there are very few springs showing at the surface, and but rarely any springy or oozy places. The water and the wetness of the numerous smaller swamps are due entirely to rains. On the higher spots, or larger high spaces, the early settlements were all made, and tillage has there been continued, with but little respite, to this time. The intermixed lower lands, or smaller swamps, were deemed worthless, and their culture was rarely attempted until within recent times. Yet, even with the imperfect superficial drainage which only is in use, these swamp lands are found to be best, and of fertility rarely exceeded anywhere. Some of this firm swamp, in Perquimons, of which Mr. J. T. Granberry's[3] estate in part is composed, and which but lately has been drained or brought under cultivation, he bought lately, at $55 the acre, unreclaimed. A highly intelligent neighbor told me that he remembered when the same land could not have been sold for 75 cents the acre, and was deemed of no value whatever for tillage.

The soils and also the subsoils vary in texture from moderately light to extremely stiff, close, impervious (now) to the descent of water, and remarkably intractable under tillage, and almost always either too wet or too dry for good ploughing, even under good farmers. Under the worst cultivators such soils are sometimes mud or mire, and sometimes clods almost as hard as brickbats. These soils are general or common in Perquimons only. Yet, on good farms, of this very difficult soil, there are seen the best (and excellent) crops of wheat, and other best crops of all the counties on the sound. The greatest drainage labors and most of the best farmers and best cultivation are also in that county; yet even there, and though many of the ditches are of great size and the drainage labors are remarkable for their extent and cost, still, almost every where, the tilled land is but partially and insufficiently drained. On much the larger portion, perhaps nineteen-twentieths of all the cultivated and even highest surface of the whole region, the drainage is much worse and still more insufficient.

V. The True Principle of Drainage for This Region and the Geological Facts on Which the Principle Is Founded

The great error of the method of drainage, general in all this region, is that the drains or ditches are designed, and only operate, to draw the superfluous and, therefore, injurious rain-water from and over the surface. The principle I would propose to substitute, is to draw off (and keep drawn off) the water which is in excess some feet below and up to the surface, and by thus removing the before constant saturation or glut of the lower earth, to permit the excess of falling rain to sink into the lower earth, and thence pass off below instead of being kept on and near the surface, as now and heretofore, until it either can flow off on the surface to ditches or is evaporated. Both the existing error and the evil effects and also the benefit of the proposed substituted plan are dependent on the geological structure of the land, and especially of its inferior beds. But, in advance of all description and reasoning as to the causes of the supposed existing phenomena and of tracing the effects in reference to draining, I will simply assume the truth of the great and all-important fact on which my plan and reasoning are founded. This fact is, that the whole of this low and flat country, at some few feet below the surface, (within the extreme limits of from 2 to 8 feet, and more generally from 3 to 5 feet,) has underlying it a bed of pure sand which, at least in all wet seasons, is glutted with water from its bottom to its top. This fact is unquestionable, and may be tested easily by every proprietor. But I have to infer, from the geological structure of the region and on reasoning, which would require too much space to state here, the further fact, that this underlying bed of water-glutted sand is nearly horizontal, but, like the overlying earth and its surface, has a gentle and general dip or declination toward the seacoast or in a southeasterly direction.

As to the general presence of the sand-bed, it is proved by every well that is dug, and not only here, but in much higher localities of the tide-water region. In the higher country, and at higher levels of surface, the sand-bed lies deeper; and also, then, generally, its upper part is dry, (or without water,) though, by digging deeper, the lower sand, there also is always found filled, (but not surcharged,) with water. A like bed of sand underlies most, or all of the bottom or low land, along the rivers in the higher tide-water counties in Virginia; and, as I infer from but limited personal observations, such sand, with much more regularity of position and operation, underlies the whole superficial layers of the great low-land region here under consideration. But in these low-lands, the sand-bed is naturally always glutted with water, which water is a source supplying moisture to the overlying earth, and also, by being already as full of water as it can be; the glutted sand-bed is an effectual barrier to the descent of more rain-water from the surface of the land. This sand-bed is, therefore, the great cause of the existing wetness of the upper beds, and surface soil, and the reason why the usual surface draining is so imperfect in operation. And the same feature offers the manner and means for effectual drainage.

Of course, very few particular facts, and in narrow spaces, have been learned from our own personal observations in this low country. But I had previously discovered the underlying and also water glutted sand-bed, (concealed from all previous knowledge, as a general fact,) below the broad bottom lands of my own farm on the Pamunkey river, (in Hanover county, Va.,) and had long studied its effects, and in reference to it, and had devised, and conducted successfully, extensive draining labor. At first, I had supposed this remarkable and then newly discovered feature to be peculiar to the particular locality of my own farming labor; but in the progress of my draining operations, and the necessary study of the whole subject, and the true principles of drainage, I came to infer, that the same feature of an underlying sand-bed, belongs to the whole of the lands of our great tide-water region, and that this sand-bed, where dipping lowest, and glutted with water, was the great cause of the evil of excessive wetness of the lowlying soils above. I felt so confident of the correctness of my deductions, that it induced me at the first time of leisure, to visit the region in question, to seek and to find the facts to conform and to sustain my theoretical views. And before my first visit to this country, I offered to a friend, residing therein, advice for the proper drainage of his farm (by seeking for and tapping the glutted sand-bed,) which he acted upon to some extent, and found therein the precise effects and all the benefit that could have been expected from his limited first operations on this new principle.

To obtain numerous evidences of the very general existence and position of the sand-bed, it was not required for me to dig or bore into the under beds, or even to seek the surface of every locality. Every farm house is supplied with water by one or more wells, and these numerous, previously, and long used wells, go far to supply all the facts required. Whether the sand-beds exists, and near enough to the surface to effect its natural drainage, may be learned usually from inquiries about the wells, their depths, and the manner and varying quantities of their supply of water. From even but a few such examples, and applying thereto my general views derived from practice and experience of draining in far distant localities, I was confirmed in the general opinions previously formed, in advance of all personal observation. The conclusions thus reached, and for which I will proceed to argue for the conviction of others, may be thus stated; that nearly all the higher and firm, as well as the lower lands, lying between the Chesapeake and Albermarle Sound, are rendered and kept too wet, not (as universally alleged,) because the soils or their under beds are of too close texture to permit the superfluous rain-water to sink, and so be discharged by percolation; but because the underlying sand-bed is already surcharged with water, and by its supplying moisture upward, renders the moist earth incapable of drinking up more water from above.

In the upper and middle ranges of the tide-water counties of Virginia, the reaching the sand-bed, and its being dry when reached, are essential conditions to the construction of a good ice-house—the dry sand bottom serving immediately to absorb, and convey away, by downward filtration, all the water formed by the melting of the ice. This is the operation of the principle of drainage of the higher beds, by the agency of a dry (or drained) upper part of the sand-bed below. It is also essential to the utility of every well, that it should be sunk through the upper and dry layer (if there be such) of the sand-bed, and into the water glutted lower part, for the purpose of its furnishing a permanent supply of water. And if, as generally, in the flat low country, the sand-bed is full of water to its top, (unless after long droughts,) and is so surcharged that the water is pressed upward, then in wells there dug, not only would water be obtained as soon as the sand-bed was reached, but the water would rise still higher, and even near to the surface of the land in very wet seasons. Thus, every well in this low country may afford evidence of the existence, height, and character of the sand-bed at its top, and also the height to which water will rise therefrom, and how near the surface of the land the upper bed must be injuriously affected by the water glut below, and whether permanently, or but for the wettest seasons. Hence, it follows, that little as has heretofore been noticed, or thought of, in regard to these important facts, and the more important deductions from them, and few as are the residents who have thought at all on these particular points, it is only necessary for farmers and thinking men to reflect upon, and apply the facts they already know, to be assured of the true principle and method of drainage for their land, which will now be more fully explained and argued.

VI. The Underlying Sand-Bed and Its Opposite Operations in Regard to Draining

Whether the underlying sand is of one continuous bed connected throughout, or broken, or separated, is not important. It is enough that it is general, and known as wanting nowhere. Neither is its general thickness known, nor is its bottom but rarely accessible or known. But it is certain that this sand-bed lies upon some lower bed, impenetrable to water from above; and which bed, in many known cases, is marl. But whatever may be the lower bed or its texture, the sand-bed itself, however open and loose in texture, if already glutted with water, is incapable of receiving more. Therefore, there is no layer of earth so impenetrable by water, as any earth, and even sand, already full of water; and in less degree, all dampness or moisture of the underlying bed of earth is so much impediment to the reception of rain-water from above. The following rough figure will serve to exhibit a profile or section of the supposed strata of the low-lands; but to render the differences of level apparent to the eye, it is necessary greatly to increase the thickness of the strata, and the rate of their dip, in the figure, exceeding the natural and actual conditions.

Suppose this figure to represent the surface soil (*a b*,) and also the inferior beds, all dipping very gradually, (and very much less than in the figure,) from northwest to southeast, or in the direction from the falls of the rivers towards the ocean. The finely dotted line, *c d*, indicates the horizontal level. The upper bed, next below the surface soil (1,) let us first suppose here to be clayey, or of close texture, and not readily permeable by water. The next below is the sand-bed, which is wholly glutted with water, or partly dry (at top,) according to its level, or dip, on the variable supply of water, and its manner of discharge. The next bed (3,) is of marl, or other impermeable earth, or otherwise, from its constant wetness, incapable of receiving more water from above.

Now, of all the excess of rain-water that falls on the whole surface of the tide-water region, (as everywhere else,) part flows off over the surface of the land, and of that which remains, part is soon or late evaporated, and part sinks into the earth as low as it can be admitted into, or absorbed by the lower earth. The greater discharge of rain-water by its flowing off will be on hilly surfaces, and soils of close and compact texture. The greater discharge by downward percolation, or filtration, will be on the most sandy or porous earth, (if dry before and to enough depth,) and the more so if on level surfaces. Whatever water is not taken off by these two modes, can be removed only by evaporation, and until so removed, the remaining excess of water must saturate the soil, if not cover it in part in stagnant pools, and, for the time, destroy its productive power, and prevent all proper tillage labors. Every transient occurrence of such wet condition must be injurious to tillage lands, and the frequent occurrence of such conditions, even if each one be transient, is enough to render even rich, arable land of very little value.

Of the rain-water that falls on the higher lands (at and above a,) and that sinks into the earth below, and which is too much to be held absorbed by the next beds, (1,) the excess must sink still lower, and go to supply or to surcharge the sand-bed (2,) below. And all the water in that bed, whether filling it wholly, or only its lower portion, would be slowly but continually pressing laterally in the direction of the dip (towards e,) to seek (and find, ultimately,) a long delayed discharge in the lower channels of rivers. Although the beds of earth may be nearly horizontal, the slightest degree of their general dipping must induce the operation stated. Thus, the supply of water to glut the sand-bed is not only increased by rain-water fallen immediately above, and over porous upper beds, (at 1,) but also another and continuous supply is pressing on laterally, derived from higher levels of the sand-bed (2,) and from rains that fell many miles distant, on the higher country. And therefore, while the upper layer of the sand-bed in the higher country, (or temporarily in the lower country,) may be left dry, (as represented above the level of the dotted line at c,) at the lower level of the same sand-bed, and at the same time it will be necessarily surcharged with water, and which not finding sufficient discharge in its gradual and slow descent along the dip of the bed, presses with all the weight of its higher lying water in every direction, and not only downward and laterally, but also upward. This is evident even to the eye. For if the water received partly on a higher and distant surface, (near to and far westward of a,) seems to keep the water in the sand-bed no higher (at any one time) than the horizontal line at c, it will still fill the whole depth of the sand-bed as descending farther eastward. As the sand-bed dips, the water confined therein (by the higher bed being but lightly permeable,) would be pressed by the weight of the higher and remote water (rising to c,) and, by a well-known law of hydrostatics, would rise as high as the line *c*, if having an upward vent. And precisely such a vent is afforded by a well, sunk at *w*, in which the water reached in the sand-bed (2) will rise to the level of *d c*, or as high as may then be the then height of the supply of water near to *c*. Thus, in nearly every well in this low land region, the water sometimes rises above the sand-bed which yielded the water; and after great falls of rain, or long continued wetness of the earth, the water supplied only by percolation, and mainly from a distance, rises much higher than usual, and, in some cases, to within one or two feet of the surface of the land.

So far, from more clear explanation, it has been supposed that the higher bed, (1,) was more or less impervious, and so served to confine in the sand-bed below its water, and greatly to resist and impede its escape by upward discharge. But if, as is more general, the higher bed (1) is of texture permeable to water, that difference does not materially vary the circumstances as to the need and manner of draining. A pervious upper bed will absorb more freely and speedily all the water that hydrostatic pressure would force upward, so as to leave much less visible results of such pressure in particular places, as in wells and deep ditches. But in either case there would be the same general evil to the upper earth and surface soil, of moisture derived from below; and the same remedy required, of discharging the injurious supply of water from its reservoir below.

To whatever height the water (proceeding from the sand-bed) can rise in the unobstructed passage afforded by a well, (or in an auger hole, bored for trial,) to the same height must there exist the force to raise the water, more slowly by filtration, but by the same hydrostatic pressure, in all the neighboring ground. The bed of earth lying over the glutted sand may be so close (in its moist condition) as to be impervious to the descent of rain-water, from the surface, which would act only by the pressure of gravity. But scarcely any earth is close enough to prevent the absorption of water, pressed upward by the much stronger force acting on the water confined below. Therefore, even when the sand-bed may be as low as six or eight feet below the surface, and a bed of unusually close texture between, the confined water may be so strongly pressed upward as to reach within two feet of the surface. In such cases injurious moisture will rise still higher, by capillary attraction, and more evidently over sandy than a close subsoil or under-bed. It is owing to this condition of things, that many spaces, without showing any standing or flowing, or even the slightest oozing water, either at the surface or in shallow ditches, are always damp and cold, produce only aquatic grasses or weeds, and exhibit every indication of wetness except the actual and usual presence of water. But after every rain, and even light rains, water will stand in puddles on such places, if level, even though the soil and subsoil are sandy and open. For moist sand is soon filled by water to repletion, and wet sand will hold water on its surface like a dish.

Thus, I infer that the whole of this low land is underlaid by a sand-bed, glutted with water to its top, and which sand-bed is generally so near the surface soil as to affect it injuriously by water from below. But even if this confined water lay too low to affect the surface earth directly, it would do it indirectly, by preventing the rain-water from sinking, and its excess being discharged by downward percolation. If the sand-bed below were dry, or always free from water from its upper twelve inches only, (as near c,) that upper layer of dry sand would serve as natural under-draining for all the upper earth. Such is the condition of things under the excellent and dry low grounds of Brandon [plantation], on James river; and such is inferred to be the case with all the similar low lands, which, though level and of stiff soil, require but little draining labors, and can dispense with all under-draining. The upper layer of the universal sand-bed, being there dry, is always ready to receive and to discharge below all water sinking from above. Thus these fine lands are under-drained by nature. And the only reason why that general under-drainage is not perfect in operation, and ample for all wants of the land, is that this dry sand is many feet (10 to 14) below the surface of the land, and the intervening beds are of clayey and compact texture. Even these impediments would not prevent the surface being generally and perfectly dry, if without any artificial drainage. But the natural draining process is too slow, and therefore the aid of some surface ditches are there needed to pass off more quickly the temporary rain-floods.

But when, instead of the upper sand being dry and so serving to drain the upper beds, the whole sand-bed is full of water, and that water is pressed upward, then all the upper beds are kept more or less wet or moist, and are thereby rendered unable to receive any more rain-water from above by filtration or percolation. The stiffest and closest clay, when dry, is full of minute fissures; if no moister than usual at some feet below a dry surface, such clay will absorb water from above and slowly pass any excess by percolation to an absorbent or receiving bed below. But earth made wet or moist by water forced upward from below, whether it be close clay or loose and course sand, can receive no more from above, and all excess of rain-water left there in pools must remain until evaporated.

We may best estimate the enormity of this evil, of the wet earth below preventing the rain-water from sinking, by the condition of the level woodland still remaining in a state of nature and without any aid from ditches. On such land, in wet seasons and usually in every winter and spring, the excess of rain-water remains and covers most of the surface, and in many cases for weeks or months together. This is universally ascribed by the proprietors and neighbors to the soil or its under-earth being too stiff and close to permit the descent of water, and this is held even where the upper bed is open and light enough for any purpose. Now let us proceed to examine the actual remedy or the drainage plan in general use, and its effects, and next the different principle of drainage and method which I propose.

VII. The Usual and General Plan of Draining and Its Radical Defects

The actual plan or system of draining which is in general and approved use in this region is very uniform in the general principle and features, and also very simple. It consists in digging numerous ditches, mostly shallow and small, merely for the purpose of collecting therein and conveying from the field so much of the excess of rain-water as will flow over the surface. These ditches are at various distances, according to the greater or less excess of wetness of the land, and they are of various degrees of imperfect effect, according to their number and depth. But on no farm is this mode of ditching effectual for drainage, and on a few only has it ever approached that desired end, where the ditches were much deeper than usual and great labor has been bestowed, though on an erroneous system.

The numerous swamps, so-called, or spaces, either broad or narrow, a little more depressed or level than the adjacent ground, serve to afford ground for outlets in deep and large ditches acting as main water-carriers through these swamps to some one of the numerous rivers or deep creeks with which the whole country is intersected. Some of these deep and main discharging ditches may severally receive the waters from two or three different farms and properties, and extend for miles before reaching the final outlet. Still, by combined effort for the common benefit, these longest ditches may be made cheaply enough for their object, and may be made deep enough to suit for any system of drainage.

Supposing that a proper outlet has been secured through which to discharge the water into the river, then each farmer next proceeds to dig the receiving smaller ditches to collect the excess of rain-water from the field. In most cases the farms are so level that the ditches may be laid off in almost any direction, and usually they are made to coincide with the cardinal points of the compass, or otherwise made parallel with, or perpendicular to some road or other straight and long outline of the field. As the most laborious, and also the most perfect draining on this plan, and only the stiffest soil, is seen in Perquimons county, the operations there will be held especially in view in the following description:

In beginning a large drainage operation, or in renewing and substituting a former irregular and imperfect laying off, the main ditch of the field or farm is first dug to discharge into some common main water-carrier, or other deep outlet. But so uniform is the general level and shape of surface, that the required main ditch can usually be made straight, and to agree in the preferred manner, with the other smaller ditches, and with the direction of the ploughing. Into the "main" and deepest ditch, (usually 3 to 4 feet deep,) and at right-angles to it, and 1,000 feet apart, the parallel "leading" ditches enter, which are 2 to 3 feet deep. Then crossing the last, and parallel to the main ditch, and 150 feet apart, (on some farms, only 125 feet,) are dug narrow "tap ditches," 18 or 20 inches deep, and which empty, at both ends, into the "leading ditches." The land is tilled in five feet beds, laid off parallel with the smallest or tap ditches. Still, all these ditches, with the narrow beds and their alleys, (or water-furrows,) are deemed insufficient to carry off the excess of rain-water, without the further aid of "hoe-furrows," which are opened first by a plough, and afterwards cleaned out by hand-hoes after every ploughing of the field, because every ploughing (or horse-tillage) fills them. These "hoe-furrows" are made across the narrow beds, at irregular distances of from 18 to 25 yards, and empty into the tap ditches. A "hoe-furrow" is made to pass through every slightest cross-depression, and wherever else deemed most necessary. Thus the alleys of the five feet beds first receive the surplus and overflowing rain-water; and so much thereof as can flow off over a level, or nearly level surface, passes out of the open ends of the alleys (from both ends) into the leading ditches, or across the beds along the hoe-furrows into the tap ditches, and thence to the leading ditches. From the latter the water passes into the broader, and deeper, main ditch, and from it to the common outlet of the farm. The hoe-furrows (or grips) are a little deeper than the alleys of the five feet beds. The alleys may be 6 or 7 inches below the crowns of the beds. This plan is, on some farms varied by the leading ditches running parallel to the main ditch; but the number of ditches and furrows, and the spaces between, are not varied.

The object of this plan, and the only possible operation of it, is to draw off the excess of rain-water mainly *over the surface*; and with all these numerous ditches and furrows, on perfectly level land, no water can flow off until it has saturated the soil, or stands above it in numerous little shallow pools; and if the field is under tillage, and has been deeply ploughed, all the ploughed layer will suck up as much rain-water as it can retain, before any surplus will begin to flow off over the surface, or by lateral and horizontal percolation, to ooze out from the soft soil into the lower furrows and ditches. Such draining at best only begins to remove the injurious excess of water from the soil, after it has effected all the drainage it can do for the time. It is true that every hour of the continuance of the water would greatly increase the first damage of the saturated soil, and that continuance the numerous drains serve to cut short and reduce, in time and in evil effect.

Some of the main ditches in Perquimons are of much greater depth and of unnecessary width at the bottom, (which should always be narrow, no matter how wide at top and how deep a ditch may be.) Mr. J. T. Granberry's main ditch is 7 to 8 feet deep; and, though without its being so designed, this ditch reached the sand-bed and tapped its glut of water. This great depth had been sought only for the different purpose of having a sufficient vent for the great quantity of surface water to be discharged from the field.

This system cuts up every field, by spade-dug ditches, into separate spaces of little more than three and a half acres each. Then bridges are required at suitable crossing places over every main and leading ditch, and also over every tap-ditch when they are crossed by a farm road or a temporary track for hauling in a crop. As many other rough wooden structures are required to give passage to water and to exclude hogs where ever a fence crosses the tap or other ditches. The labor necessary to dig and keep open all these ditches, with all the other accompaniments and the increased labor of tillage, &c., among these open ditches must be enormous. It would not be much more costly, and would return much more nett profit to adopt, instead, the modern English system of deep and covered under-draining—which system, after all, is but the drainage of surface-water, derived from rains, by downward filtration, and as soon as may be effected after the rain has fallen on the surface in excess.

This plan of draining by numerous ditches separating and surrounding small rectangular spaces was first used on the low (marsh) rice-lands of South Carolina, where it was not inconvenient for tillage, inasmuch as no ploughing or other team-labor was practicable on the soft and miry soil. Thence the same system was transferred to much of the high and firm land under cotton culture, but which needed some attention to drainage. Such ditching was practiced as late as 1843, on much land in Charleston district which scarcely needed a ditch (dug by the spade) any where. But there, while these frequent ditches were deemed indispensable by many planters, they were also deemed so great an impediment to the plough that that implement was excluded therefrom and these fields were cultivated by handlabor entirely. In Perquimons full use is made of the plough despite of the many obstructing ditches. And it has not been very long since crossploughing also was in use among these many ditches—the corn rows being laid off and ploughed across as well as lengthwise of the long and narrow rectangles.[4] Of course the culture then must have been flat or without beds and intervening alleys, preserved throughout the years' tillage, as since and now.

VIII. Evidences or Illustrations of the Existing Injuries from Superfluous Water, and of the Proper Means for Relief

The plan or principle on which I would propose to drain the lands of this low country is very different from what has heretofore been unusually aimed at, and, but partially effected. Instead of removing the excess of water by passing it off *over the surface* through numerous shallow and open tap-ditches, I would, by a few deep and mostly covered drains, tap the glutted sand-bed below, and thus, as much as practicable, lessen or entirely abate the previous upward pressure and direction of the confined water, and thereby relieving the upper bed of earth of its present supply of moisture from below, make it dry and permeable, and so permit, for the future, the excess of rain-water to sink into the drained upper bed, and be thus drawn off by percolation to the still lower sand-bed, (then empty enough at top to receive such temporary additions,) and thence the water to pass along the dip of the sand-bed, and far beneath the surface of the land, to the nearest deep stream or other place of discharge.

It is admitted that, except as to my own limited operations and experience, on a single farm, (Marlbourne,) there is almost no such practical proof of the effects here anticipated in regard to this great low-land region, of which so little is well known to me. But recent, and few, and limited as have been my means for personal examination and investigation in this region, there can be little doubt of the general existence of the one important natural feature on which my plan and reasoning rest, viz: the under-lying and water-glutted sand-bed, having a general, very slight, continuous dip. If this is the general and natural condition of the land, and if it is a sufficient cause for its present wetness, then it follows that the true principle of drainage, which sound theory would direct, is to draw the water from the *bottom*, and not from the *top*, as is the only function of shallow ditches. It may be, in some few localities, that the glutted sand-bed lies too low to be reached by ditches without too great labor and expense. But even such objections to the practical operations will not invalidate the correctness of the theory. And such good objections to practice probably exist in but few cases of limited localities.

It is manifest, to the least consideration, that the usual and universally approved plan and procedure cannot drain this land. As to the moisture infiltrating from the glut below, or driven upward by hydrostatic pressure, or drawn still higher and diffused as mere dampness by capillary attraction, it is obvious that this moisture cannot be lessened by any number of ditches in the upper earth. As to the excess of rain-water, when remaining separate on the surface, some of it will flow off in shallow ditches. But none will so pass off, from a level surface until the excess of water stands in small pools. Nor can any of the surplus water escape by filtrating laterally through the soil until the soil or upper earth has drunk up more rain-water than it can retain. These conditions of extremely wet earth, (and the more if of recently and deeply ploughed land,) must exist before the present system of drainage can even begin to act, and must still remain in force after the ditches have ceased to draw from the land that portion of the water which cannot be held absorbed. All the still remaining water, (and enough for the time to convert tilled soil to mire,) will be removed only by evaporation, as none can sink into the earth below in its present and usual wet state caused by the glut of water in the sand-bed, and the moisture always rising therefrom.

The best farmers seeing the imperfect operation of this plan of draining, have sought the desired improvement in digging all their ditches deeper than usual. But, unless such deepening reached and tapped the sand-bed, the deeper ditches could not gather any water from below, and could convey no more from the surface of the land than would be done by shallower ditches in somewhat longer time.

IX. *The Upper Beds Always Permeable if Drained*

But even if it be conceded to my argument that the sand-bed could be tapped, and the previous upper layer of its water be drawn off and kept permanently lowered, it would still be denied by most of the farmers that the rain-water can then sink through the earth. This denial would be founded on the supposed impervious texture of the intervening bed of earth. This belief of the under earth being impermeable to water is not only general in Perquimons, (and with much color of truth there,) where the upper earth is extremely close and stiff, and in some places eight feet or more in thickness, but also in Princess Anne and Norfolk counties, where the soil and under earth are abundantly porous, and not generally more than four feet thick.

Further, the immense quantity of rain-water which remains long, and covers much of the surface on the forest land in its natural condition, and which water passes off where ditches have been dug, makes it seem incredible that even half of all this water could sink through the earth below. It is also a prevailing belief that there is more rain in this region than general. I presume that no more rain falls from the clouds, but as very little of the excess of rain-water sinks into the earth, (because of its wetness below,) there is far more of the surplus rain-water to be removed and discharged by ditches than in other localities. In some of the nearly as level but higher lands of parts of the Southampton and Surry [Counties], in Virginia, scarcely a ditch is required, and there is no evil of rain-water remaining on the surface. There, in furnishing a pervious soil and sub-soil and dry under-beds, nature has effectually under-drained such lands, and in so doing has enabled most of the surplus rain-water to disappear by downward filtration. The great quantity of rain-water in the low-lands which passes off in the ditches is owing to the small absorbing power of the always wet lower earth, and, in less degree, of the upper also.

X. *Examples of the Effects of the True Principle of Drainage, in Both Artificial and Natural Operations*

Though there has been very little practice in this region on the plan of tapping and drawing off the confined water of the inferior sand-bed, and almost none by design, there still have been some such operations, and with marked beneficial results. Mr. J. T. Granberry, in Perquimons, and Mr. E. H. Herbert, in Princess Anne, tapped the water of the sand-bed when they anticipated nothing of the important effect, and merely designed to make unusually large and deep ditches. Mr. W. Sayre,[5] then of Norfolk county, acting on my general views and advice, given to him before I had seen his land, or even any part of the region in question, sought for and found the wet sand-bed at four to five feet deep, and to which no ditch on his farm or near to it had before penetrated. He deepened the greater length of his general outside ditch to the sand, and found great increased draining benefit therefrom in the single year which he continued to own and reside on the farm. One of the effects could scarcely be mistaken. In the summer after the first opening of this deep encircling ditch to the sand-bed, the well, half a mile distant from the ditch, ceased to supply water, and continued thus nearly dry until in the following winter. This well, (or another very close by,) had always before, and as far back as known, yielded water abundantly, and through the dryest seasons. The subsequent and long failure must have been caused by the cutting off, by the deep outside ditch, the supply to the well of water from the sand-bed. It is difficult to appreciate such slow and gradual effects, or to know always to what particular courses to ascribe them. Such effects from this mode of drainage may be slowly increasing for years before reaching their maximum of operation.

But on this principle there are many other and great drainage operations which nature has executed, and which show the beneficial results that are here promised. Every river or smaller deep water channel in this low-land is, in effect, a deep drain cut into the glutted sand-bed, and which cut or tapping has been operating to draw off the neighboring confined water, and to prevent its upward pressure so far as circumstances permitted. Along the sides of every river and deep branch the bordering lands, for half a mile or more in breadth, are much drier than any other adjacent lands of equal elevation and like surface. This is the case as in Durant's Neck, where the land is very level, and also lower than is usual for the firmest soil. This is the long peninsula of good land lying between Perquimons and Little river[s], and extending to Albemarle Sound.

The depressed shore of a river does not serve the better to drain bordering land because the river is a mile or more in width. A covered drain, having but a four-inch pipe or passage for water, if serving to reduce and convey away all the excess of under-water, and to prevent its previous upward pressure, and so leave the upper layer of the sand-bed dry, would, for draining effect, serve all the purposes of the widest river of no greater draining depth. If the natural depression for the river's passage serves to drain by lateral percolation half a mile width of the bordering land, a deep artificial drain sunk a foot or two into the sand-bed, and whether open or covered, may be expected to do as much. And if so, deep parallel drains a mile apart perhaps might drain the intermediate land. And such drains, even if ten feet deep and covered, would still be made and kept at less cost than the never-ceasing trouble of the numerous shallow and open ditches in Perquimons. But in most other places, as Princess Anne and Norfolk counties, the glutted sand-bed is not usually more than four feet below the surface, and drains sunk into the sand, and if four or even eight of them to the mile of width or cross-distance, would not be very costly, and could scarcely fail of their object.

XI. Draining Vertically by Bore-Holes

Where the water is closely confined in the sand-bed by the compact texture of the wet overlying earth, and the upward pressure of the confined water is considerable, (because of the quantity, or height, or weight of the water at the higher sources,) a portion of the water may be drawn higher than the top of the sand-bed by use of the auger. As in most of the wells the water rises to more or less height above the top of the sand, so it would rise as high in the holes bored by an inch-auger. And if the main or discharging ditches were sunk but a few inches lower, then the water could be thus drawn up in holes bored in such ditch, the water rising through the boring would continue to flow off along the bottom of the ditch. In such cases, the holes, if found operative, should be bored every thirty to fifty yards in a new ditch, as some will not act at all. Each such bore, when acting to bring up a continued stream, is an artificial "boiling spring." And if there is sufficient quantity and force of the water thus rising, there is no more reason why the artificial boiling spring shall be obstructed and its flow stopped than a natural one.

XII. The Presence of Quick-Sand Both as an Impediment or an Aid to Effectual Draining

It was by such borings (commenced for a very different object) that I first discovered the general existence and the properties of the water-glutted sand-bed on my own farm, and by them drew up and passed off water in considerable quantity before my main ditch had been sunk within two feet of the sand-bed. But if it is practicable and safe to go deeper with the spade, this vertical draining, in open ditches, should be but a temporary expedient, as it was in my own case. If the water will rise, say two feet in such bore-holes, to the then bottom of an open ditch, it will operate partially to reduce the glut of water below and prevent so much of its upward pressure. But the reduction will not be of any water that cannot force its passage so high. The greatest value of the fact of thus draining up water by boring, is the sure indication it affords of the still greater success of a future deeper digging of the ditch. If water thus rises to the height of two feet, it will rise with much more force and longer continuance if the ditch is sunk deeper and the water has so much less height to rise. If by still later and deeper digging the ditch is sunk into the sand, then there will no longer be vertical or boiling springs, but, instead, water oozing or flowing in laterally from the upper sand and along the whole line of such digging. Of course, and the more if the sand is very fine, such continuous opening is better than any number of auger holes, even if the bores should always continue open and discharging.

The inability to execute so extensive and costly an operation at once compelled me to deepen my main ditch at different times and in several successive years. But there is another reason for such gradual deepening, which will probably be found to operate in all diggings into the sand-bed in this low country. It is most likely that this water-glutted bed is every where a "quick-sand," almost semi-fluid, and which, as soon as dug into, will flow in from the sides and fill with sand the deeper excavation. And if the digging is persisted in it will cause caving or falling in of the solid and dry upper margins of the ditch, so that any effectual or permanent deepening at that time will be impracticable. If quick-sand is the greatest impediment to continued and successful deepening of the digging, its presence is also the surest proof of the necessity for the work and the best surety for its final and complete success. Quick-sand is nothing but a very pure and loose sand of which all the interstices are glutted with water. There is no coherence of the different particles of such sand, and the water contained therein is nearly as much in bulk as the solid matter of the sand itself, and when drained and passing off the water is continually renewed by lateral supply from more or less remote and higher sources. Hence quick-sand is semi-fluid, and flows in almost as freely as water, fills every lower cavity of an open ditch, and is like to enter every crevice of the filling material of a covered drain, and finally to choke the narrow conduit. Nothing can be worse than quick-sand to oppose the immediate and complete excavation of a ditch, whether to be covered or left open. But delay and time afford the remedy. When quick-sand is reached the digging should at first go no deeper than its surface, or no deeper into the sand than may be without causing damage. Then the before confined water, which rendered the sand "quick" or semi-fluid, will find a discharge into the ditch. The previous upward pressure will be removed. Later, the water will subside, leaving free the upper sand, thus drained into the ditch, and as low as the level of the discharge. In a year after the first operation, the then bottom of the ditch will no longer be of quick-sand, as at first, but will have become firm, and may then be deepened some six or eight inches more, before reaching what is still quick-sand below. Thus so much deeper and fuller discharge is given to the water, and so much more of the quantity removed, that thereby another layer of the then highest quick-sand is gradually converted to dryer and firm sand, and which may also be subsequently taken out safely by the spade. In this manner, and easily, and with best effects, I have, in three successive years, gained two feet of depth below the original surface of a bad quick-sand, in which at first I could not keep open the shallowest permanent passage. If all the glutted sand-bed of the low country (as inferred) is also of quick-sand, in like manner it may at first be barely tapped by ditching, and afterwards, and gradually, be dug into deeper, until all the injurious excess of under-water has been reduced and removed.

XIII. Tests by Which to Judge in Advance of the Expediency or Success of Desired Draining Operations, and Illustrations of Effects

Such is my view of the cause of the general wetness of this low-land region, and such the proposed remedy. If the principle is sound and the deduction true it is enough for my argument, and also for every extreme applications of the theory in practice. But it is not for me, slightly informed of particular facts and localities as I am by personal observation, to offer particular directions for practical operations, or to state the natural and various conditions of different localities, which may either invite or discouage and forbid efforts to drain by means of reaching the deep-seated sources of the injurious waters. In many or most localities of this great low-land region the proposed means may be used both cheaply and profitably. In others, owing to the greater depth of digging necessary, the operation, though equally sure of success, might be of more cost than profit. Every judicious farmer acquainted with the local details can best determine as to the applicability of my general plan to his own farm and vicinity. But there are certain indications and preliminary tests of the need for and probable success of such undertakings, which each farmer should consult in advance. These will now be mentioned.

The shallow wells on every farm will have shown whether a sand-bed has been reached, whether its being tapped brought up water, and at what height above the sand, if any the water stands permanently, and how much higher after winter, or the wettest seasons. These facts would serve to show how high the water may be drawn up by borings, and how much below that height it may be sunk by deep ditching. Thus, any depth of ditching below the highest temporary rising of the water, in wells or bore-holes, would do *some* good in draining off or reducing the glut below, and its upward pressure, though such benefit might be but for the wettest seasons. But the deeper the digging the greater would be the reduction of the hurtful excess of water. And the remedy would not be complete, until the main ditches were sunk into the sand-bed, so as to take off from the adjacent ground, all the former upward pressure of the under-water, and also render the upper layer of the sand-bed dry, and therefore capable of freely imbibing new supplies of rain-water infiltrated from above.

Next, as to the assumed permeability to water of the upper bed of earth. It has been admitted that the upper beds, even if of the most sandy and loose texture, if full of water below, are impermeable to more water standing on the surface. But if such wet earth be deprived of all superfluous moisture, (as by any proper draining,) then, what was impervious before, may become as pervious as desirable. Every one has observed such change in clay, when dug into, and the bottom of the excavation left exposed to a drying atmosphere. Of course, such extent of drying, and the consequent great opening of fissures, is not to be looked for under the covering earth. But in long droughts, earth not affected by under-water, will become as dry as dust for four feet or more below the surface. This is often seen in the digging of graves in summer; while in that dry condition there must be formed innumerable small pores and fissures, caused by contraction, in the most compact earth, through which water would freely sink, and in great quantity, and as low as the earth had thus dried, and fissures been formed. And these fissures could not be again entirely closed by wetness and expansion of the earth, so as to exclude all percolation of water. It is not for me to assert that there will be enough of these fissures, and reaching to sufficient depth, to serve to carry down by percolation *all* the excess of rain-water, even when gradually falling on the earth. But there can be no question that water will be so absorbed, and conveyed away in great quantity, in a soil with under-beds thus drained, when the same earth, before being drained, would have been incapable of absorbing any water below the quickly saturated surface soil.

For the good effect and success of the plan of draining the earth from below, it is not necessary that all or even a large proportion of the water in the sand-bed shall be so drained off. It may be that the bed is twenty feet thick. However thick the bed, its being full of water and surcharged, (proved by the water pressing upward,) shows that the supply of water from the higher parts of the country is greater than the sand-bed has openings for its lateral discharge. Thus, suppose the whole natural discharge of the sand-bed, into rivers and other outlets, and by evaporation, to be in volume, as 19, and the supply of water from rains, and from the more elevated and distant parts of the bed, to be as 20, then it is seen that the excess of supply of 1 part can only be removed by being forced upward through the earth. This is the water that operates injuriously, directly, by causing wetness to the under-earth, and indirectly, by preventing the excess of rain-water from being discharged by sinking. Then, if by tapping the sand-bed, this twentieth part of the water only is removed, the whole upward pressure, and the surcharge is prevented. But further, if by deeper draining the still full (but not then over-gorged) sand-bed has its water drawn off and lowered only one foot of its 20, or more of supposed depth, that upper foot of sand, thus made dry, will serve as under-draining (or absorbent) material for all the upper earth, and may receive and continually pass off all the surplus rain-water that may therefore fall on the surface. Such is the fortunate natural condition of the best low-ground farmers on the lower James river before adverted to—best not so much for their great natural fertility, and good constitution, valuable as these are, as because they are thus under-drained by nature. The upper layer of the sand-bed under these lands, is always dry for some feet down. This dry layer, though some twelve feet or more below earth of clayey texture, is the true cause of the usual dry condition of those soils. And although the wells reach water in abundance a few feet lower in the sand, that water has no upward pressure, and cannot damage the beds of earth and soil. In these cases the natural means for the lateral discharge of water from the sand-bed, (in its high level,) are greater than needed from the quantity supplied. Therefore, the higher layer of the sand-bed is kept free from water, and always ready to receive, and convey still lower, any new and temporary supply from the upper beds and soil. If, on the contrary, the average supply of water had ever so little exceeded the means for average discharge; this upper layer of sand would have been always over-gorged with water, and the surface would suffer with wetness, as do the lowlands on the Pamunkey river, and all this great low-land region here under consideration.

Though wet earth is perfectly impervious to the entrance and passage by percolation of more water from the surface, (pressing downward, and by its own weight only,) I doubt whether any earth in the tide-water region is impervious, if previously drained, at least, none such has occurred in my extensive draining labors and experience. Much soil is made more impervious by having been ploughed or tilled when wet. This operation approaches, in effect, to what is called "puddling," or kneading wet clay, or loam, which is done for the purpose of closing all the pores, and making the earth impervious to water, such, in the greatest perfection, is the working of clay for pottery, and in less degree, for making tiles and bricks. Hence it is that deep and proper ploughing, introduced on land before often ploughed wet, and always shallow, has well-known draining effect, because the "puddled" and impervious pan is broken up, and the rain-water then permitted to sink through the natural fissures of the lower earth.

XIV. Some of the Farming Practices of the Low-Lands—Defects and Proposed Improvements—Rotations of Crops—Pea-Fallow, and Narrow, and Broad-Bed Tillage

In my hasty journeys through this country, though diligently engaged in taking general and superficial views, I had but little opportunity to observe extensively, or to examine the details of farming. Therefore, nothing like minute description will be attempted, and only general remarks offered on some of the most striking advantages and capabilities of the lands, and defects of their culture.

The early settlements were made on the dryest places, and on most of these, tillage has been continued almost incessantly, from the first settlement to recent, or to the present time. Under such treatment, and with the necessary, or at least certain, and frequent wet ploughing of land always too wet in winter and spring, it is surprising that fields so abused have not become poorer than they are. I saw none that were so unproductive as the poorest fields of the higher tide-water counties in Virginia, which have not been marled or limed, or as all such most exhausted lands were before marling and liming were begun; and wherever the formerly most reduced lands have latterly been occupied by good farmers, they have been greatly and rapidly improved. Sundry such cases are to be seen, and especially in Perquimons county. The oldest tilled lands are here referred to. The greatest recent improvements have been the bringing under culture the extensive firm swamp lands which have lost little or nothing of their original and great fertility.

On the farms of Messrs. Francis Nixon[6] and J. T. Granberry, I saw the manner in which these swamp lands are brought under cultivation. The large trees, not needed for timber or fuel are belted and so killed. The heavy forest growth is mostly of gum, poplar, oak, and large swamp pine, (used for naval timber,) some of the latter of great size. The smaller growth is cut down more than once, and mostly dies. The land is used for grazing, until the roots are enough rotted to permit ditching and ploughing. This will be in about five years after the belting of the trees. Then the principal ditches are dug on the plan before described, and as they are to remain, except that when encountering a very large tree in the spring, (or before) the smaller ditches are also cut, and the land ploughed and planted in corn.

There is no marl in this region, except at a few exposures of small extent—or rather, the marl lies too deep to be accessible. Some marl has been excavated and used in Princess Anne. There are extensive Indian banks of muscle-shells on the borders of the Chowan river; and in Currituck, an Indian bank of oyster-shells stretches almost continuously for forty miles along the eastern margin of the sound. There are also in shallow waters of the sounds immense beds of oyster-shells, in places where the animals lived, before being killed by the water becoming fresh. So there is no want of material for calcareous manuring, independent of the supplies of lime and of shells, available from the waters of the Chesapeake. Some of the Indian bank-shells have been used, and more lime, and to good effect, as reported, and better than ought to be expected on land not well drained. Next after supplying the first necessity, draining, liming would be especially beneficial to all the lands of this region. Besides other reasons, and benefits to be gained, lime applied on the new and rich lands would serve the better to preserve their fertility; and, on the poorest lands, it will enable the most speedy and complete acquiring of fertility. But the best effects from lime can be counted on only on land previously well drained, or not needing draining.

The great crop of the North Carolina counties is corn. Next to this, and especially in Perquimons, is wheat. These two are the only great crops for market. The lands generally, if not suffering much from wetness, produce corn well. On the new clearings of firm swamp lands, ditched well on the ordinary plan, fifty bushels to the acre may be made. I saw a small field of wheat in Princess Anne, (where that crop is rarely attempted, and never on large spaces,) and several large fields in Perquimons, that in growth equalled what I had just before seen on some of the best lands on James river. There is no better land for the growth of wheat than the soils of close and medium texture here. But the imperfect draining of the fields must prevent the product and quality of the grain being in proportion to the growth of straw; and, moreover, the humid air of the whole region, (caused mainly by the general want of draining,) makes the wheat crops more liable to be diseased with rust.

It was with much surprise, some years ago, that I heard that the best and largest crops of wheat in Perquimons, and in some other parts of this region, were still reaped by the sickle, or reap-hook. This primitive mode of harvesting, which is older than the days of the patriarch Jacob, and which formerly was general in the United States, as it still is in Europe, I had supposed had everywhere, in this country, been substituted by the more expeditious scythe and cradle, if not by the still more modern reaping machine. And when first informed of the ancient usage remaining here, I had erroneously inferred that it indicated very slow progress in agricultural knowledge and improvement. But, when on my visit, while finding this practice far more extended than my previous idea of it, I also heard reasons in its defence, which seem to maintain its good economy. Neither is this practice confined to small crops. The best farmers and largest wheat growers, who sometimes make crops of more than five thousand bushels, reap them with the sickle. I knew that, by this mode, there may be avoided much of the great waste of wheat that is usually made by cradling; but had supposed that the slower operation of the sickle, and the high prices of harvest labor, and the scarcity of laborers at any price, had caused this implement to be abandoned everywhere in the United States, except for spots of rank and tangled wheat, or on steep hill-sides. Even for these latter circumstances, in which the proper use of the sickle would always be preferable, I have not been able to resort to it, and they would make awkward and very slow work. But in this district, the regular use of the sickle has never been abandoned, or suspended, and, therefore, the laborers are expert; and in a heavy growth of wheat, a good hand, with the sickle, can reap more wheat than he could, on the same ground, with the cradle, besides saving much more of what is cut down. The difference of waste will more than pay the difference of amount of labor and greater expense through a crop. Further, by using the sickle, and cutting as high as can be to save the wheat, most of the tall straw is left standing as stubble in the field, which is the cheapest, and as good a disposition as can be made of it for manuring the land, and makes a vast saving of labor in the hauling, thrashing, and stacking, compared to the handling of all the greater length of straw, as usually cut by the scythe and cradle, or by a reaping machine. But, if admitting that the reaping of a heavy growth of wheat by the sickle is preferable, still, in a merely agricultural country it could not be done, for want of the additional force of hands which this process certainly requires. But in the peculiar condition of this district, this objection does not apply. There are so great a number of laborers employed in cutting timber, and in the fisheries, that there are enough, for the higher wages of harvest, to supply the then extraordinary demand for labor on every wheat farm.

Light growths of wheat are often reaped by cradling; and where both modes are thus in use together, the more extensive use of the sickle is, in itself, good evidence of the heavy crops of wheat raised here by good farmers, and on good land. Perquimons has generally stiff soil, and is much the best wheat producing part of this region, (not including the Roanoke bottom.) In Pasquotank the lands are also good, but lighter, and better for corn. Those of Camden and Currituck are inferior in value of soil and agricultural products, and also as to improved farming. Currituck, especially, is so intersected by navigable waters, and bounded by the sound and the ocean, that the labors or pursuits of the residents are all more or less connected with the water and its products.

Except corn and wheat, there is scarcely a crop of large culture raised for market in the North Carolina counties. Cotton, which is so universally and extensively cultivated in the nearest higher [westerly] counties of North Carolina, and even to some extent in those of Virginia, is not attempted here, as a crop, for market. The general prevalence of wet soil is a sufficient cause for the absence of this crop. Oats, and especially hay, would be good crops for this humid climate and the fodder and shucks of corn serving in the place of hay, as everywhere in our corn-growing country. Yet vessel loads of coarse and mean hay, from the northern States, are continually brought here for the use of the towns, and for the teams of the lumberers working in the swamp forests. There is no better country for grass east of the mountains. On the farm of Edward H. Herbert, Esq., Princess Anne, on a large space, and elsewhere in Norfolk county, in small lots, I saw dry meadows of orchard grass and clover that would have been deemed good for the best grass districts, and which well attested the value and good drainage of the fields on which these crops grew.

In the counties in Virginia, where near to Norfolk, and with easy access by the regular steamers to the great northern cities, "truck" farming, or cultivating green vegetables and fruit for sale, is the sole business on sundry of the most valuable farms, and it enters more or less into the culture of many others. This business is carried on exclusively, largely, and successfully in Norfolk county, on river farms only, and within a few miles of the wharves. The limitation to these localities is compelled, first, from ready access to the steam-vessels, and also because only in close neighborhood to a considerable town can numerous laborers be hired whenever wanted for gathering vegetables and fruits, which requires, rarely, many hands, and for short and uncertain lengths of time. This kind of farming is the most perfect in all its operations, the most costly in money and labor, and the most productive, not only in the gross returns, but in net profit; and, as reported, it is the only kind of farming in the county that is well conducted. It is not long since this "truck farming" has been established on any thing like its present important position; and in that time, the lands near Norfolk and Portsmouth, suitable for this business, and so used, have increased, in market value and price, from 500 to 1,000 per cent.

This market gardening, or "truck-farming," in these large operations, is a peculiar and remarkable branch of agriculture, which well deserves thorough examination, and more full report, than this slight notice. It is an important and admirable kind of what in England is called "high farming," requiring great expenses, but returning so much the larger profits. Compared to nearly all other farming of the surrounding and neighboring lands, the "truck" farms appear like an oasis in a desert. The quantity and the cost of manures applied on these farms, and the magnitude of other expenses, and still more the great returns of the products and profits would be astonishing, if not appear incredible, to a stranger. Still, this business is the most laborious employment of a proprietor, exacting unceasing attention, care, and anxiety, for every hour. Nothing short of untiring industry, care, and also good judgment, can attain success and its great rewards; and even all these will not always prevent heavy losses. The business is precarious, and subject to great changes and hazards, and losses, which no industry or care can guard against. A single severe frost, at an unusual time, may destroy a valuable crop, for which all the expenses have been incurred, except for the gathering and shipping; and which loss may reduce the net receipts expected by thousands of dollars.

In the Virginia counties the required drainage and culture are of much easier executions than in Perquimons, and yet both are more negligently performed. No where does there seem to be any regular system of rotation of crops. This essential part of good farming is neglected every where by poor and bad farmers. The most energetic and successful cultivators and improvers here have been so much occupied in the heavy labors of clearing and draining their new and rich swamp lands, that they had no opportunity to use any newly cleared lands, for which, for some years, regular rotation would not be required, and would even be improper. But this circumstance and the continued additions of new surface to the tilled land should not prevent the older and poorer land being kept under a proper rotation, or at least under a proper succession of crops. And the neglect is the more reprehensible and strange, inasmuch as the farmers of this region possess peculiar facilities for rotations in the pea-crop, and a climate admirably adapted to its growth. The limited territory on which both the pea and the wheat crop can grow well, (the one suiting so well to prepare for and aid the growth of the other,) I deem the most favored of all agricultural regions. Still more strange appeared to me the general neglect of peas as a manuring crop in this region, from some of the best farmers of which I obtained most of my early instruction as to this particular value of the pea crop. Yet this great means for improvement, on most farms, seems to be but little used or appreciated. It is true, that peas are planted, as a secondary crop, in every field of corn, and the returns are highly valued. But this pea-crop, except so much as is gathered for seed or for sale, is generally eaten on the ground by the hogs design[at]ed for slaughter, (greatly indeed for benefit in that respect,) so that very little of the crop, except the roots and stems, go to manure the land. I heard of no separate crops of broad-cast peas, (or "pea-fallow,") to prepare and manure for a succeeding wheat crop, the most valuable use to which the pea-crop can be applied. It is a frequent practice here for the land in corn (and secondary peas) not to be sown in wheat the autumn of the same year, (as is usual in lower Virginia,) but for the field to remain until the autumn of the following year, and then to be sown in wheat. This practice leaves the field idle and useless all the spring and summer, when in that time it might be sown in peas, and bring a manuring and cleansing crop to precede the wheat, without any loss of time or of land. This is a regular part of my own established rotation, and, as supposed, its best feature, though my more northern position and shorter warm season render the pea-crop much less productive and beneficial than in this more favored region. Still more than this omission, another is common and as reprehensible. Wheat, in some cases, is made to follow wheat in two successive years. If, in such cases, there was merely interposed between these two crops a broad-cast crop of peas, (for which there is plenty of growing time,) that addition only would serve to substitute a cleansing, enriching, and judicious succession of crops, for one that is inexcusable and abominable. Clover is made on most of the good farms of Perquimons, and used as a preparing (or fallow) crop for wheat. With the superior facilities for the best growth of peas, if I were farming in this region, I should much prefer pea-fallow to clover-fallow to precede wheat.

The reason offered for the total omission of pea-fallow is the great and engrossing tillage labors required for the great crops of corn, and also for the wheat harvest, both of which occur with and include the very time in which the land for broad-cast peas should be ploughed and sown. This is true, and a sufficient reason, if it is necessary to plant in corn as much land as the laboring force can cultivate. But it would be much better to secure the great benefit of a manuring pea-crop to precede wheat, by the (temporary) sacrifice of omitting to plant as much corn as would release enough labor for the additional pea-crop. This sacrifice was a necessary incident of my own change (in 1848) of the five-shift rotation, without pea-fallow, to the six shift, with one entire field under broad-cast peas. The fields of both corn and wheat, by this change, were reduced, severally, to five-sixths of their previous size. Yet the wheat crops have continued since to increase, on the general average, and to exceed more and more the previous entire product, and so have the corn-crops, except in the first year only of the reduced extent of cultivation. Yet the advantages of manuring by the pea-crop in my localities and climate are very inferior to those of this region of North Carolina.

While the many firm swamps remained generally under forest, these lands afforded excellent "range" for live-stock, or a great quantity of food, especially for cattle and hogs. But this benefit, (if it was one,) has almost ceased in the best cultivated parts of the counties on the sound. Such is Durant's Neck, the narrow and level and very low peninsula which stretches for twelve miles between Perquimons and Little river to Albemarle sound. This land, being but a few miles wide anywhere, and bounded nearly around by these deep waters, is in consequence better drained, naturally, than the interior lands, and is very productive. Nearly all this "neck" is enclosed, and an unusually large proportion of the whole is under tillage, and there is scarcely any unenclosed forest or waste land for ranging live stock, and none that affords any grazing profit. I know no place where it would be so profitable to dispense with fences, as is done, by mutual agreement, by the proprietors of three several neighborhoods in Prince George county, Virginia, each including from 4,000 to 8,000 acres, and making from 10 to 15 farms and separate properties.[7] If the cultivators of Durant's Neck would do the like they would only have to make one short and straight fence to enclose all their fine farms, and save all the cost of their present useless fences. Yet every farm and field is now separately fenced in, and some of the proprietors have no materials for fencing, and buy, and transport from a distance, all their rails. This locality, more strongly than any other, shows the absurdity of our fence laws, and also the strength and long vitality of old habits and opinions, when the former good reasons for them have long ceased to exist. If the live-stock were reduced in numbers to one-fourth, and these were well kept, by being herded within the farms, one cow would yield as much profit as four do now. And when the grazing stocks were so lessened in number there would be much surplus grass left to manure the pasture on other land. While three-fourths of all the present fencing might be dispensed with, the other fourth would serve to make a sufficient pasture enclosure for every farm. For nothing in geometry, is more clearly demonstrable than the proposition that it will require greatly less length of enclosure to fence in the cattle of any well cleared and settled section of the country, than to fence in all the fields and crops to protect them from the cattle if left at large. One-fourth of the present fencing in Durant's Neck would suffice not only to make on every farm a proper pasture enclosure, but also the general and joint barrier fence against all other people's stock. Most of the farmers in Prince George, who have joined in these arrangements, if not situated on the border, have no fence except the pens in which to confine the animals at night. But this extreme course is not true economy.

In Princess Anne, there still remains so much uncleared and swamp land, that the leaving cattle to range at large is deemed very profitable to the owners, and perhaps, in general, it is there, more an offset to the expenses of fences, under our fence law, than in any other county of lower Virginia. The open swamps bear reeds in great quantity, which afford abundant and excellent food for cattle that have become wild, and are made use of when wanted for beef, only by being hunted and shot. These wild cattle would be very profitable to their owners, as they require neither food nor attention, except that they are as much at the disposal of every other person who may be inclined to shoot and steal them.

It becomes a slight observer of a newly seen agricultural district of novel and peculiar character, to be diffident of his own opinions thereon, and more especially, when they are in opposition to those of the judicious and experienced resident farmers. One of such subjects I will mention, though without any view of urging the superior value of my opinions and practice, in this respect on my friends in this region, who unanimously and strongly protested against them at least for their lands. Their experience of facts, in contradiction, certainly deserves more to be respected, than my theoretical views as to this region, even though they have been sustained by the results of my own practice and experience elsewhere.

As stated before, the tillage generally, and on the best managed farms, is in narrow beds (five feet,) for corn, and the same size is preserved for wheat. The beds are reversed for every crop, both of corn and wheat. I will not here repeat my objections to this narrow bed tillage, nor my reasons for preferring (where any are necessary) beds of twenty-five or more feet in width. These views have been stated and argued at length in different former publications. (The latest and fullest articles on tillage in broad beds, and also on draining in general, are in "Essays and Notes on Agriculture, 1855.")[8] I will only say here, that all the reasons for preparing wide beds for low and flat lands generally, apply with greater force to the lands of this region, and especially in Perquimons, because they are of more regular level, and with fewer alternations of slight depressions and elevations, than any other low-lands within my knowledge. The best farmers here, with whom I have argued this question, object on various grounds to my broad-beds, but especially, because their frequent cross "hoe-furrows" are deemed indispensable, and if the broad and higher beds, and their deeper alleys were in use, the "hoe-furrows" would have to be made still deeper, and require more labor to dig, and to renew after every ploughing or horse-tillage, and be even inconvenient for the ploughs to cross. This objection would be valid, if indeed it would be necessary (with the broad-beds, and deeper alleys) to retain the hoe-furrows; but this necessity I doubt. For with so much higher beds and deeper alleys between them, on land scarcely varying from a level, or from a regular and gentle slope, I think that the deeper alleys would substitute the hoe-furrows, and render them superfluous, except where a cross depression of surface required a particular cross grip. In my own practice, on the Pamunkey flats, the surface is much more irregular, yet there are no grips kept across the beds, except along the cross depressions. If the inequalities of surface level were as rare as on the Perquimons lands, my cross grips would be fewer and less necessary than they are.

But if my plan of broad beds would suit this region, there might still be added thereto another improvement, which I commenced using in 1855, and which has been continued since on the Marlbourne farm, with increasing confidence and approval. Without taking time here to describe and recommend the operation in general on the different circumstances of my own farm and practice, I will merely apply the plan to the present existing divisions and ditches of the Perquimons lands.⁹ We will suppose that these present ditches are all necessary and proper to be retained, though such is not my opinion, if a different system of drainage were in use. Then suppose merely the change that each of the rectangular enclosed spaces of 150 feet wide, instead of being as now in thirty beds of five feet wide, was ploughed into six beds, each of twenty-five feet width. After two or three years of ploughing, and tillage, and gathering of these wide beds separately, they would be as high, and their intermediate alleys as deep as desirable. Then, instead of continuing to plough each bed separately, the first furrow should be cut alongside of the central alley, and turning the slice into it. This furrow should begin and end at 75 or 80 feet distance from the ends of the rectangular "slip," or at (or less than) the same distance of the central alley from the sides of the slip. Turning the plough at that distance, another furrow should be cut alongside, and throwing the slice to the first, thus making a "list" in the former central alley. So the ploughing would proceed around this first list, cutting across the ends as well as along the sides, and throwing every furrow-slice towards the centre of the ploughing. This ploughing, though flush, and cutting across the ends as well as along the beds, and with no regard paid to the alleys, would scarcely alter the outline of the previous surface, and would not lessen the height of the crowns of the beds or the depth of the alleys, except the central alley, which would in time be filled, and would not then be needed. The outside furrows would just reach the encircling ditches of the "slip," turning the depth and width of a furrow-slice from each at every repetition of such ploughing. One or two furrows run along each of the old alleys, after the flush ploughing, would clean them out and put the broad beds in their original shape, and they would be more thoroughly broken by this mode of ploughing. Every successive ploughing of the land to prepare for any crop should be done in like manner. The tendency and operation would be to raise the central part of each rectangular division so ploughed around, and to lower and slope the sides and ends, or margins, next to the surrounding ditches. After a few such ploughings the shallow tap ditches would be to the eye almost obliterated, or changed to mere ploughed alleys or grips. Yet, in fact, they might be deeper than before, and would certainly be more operative for surface drainage than before. The preserving and cleaning out of these "tap-ditches," instead of requiring spades and shovels, would therefore be as well done by the last finishing furrows of the plough. These ditches would no longer prevent any obstruction to the crossing of ploughs, or partially loaded carts. If desired, (and it might be even desirable in future time,) the corn-rows and their ploughing, in narrow beds, might be directed across the beds and tap-ditches. Further, the end margins of the "slip" being equally depressed, and sloping to the edges of the larger leading ditches, these would be much more easily crossed by teams, and fewer and smaller bridges would be required. Thus, in the course of time, each separate "slip" would be converted to one broad bed of a 150 feet wide, and gently rounding surface, and 1,000 feet long, (the present dimensions of the separate divisions,) with sloped margins and ditches between deeper than before, yet presenting either little obstruction, or none, to the crossing of ploughs and teams.

Part Four

Valedictory

13 Toward the Managed Landscape
An Address on the Opposite Results of Exhausting and Fertilizing Systems of Agriculture (1852/1860)

The particular object of the address which will now be read, is to exhibit in full, and place in contrast, the opposite results on a country and people, of *exhausting and improving systems of Agriculture.*

In every feeling and opinion there is no more true and zealous Southerner than myself. I have long studied the domestic life and institutions, and social and moral condition of the people of the slave-holding States, and in every important respect, I may truly say, that I concur with, approve, and sympathize with yourselves on these subjects. Yet it is my present design and business not to treat of our many points of perfect agreement of opinion, but of the few of difference; not to speak of your laudable works, but your errors; and to apply to the planters of South Carolina, censure where deserved as readily as I would applaud them in other respects, which have no relation to my present general subject. Even in the general system of southern agriculture, in which there is so much to condemn, I cannot but admire the energy and intelligence exercised by the cultivators to attain the object usually sought—which is to draw from the land the greatest *immediate* production and profit. If their object were instead, as it ought to be, the greatest *continued* products and profits, and that object were pursued with as much ability, the people of South Carolina would soon stand in as exalted a position of agricultural success, as now and heretofore, for social and moral qualities, as men and citizens. Even for the few years which have passed since I investigated and reported upon your abundant resources for fertilization, and urged their use,[1] if these means had been properly applied, already the agricultural production of half the arable lands of the State might have been increased full fifty per cent. I may dare to express this opinion, inasmuch as on a newly purchased farm, I have myself more than tripled that amount of increase by the means recommended, and within the same short time since uttering the precepts for the like improvement here.

The great error of southern agriculture is the general practice of exhausting culture—the almost universal deterioration of the productive power of the soil—which power is the main and essential foundation of all agricultural wealth. The merchant, or manufacturer, who was using (without replacing) any part of his capital to swell his yearly income—or the ship-owner, who used as profit all his receipts from freight, allowing nothing for repairs, or deterioration of capital—would be accounted by all as in the sure road to bankruptcy. The joint-stock company that should (in good faith, as many have done by designed fraud.) annually pay out something of what ought to be its reserved fund, or of its actual capital, to add so much to the dividends, would soon reach the point of being obliged to reduce the dividends below the original fair rate, and, in enough time, all the capital would be so absorbed. Yet this unprofitable procedure, which would be deemed the most marvelous folly in regard to any other kind of capital invested, is precisely that which is still generally pursued by the cultivators of the soil in all the cotton producing States, and which prevailed as generally, and much longer in my own country, and which, even now, is more usual there than the opposite course of fertilizing culture. The recuperative powers of nature are indeed continually operating, and to great effect, to repair the waste of fertility caused by the destructive industry of man, and but for this natural and imperfect remedy, all these Southern States, and most of the Northern likewise, would be already barren deserts, in which agricultural labours would be hopeless of reward, and civilized men could not exist.

Let me not be understood as extending censure to all southern agriculture, and charging this great defect as being universal. It is truly very general—but there are numerous exceptions, of which it is not my purpose to treat. My present business is with the errors and defects of southern agriculture, and not with its points of admitted excellence—as, for example, the elaborate system of rice culture, and, for other tillage, the very general and commendable attention paid to the collection of materials for putrescent manures. Nothing has appeared to me more remarkable in the agriculture of this region, than the close neighbourhood, (often, indeed, seen on the same property,) of the best husbandry, in some respects, and almost the worst in most others.

The great error of exhausting the fertility of the soil is not peculiar to cotton culture, or to the Southern States. It belongs, from necessity, to the agriculture of every newly settled country, and especially where the land before being brought under tillage, was in the forest state. When first settled upon, forest land costs almost nothing, and labour is scarce and dear. Even if labour is more abundant, it still will be long before enough land can be cleared to allow changes of culture and rest to the fields; and for some years after each new clearing, it would be even beneficial to continue the tillage of corn, tobacco or cotton, so as effectually to kill all remains of the forest growth. But as soon as enough land can be brought under culture, and has been put in clean condition, so as to allow space for change of crops and due respite from continued tillage, the previous exhausting course will no longer be best even for early profit. Even in a new country, while land is yet fertile, it is cheaper to preserve that fertility from any exhaustion, than it is to reduce it considerably. And in an older agricultural country, like South Carolina, having abundant resources in marl and lime for improving fertility, it would be much cheaper, and more profitable, to improve an acre of before exhausted land, than it is to clear and bring under culture an acre of ordinary land from the forest state, allowing that both pieces are to be brought to the same power and rate of production.

New settlers are not censurable for beginning this exhausting culture. But they and their successors are not the less condemnable for continuing it after the circumstances which justified it have ceased. The system was first begun in Eastern Virginia, because it was the first settled part of the present United States, and it continued to prevail almost universally, until since the course of my adult life began, and only has partially ceased since, because the country was nearly reduced to barrenness and the proprietors to ruin. From this erroneous policy, so long pursued in Virginia, and the manifest and well known disastrous results in the general and seemingly desperate sterility of the older settled portion of the State, the younger Southern States might have taken warning, and have learned to profit by the woeful and costly experience of others. But it seems that every agricultural community must and will run the same race of exhausting culture, and impoverishment of land and its cultivators, before being convinced of the propriety of commencing an opposite course—after the best means and facilities for making that beneficial change have been greatly impaired by the lapse of time, and progress of waste of fertility—if, indeed, these means are not then irretrievably forfeited.

If, at this time, the work of improvement, with the aid of marl and lime, were properly begun and prosecuted, there would be found here incalculable advantages over those of the pioneers in the like work in Virginia. These advantages would be—first, a tenfold better supply of far richer and cheaper marl than is found in Virginia; second, much more remaining organic matter, or original fertility of the soil, as yet unexhausted; third, full information to be obtained of the operations and opinions of thousands of experienced and successful marlers to refer to, of which advantage there was almost nothing existing thirty years ago. In South Carolina more marling could now be done in a year, and in a proper manner, than was done in Virginia for the first twenty years; and, though judging merely by analogy, I infer that the benefit would be fully as great in this region as in my own.

And now I will state, from unquestionable official documents, something of what has been effected in Virginia, not merely in cases of particular farms, and those entirely marled, which might show tripled or quadrupled products and market returns, and tenfold *intrinsic* value, compared to their former low condition, but cases showing the bearing of the comparatively few marled and limed farms on the aggregate assessed value of all lands in lower Virginia, and upon the receipts of land tax from the same, although not one-twentieth part of the whole tide-water district has yet been improved in fertility, or is in the least degree better (and, probably, the great remainder is much poorer) than when the marling of other lands first began to raise the general average of assessed values throughout this whole district.

It appears, from the latest state assessment of lands in Virginia, for 1850, that the actual increase of value in the tidewater district only, since 1838, the previous assessment, was more than seventeen millions of dollars. On this valuation, and at the same rate of taxation, there is more than $17,000 increase of land tax alone accruing annually to the state treasury. It is obvious that any increased value of lands, caused by their increased production, would necessarily require an increase of labour and of farming stock, and would produce proportional increase of general wealth of the improvers, and would add other receipts from taxes in proportion—all serving still more to augment the public revenue.

The recent addition to the aggregate value of lands in Eastern Virginia, is admitted to be the effect of agricultural improvements; and that more than all the nett increase is due to marling and liming only, would be equally evident, if I could here adduce the proofs, as I have done elsewhere.[2] Further; though 1838 was the date of the earliest assessment made after marling and liming had begun to increase aggregate production and value of lands, it is an unquestionable fact that the general impoverishment had been greater, and values much lower, about 1828. And if this earlier time and greatest depression had been marked by an assessment then made, the full increased value of lands from that time, would have appeared at least $30,000,000 in 1850, instead of seventeen and a quarter millions, counting from the already partially advanced improvement and enhanced values of 1838. However, even if these, my deductions and estimates, go for nothing, there will still remain the proof, by official documents, of the actual increase of the value of lands in twelve years, of seventeen and a quarter millions, or nearly one and a half millions yearly.

Now, bear in mind that these are not the results of the improving of all the tide-water region, nor all of its much smaller arable portion, but, probably, of not more than one-twentieth of the cultivated land. All the remainder, if uncultivated, is stationary; and if cultivated, is generally in a continued course of exhaustion; and the small quantity of enriched land had first to make up for all deficiencies of the impoverished, and lessenings of production throughout the whole tide-water district, and after all such deductions, still exhibited a clear surplus of seventeen and a quarter millions of increased aggregate value. This is the result of but the beginning, and a very recent beginning of measures for improvement, executed in every case imperfectly, often injudiciously, and sometimes injuriously, altogether on less than one-twentieth of the space on which calcareous manures are available. The great omitted space will hereafter be fertilized in the same manner. Then the actual increase of value of lands, founded on increased production, will be counted by hundreds of millions of dollars. And this anticipated enormous amount of fertility and capital to be created, might have been even now in possession, if our improvements by calcareous manures had been begun thirty years earlier, instead of there having been continued through all that time, the progress of wasting and destroying the remaining powers of the soil. South Carolina began exhausting culture much later, and is now full fifty years less advanced towards the lowest depth of that descent which we had nearly completed. If that future of fifty years of continued exhaustion could now be cut off, and the improvement of lower South Carolina by calcareous manures could be at once begun and continued, the loss of at least one hundred millions of dollars of now remaining value would be saved, and a gain of three hundred millions from improvement would be reached sooner by the same fifty years.

This would be better, by all the great value, than even the following out precisely the first sinking and now rising course of lower Virginia. In that region, the cultivators waited until the fertility of the land had so nearly expired, that it was supposed to be in *articulo mortis*—at the last gasp—before the work of resuscitation was begun.

The comparative results of the opposite systems of improving and exhausting cultivation may be thus illustrated. Suppose a certain investment of capital will yield twenty per cent of present annual interest, or nett products, and two persons invest equal amounts in the business. The more provident one draws and spends but fifteen per cent annually of his income, and leaves the remaining five per cent to accumulate and to be added to his interest bearing capital. The other proprietor draws each year, and spends all of the certain and annual average returns of his capital, and five per cent more of the capital stock itself. He reasons (may I say it?) like many cotton planters, and infers that so small a detraction from his capital will do no harm, as he will have so much the more of quick returns for immediate use or re-investment. In less than twenty years, one of these individuals will have doubled his original capital, and also his twenty per cent income, and the other will have exhausted his entire fund.

But it may be said, (as alleged in regard to the squanderers of fertility,) that as the latter person had received so much more of annual returns at first, he might have re-invested and thus have retained his over-draughts of annual products. If a planter—and, of course, his over-draughts had been from the fertility of his land—he might have bought another plantation, to work and to wear out in like manner. But even if so, wherein would be the gain? He would have had the disadvantages of a change of investment, of removal, and making a new settlement. But where one man would so save and re-invest his over-draughts from his capital, two others would use, or, perhaps, spend theirs, as if so much actual clear profit or permanent income. When the land is utterly worn out, and the total capital of fertility wasted, (or the small remnant is incapable of paying the expenses of farther cultivation,) it will most generally be found that the channels into which the early full streams of income flowed, are then as dry as the sources.

I do not mean that it necessarily follows that the planter who exhausts his land also lessens his general wealth. Would that it were so. For, then, such certain and immediate retribution would speedily stop the whole course of wrong doing and prevent all the consequent evils. It may be rarely, and it might be never the case, that the exhauster of land becomes absolutely poorer during the operation. He will have helped to impoverish his country, and to ruin it finally, (by the same general policy being continued,) he will have destroyed as much of God's bounties as the wasted fertility, if remaining, would have supplied forever, and as many human beings as those supplies would have supported, will be prevented from existing. And yet the mighty destroyer may have increased his own wealth. Nevertheless, he does not escape his own, and even the largest share of the general loss he has caused. While thus destroying, say $20,000 worth of fertility, the planter, by the exercise of industry, economy and talent in other departments of his business, or from other resources, may have grown richer by $10,000. But if, as I believe is always true, it is as cheap and profitable to save as to waste fertility, in the whole term of culture, then the planter, in this case, might have gained in all $30,000 of capital, if he had saved, instead of wasting, the original productive power of his land.

Even if admitting the common fallacy which prevails in every newly settled country, that is profitable to each individual cultivator to wear out his land, still, by his doing so, and all his fellow proprietors doing the like, while each one might be adding to his individual wealth, the joint labours of all would be exhausting common stock of wealth, and greatly impairing the common welfare and interest of all. The average life of a man is long enough to reduce the fertility of his cultivated land to one-half, or less. Thus, one generation of exhausting cultivators, if working together, would reduce their country to one-half of its former production, and, in proportion, would be reduced the general income, wealth and means of living, population and the products of taxation, and, in time, would as much decline the measure of moral, intellectual and social advantages, the political power and military strength of the commonwealth. The destructive operations of the exhausting cultivator have most important influence far beyond his own lands and his own personal interests. He reduces the wealth and population of his country and the world, and obstructs the progress and benefits of education, the social virtues, and even moral and religious culture. For upon the productions of the earth depends more or less the measure to be obtained, by the people of any country, of these and all other blessings which a community can enjoy. There is, however, one very numerous class of exceptions to this general rule, which is, when an agricultural people, or interest, is tributary to some other people or interest, whether foreign or at home. Such exceptions are presented in different modes, by the agriculture of Cuba being tributary to Spain, of many other countries to their own despotic and oppressive home governments; and of the southern states of this confederacy, to greater or less extent, to different pauper and plundering interests of the northern states, which, through legislative enactments, have been mainly fostered and supported by levying tribute upon southern agriculture and industry.

The reason why such woeful results of impoverishment of lands, as have been stated, are not seen to follow the causes, and speedily, is that the causes are not all in action at once and in equal progress. The labours of exhausting culture, also, are necessarily suspended, as each of the cultivators' fields is successively worn out. And when tillage so ceases, and any space is thus left at rest, Nature immediately goes to work to recruit and replace as much as possible of the wasted fertility, until another destroyer, after many years, shall return again to waste, and in much shorter time than before, the smaller stock of fertility so renewed. Thus, the whole territory so scourged, is not destroyed at one operation. But though these changes and partial recoveries are continually, to some extent, counteracting the labours for destruction, still the latter work is in general progress. It may require (as it did in my native region) more than two hundred years from the first settlement, to reach the lowest degradation. But that final result is not the less certainly to be produced by the continued action of the causes. I have witnessed at home, nearly the last stage of decline. But I have also witnessed, subsequently, and over large spaces, more than the complete resuscitation of the land, and great improvement in almost every respect, not only to individual, but to public interests; not only in regard to fertility and wealth, but also in mental, moral and social improvement.

Inasmuch as my remarks would seem to ascribe the most exhausting system of cultivation especially to the slave-holding states, the enemies of the institution of slavery might cite my opinions, if without the explanation which will now be offered, as indicating that slave labour and exhausting tillage were necessarily connected as cause and effect. I readily admit that our slave labour has served greatly to facilitate our exhausting cultivation; but only because it is a great facility—far superior to any found in the non-slave-holding states—for all agricultural operations. Of course, if our operations are exhausting of fertility, then certainly our command of cheaper and more abundant labour enables us to do the work of exhaustion, as well as all other work, more rapidly and effectually. But if directed to improving, instead of destroying fertility, then this great and valuable aid of slave-labour will as much more advance improvement, as it has generally heretofore advanced exhaustion. The enunciation of this proposition is perhaps enough. But if any, from prejudice, should deny or doubt its truth, they may see the practical proofs on all the most improved and profitable farms of Lower and Middle Virginia. On the lands of our best improvers and farmers, such as Richard Sampson, Hill Carter, John A. Selden, William B. Harrison, Willoughby Newton, and many others, slave-labour is used not only exclusively and in larger than usual proportion, (because more required on very productive land,) but is deemed indispensable to the greatest profits, and operating to produce more increase of fertility, and more agricultural profit, than can be exhibited from any purely agricultural labours and capital north of Mason and Dixon's line.

There is another stronger reason for the greater exhausting effects of Southern agriculture, and, therefore, of tillage by slave-labour. The great crops of all the slave-holding States, and especially of the more Southern—corn, tobacco, and cotton—are all tilled crops. The frequent turning and loosening of the earth by the plow and hoe—and far more, when continued without intermission year after year—advance the decomposition and waste of all organic matter, and expose the soil of all but the most level surfaces to destructive washing by rains—and rains the more heavy and destructive in power, in proportion as approaching the South. The Northern farmer is guarded from the worst of these results, not because he uses free-labour, but because his labour is so scarce and dear that he uses as little as possible for his purposes. Besides this consideration, his climate is more suitable to grass than to grain, and his other large crops are much more generally broad-cast than tilled. These are sufficient causes why, in general, the culture of land in the Northern States should be less exhausting than in the Southern, without detracting anything from the superior advantages which we of the South enjoy, in the use of African slave-labour.

At the risk of uttering what may be deemed trite or superfluous to many of those who now honour me by their attention I beg leave to state concisely, the fundamental laws, as I conceive them to be, of supply and exhaustion of fertilizing matters to soils, and aliment to plants.

All vegetable growth is supported, for a small part, by the alimentary principles in the soil, (or by what we understand as its fertility,) and partly, and for much the larger portion, by matters supplied, either directly or indirectly, from the atmosphere. More than nine-tenths usually of the substance of every plant is composed of the same four elements, three of which, oxygen, nitrogen, and carbon, compose the whole atmosphere. The fourth, hydrogen, is one of the constituent parts of water; and, also, as a part of the dissolved water, hydrogen is always present in the atmosphere, and in great quantity. Thus, all these principal elements of plants are superabundant, and always surrounding every growing plant; and from the atmosphere, (or through water in the soil,) very much the larger portion of these joint supplies is furnished to plants; and so it is of each particular element, except nitrogen; much the smallest ingredient, and yet the richest and most important of all organic manuring substances, and of all plants. This, for the greater part, if not all its small share in plants, it seems is not generally derived even partially from the air, though so abundant therein, but from the soil, or from organic manures given to the soil.

But though bountiful nature has offered these chief alimentary principles and ingredients of vegetable growth in as inexhaustible profusion as the atmosphere itself, which they compose, still their availability and beneficial use for plants are limited, in some measure to man's labours and care to secure their benefits. Thus, for illustration, suppose the natural supplies of food for plants furnished by the atmosphere to be three-fourths of all received, and that one-fourth only of the growth of any crop is derived from the soil and its fertility. Still, a strict proportion between the amount of supplies from these two different sources, does not the less exist. If the cultivator's land, at any one time, from its natural or acquired fertility, affords to the growing crop alimentary principles of value to be designated as five, there will be added thereto other alimentary parts, equal to fifteen in value, from the atmosphere. The crop will be made up of, and will contain, the whole twenty parts, of which five only were derived from, and served to reduce, by so much, the fertility of the soil. These proportions are stated merely for illustration, and, of course, are inaccurate. But the theory or principle is correct; and the law of fertilization and exhaustion, thence deduced, is as certainly sound.

Then, upon these premises, there is taken from the land, for the support of the crop, but one-fourth of the aliment derived from all sources for that purpose. And, if no other causes of destruction of fertility were in operation, one green or manuring crop, (wholly given to the land, and wholly used as manure,) would supply to the field as much of alimentary or fertilizing matter as would be drawn thence by three other crops, removed for consumption or sale. But in practice there *are* usually at work important agencies for destruction of fertility, besides the mere supply of aliment to growing crops. Such agencies are, the washing off of soluble parts, and even the soil itself, by heavy rains, the hastening of decomposition and waste of organic matter, by frequent tillage processes and changes of exposure—

and plowing or other working of land when too wet, either from rain or want of drainage. Also, a cover of weeds left to rot on the surface, or any crop plowed under, green or dry as manure, is subject to more or less waste of its alimentary principles, in the course of the ensuing decomposition. Therefore it is nearer the facts, that two years' crops or culture, for market or removal, would require one year's growth of some manuring crop to replace and to maintain undiminished, or increasing the productive power of the field. The poorest and also the cheapest of such manuring crops will be the natural or "volunteer" growth of weeds on land left uncultivated, and not grazed; and the best of all will be furnished in the whole product of a broad-cast sown and entire crop of your own most fertilizing and valuable field peas.

Thus, of each manuring crop, (as of all others,) or of the fertilizing matter thus given to the land, the cultivator has contributed but five parts from the land, or its previous manuring, and the atmosphere has supplied fifteen parts. If, then, the cultivator by still more increasing his own contributions, will give ten parts of alimentary matter to the land and crop, there will be added thereto from the atmosphere in the same two-fold proportion, or thirty parts, and the whole new productive power will be equal to forty. And if the soil is fitted by its natural constitution, or the artificial change induced by calcareous application, to fix and retain this double supply of organic matter, the land will not only be made, but will remain, as of much increased fertility, under the subsequent like course of receiving one year's product for manure, for every two other crops removed. But, on the other hand, if more exhausting culture had been allowed, instead of either increased or maintained production—or if the crops take away more organic matter than nature's three-fold contributions will replace—then a downward progress must begin, and will proceed, whether slowly or quickly, to extreme poverty of the land, its profitless cultivation, and final abandonment. In this, the more usual case, the cultivator's contributions of aliment, (obtained from the soil,) are reduced from the former value, designated as five, first to four, and next successively to three, two, and finally less than one; and nature keeps equal pace in reducing her proportional supplies, from fifteen, first to twelve, and so on to nine and six, and less than three parts. So the strongest inducement is offered to enrich, rather than exhaust the soil. For whatever amount of fertility the cultivator shall bestow, or whatever abstraction from a previous rate of supply he shall make, either the gain or the loss will be tripled in the account of supplies from the atmosphere, furnished or withheld by nature.

In another and more practical point of view, the loss incurred by exhausting culture may be plainly exhibited. According to my views, (elsewhere fully stated,)[3] soils supposed to be properly constituted as to mineral ingredients, do not demand for the maintaining and increasing of their rate of production, more than the resting or the growth of two years in every five, mainly to be left on the land as manure. These are the proportions of the five-field rotation, now extensively used on the most improving parts of Virginia. And one of these two years the field is grazed, so that parts of its growth of grass is consumed, instead of remaining on the field for manure. To meet the same demands, the more Southern planter might leave his field to be covered by its growth of weeds, or natural grasses, one year, (and also to be grazed,) and a broad-cast crop of pea-vines to be plowed under in another for every three crops of grain and cotton. But the ready answer to this, (and I have heard it many times,) is, "What! lose two crops in every five years? I cannot afford to lose even one." It may be that the planter is so diligent and careful in collecting materials for prepared manure, that he can extend a thin and poor application, and in the drills only, over nearly half his cotton field; and perhaps he persuades himself that this application will obviate the necessity for the rest and manuring crops to the land. The result will not fulfil this expectation. But even if it could, the manuring thus given directly by the labour of the planter, is more costly than if he would allow time and opportunity for nature to help to manure for him—whether alone, or still better if aided by preparing for and sowing the native pea, to the production of which your climate is so eminently favourable. All the accumulations of leaves raked from the poor pine forest, with the slight additional value which may be derived from the otherwise profitless maintenance of poor cattle, will supply less of food to plants, and at greater cost, than would be furnished by an unmixed growth of peas, all left to serve as manure.

The native or Southern pea, (as it ought to be called,) of such general and extensive culture in this and other Southern States, is the most valuable of manuring crops, and also offers great and peculiar advantages as a rotation crop. The seeds, (in common with other peas and beans,) are more nutritious as food, for man and beast, than any of the cereal grains. The other parts of the plant furnish the best and most palatable provender for beasts. The crop may be so well made, in your climate, as secondary growth under corn, that it is never allowed to be a primary crop, or to have entire possession of the land. It will grow broad-cast, and either in that way, and still better if tilled, is an admirable cleansing growth. It is even better than clover as a preparing and manuring crop for wheat. In one or other of the various modes in which the pea-crop may be produced, it may be made to suit well in a rotation with any other crops. Though for a long time I had believed in some of the great advantages of the pea crop, and had even commenced its culture as a manuring crop, and on a large scale, it was not until I afterwards saw the culture, growth and uses in South Carolina, that I learned to estimate its value properly, and perhaps more fully than is done by any who, in this State avail themselves so largely of some of its benefits. Since then, I have made the crop a most important member of my rotation; and its culture, as a manuring crop has now become general in my neighbourhood, and is rapidly extending to more distant places. If all the advantages offered by this crop were fully appreciated and availed of, the possession of this plant in your climate would be one of the greatest agricultural blessings of this and more Southern States. For my individual share of this benefit, stinted as it is by our colder climate, I estimate it as adding, at least, one thousand bushels of wheat annually to my crop.

From this digression to a particular branch, I will now return to the general subject, of the neglect of rest and manuring crops, for land.

The incessant cultivator does not the less rest, and lose the use of his land, by refusing any cessation of tillage so long as he can avoid it. If such cultivators manure so abundantly that there is no general decline of production, then they do not come under my past remarks and censure. If there be any such, I will only say of their mode of maintaining fertility, that it is less effectual and more costly, than if aided and substituted in part by manuring crops and a judicious rotation of crops. But as to many other planters, who, whether slowly or rapidly, are certainly impoverishing their lands, they will, at some future period, be compelled to allow a greater proportion of time for the land to rest, and to greater disadvantage, and less profit, than if allowing regularly either one year in three or two in five. Suppose the land to yield cotton, (or sometimes corn,) continuously for thirty, or even forty years—or, with much manuring, sixty years. In such cases, it is true, there were as many crops obtained as the land was kept years for tillage. But after the first few years, the products were declining; and for the last five or ten years, on the general average, they scarcely paid more than the expenses of cultivation. The crops also suffered during the whole time the evils of a want of rotation, and the land of want of change of condition. At the close, the land must be turned out to rest, because manifestly, not worth longer cropping. This compelled cessation and rest will continue for twenty, thirty, or forty years, when the land will be again cleared of its second (or perhaps its third) growth of trees; and with this and other extra labors, will be again brought under continued tillage, to be again, and much more speedily, exhausted of its smaller recovered amount of productive power. In this manner, though at long intervals, more than the full proportion of rest, required by an improving system of rotation, is given to the land, and enforced by its exhaustion; and the manner is such as to make the least return of benefit for the greatest expense incurred for the respite of the land from cultivation.

My former engagement in South Carolina, and the then especial object of my investigations and labours, served to make me better acquainted with a large portion of your territory than any other as extensive elsewhere. From that acquaintance was derived the opinion, which I have since asserted and still maintain, that no other as extensive region, known to me, possesses half as great advantages and resources for agricultural improvements, or more needs the employment of these means. The proper and full use of your wonderfully abundant, rich and easily accessible marl, and the recent shells and other marine remains, offer the best principal and indispensable means of fertilization, and which are available for half your territory. Another great resource, and almost as much neglected, is presented in your great inland swamps, now only wide-spread seed-beds of disease, pestilence and death; and which, by drainage, with certainty and great profit, might be converted into dry fields of exuberant fertility. It is true, that existing legal obstacles oppose these extensive plans for drainage; but these difficulties might be removed by wise legislation, with great benefit to the interests of all concerned—and improvements might be permitted and invited which would render these now worthless and pestilential swamps as fruitful as the celebrated borders of the Po.

The draining of the inland swamps of rich alluvial soil, together with the general application of marl to these and also to the now cultivated higher ground, would go far to remove the long prevailing unhealthiness to which Lower South Carolina is subject, and which is the only important evil which is not entirely in the power of the inhabitants to remedy. I will not presume to say how far this great evil may be lessened by these works of industry and improvement. But, when so much of your country consists of low and wet swamp, and of partially wet, higher lands, and all easy to be drained, it does not seem over-sanguine to suppose, that, with such drainage and the general extension of the also sanitary operation of marling and liming, the country would be as much improved in healthiness, as in fertility. Such change to greater healthiness has been most marked in my own country, in the extensively marled neighbourhoods, even where there has been no considerable draining operation executed or required. This improvement of health, is ascribed by all who have experienced the beneficial change, mainly to the sanitary influence of the now calcareous soil.

Your extensive and rich river swamp lands offer another great object for improvement, and increase of agricultural profit and wealth. Even "sandy pine barrens," now unfit for tillage, or for any useful production, other than the magnificent pine forests which cover them, if made calcareous and put under Bermuda grass, (the curse of tillage lands so infested) would be made and valuable land for pasturage, as the equally barren chalk downs of England.

Your high lands are mostly level, or of gently undulating surface, and easy to till, and the soils generally well suited to your great staple crops, corn and cotton. The navigable rivers which pervade Lower South Carolina, in their number and character, present a remarkable geographical feature, as singular as it is valuable. The main canals required for extensive drainage of the inland swamps, would be so many additions to the existing navigable highways. So low are the intervening swamp lands, that nearly all the deep navigable rivers, might be connected by canals of level or nearly level water; and in that respect, Lower South Carolina might possess the peculiar facilities of Holland for extensive inland navigation. These connecting canals, by diverting some of the superfluous supply of fresh waters of some rivers, to others where it is deficient, might perhaps serve to extend greatly the present area of tide covered land, capable of being flooded for rice culture. If such canals, mainly for drainage, but serving also for navigation, were made to connect the Edisto with the Ashley, the Cooper and the Santee, there would be another incidental advantage as remarkable as it would be valuable. The excavation of the canals through the great swamps, (and certainly between those stretching from the Ashley nearly to the Santee,) would generally penetrate into marl of the richest quality, lying a few feet below the surface of the swamps. If duly appreciated, this rich calcarious earth, to be used as manure, would go far to reimburse the costs of the excavation; and if used for lime-burning, would furnish good lime, and at one-third of the price of that for which South Carolina has paid and continues to pay millions of dollars to the lime-burners of New England. This voluntary tribute, at least, which is one of so many unnecessarily paid by the South to the North, might be ended to the immediate and great profit of both the sellers and the buyers of the substituted lime, made of the abundant, cheap and excellent native material. The buying of Northern lime by South Carolina and Georgia, is as unprofitable and as absurd a procedure as the usage of importing Northern hay. But of these and of many similar things, we of the South have no right to blame any but ourselves. All of the commodities which we import from the Northern States, and which might be more cheaply provided at home, serve indeed to make up an enormous amount of annual tribute. But this part of our general burden is fairly and properly levied by northern enterprise and industry upon southern listlessness and indolence. Very different, however, is the case as to the far greater proportion of the general amount of tribute paid by southern to northern interests—from which we have no defence, because government induces and enforces the payment, by the legislative machinery of protecting duties and the indirect bounty system. But I am straying from my designed subject, the improvement of southern agriculture to its governmental and political oppression.

Putting aside all speculative and untried subjects and modes of improvement—and counting upon nothing more than the proper use of your calcareous manures and judicious tillage, and the early results of both—and supposing that your county should so benefitted only in the same degree as has been the small portion of mine already marled or limed—the most moderate estimate of the agricultural values so to be created would now appear to you to be so greatly exaggerated as to be altogether incredible. But however much I would desire to avoid the position of a discredited witness, I will not be restrained by that fear from stating general results, which are notorious in Virginia, and to sustain the truth of which, thousands of particular facts can be adduced. These results, susceptible of clear proof, or exhibited by official documents, are that thousands of farms have been doubled or tripled, and some quadrupled in production, and the general wealth of their proprietors as much increased—the assessed values of marled lands increased by many millions of dollars, while those of similar lands, not so treated, have continued to decline as all did before; and the treasury of the commonwealth is already benefitted by many thousands of dollars received annually from the counties containing these improved lands, and derived from them, while the revenue from lands of the neighbouring and before similar counties, is still decreasing.

So far, I have spoken as to benefits which have already occurred, and which are unquestionable, and which have been derived from resources and facilities for improvement not to be compared in amount and value with those of South Carolina. I have elsewhere estimated the possible future and full fruition of this system of improvement, in Lower Virginia only, at five hundred millions of dollars of increased pecuniary value of capital thereby to be created. The full employment of your much great[er] resources, which have been named but not estimated, would be so much more in addition.

But agricultural production and pecuniary values are not the only or the greatest gains; and though others rest upon opinion only, and are incapable of being measured, their existence and their value are not the less acknowledged by all judicious observers, in our country most improved in agricultural production by calcareous manures. The improvement of health has been mentioned; the improvement of economical and social habits, morals and refinement, and better education for the growing generation, have been sure consequences of greatly increased and enduring agricultural profits; and these moral results will hereafter be increased, in full proportion to the physical and industrial producing causes. Population, though a later effect, is already sensibly advanced by these agricultural causes. The strength, physical, intellectual and moral, as well as the wealth and revenue of the commonwealth of Virginia, will soon derive new and great increase from the growing improvement of that one and smallest of the great divisions of her territory, which was the poorest by natural constitution—still more the poorest by long exhausting tillage—its best population gone, or going away, and the remaining portion sinking into apathy and degradation, and having no hope left, except that which was almost universally entertained of fleeing from the ruined country, and renewing the like work of destruction on the fertile lands of the far west. Terms of reproach and contempt, (once not undeserved,) have been so long and so freely bestowed on this tide-water region of Virginia, and had become so fixed by use, that it will be long before they will cease to be deemed applicable; or before many persons who now know this region, only by the memory of former report, will learn that it is not altogether a land of galled and gullied slopes, or broomsedge-covered fields, over whose impoverished and dwindling population, indolence and malarious disease contend for mastery.

From these matters, referred to for proof or illustration, I return to my main subject, more immediately connected with, and more likely to be interesting to my auditors.

There is not one of the industrial classes of mankind, more estimable for private worth and social virtues, than the landholders and cultivators of the Southern States. With them, unbounded hospitality is so universal, that it is not a distinguishing virtue—and, in truth, this virtue has been carried to such excess, as to become a vicious tendency. Honourable, high-minded, kindly in feeling and action, both to neighbours and to strangers—ready to sacrifice self-interest for the public weal—such are ordinary qualities and characteristics of southern planters. Many of the most intelligent men of this generally intelligent class, are ready enough to accept and to apply to themselves and their fellow-planters, the name of "land-killers." But while thus admitting, or even assuming this term of jocose reproach, they have not deemed as censurable or injurious, their conduct on which this reproach was predicated. They have regarded their "land-killing" policy and practice merely as affecting their own personal and individual interests—and if judged by their continued action, they must believe that their interests are thereby best promoted. Their error, in regard to their own interests, great as may be, is incomparably less than the mistake as to other and general interests not being thus affected. As I have already admitted, individuals may acquire wealth by this system of impoverishing culture, though the amount of accumulation is still much abated by the attendant waste of fertility. But with the impoverishment of its soil, a country, a people, must necessarily and equally be impoverished. Individual planters may desert the fields they have exhausted in South Carolina, and find new and fertile lands to exhaust in Alabama. And when the like work of waste and desolation is completed in Alabama, the spoilers, (whether with or without retaining a portion of the spoils,) may still proceed to Texas or to California. But South Carolina and Alabama, must, nevertheless, suffer and pay the full penalty of all the impoverishment so produced. The people who remain to constitute these States respectively, as communities, are not spared one tittle of the enormous evils produced—not only those of their own destructive labours, but of all the like and previous labours of their fellow citizens and predecessors who had fled from the ruin which they had helped to produce. And these evils to the community and to posterity, greater than could be effected by the most powerful and malignant foreign enemies of any county, are the regular and deliberate work of benevolent and intelligent men, of worthy citizens, and true lovers of their country.

I will not pursue this uninviting theme to its end—that lowest depression which surely awaits every country and people subjected to the effects of the "land-killing" policy. The actual extent of the progress toward that end, throughout the Southern States, ought to be sufficiently appalling, to induce a thorough change of procedure and reformation of the agricultural system of the South.

In addition to all increase of the other benefits of agricultural improvement which have been cited—pecuniary, social, intellectual and moral—there would be an equal increase of political power, both at home and abroad, which at this and the near approaching time, would be especially important to the well being and the defence of the Southern States, and the preservation of their yet remaining rights, and always vital interests. If Virginia, South Carolina, and the other older slave-holding States, had never been reduced in productiveness, but, on the contrary, had been improved according to their capacity, they would have retained nearly all the population they have lost by emigration, and that retained population, with its increase, would have given them more than a doubled number of representatives in the Congress of the United States. This greater strength would have afforded abundant legislative safeguards against the plunderings and oppressions of tariffs to protect Northern interests—compromises (so-called) to swell Northern power—pension and bounty laws for the same purposes—and all such acts to the injury of the South, effected by the greater legislative strength of the now more powerful, and to us, the hostile and predatory States of the confederacy. Even after Virginia, with more than Esau-like fatuity, had sacrificed her magnificent north-western territory, which now constitutes five great and fertile States, (and a surplus to make, by legislative fraud, a large part of the sixth State[4] [Minnesota],) and all of which are now among the most hostile to the rights of the people of the South—if Virginia had merely retained and improved the fertility of her present reduced surface, her people would not have removed. Their descendants would now be south of the Ohio, ready and able to maintain the rights of the Southern States, instead of a large proportion, as now, serving to swell the numbers, and give efficient power to our most malignant enemies. The loss of both political and military strength, to Virginia and South Carolina, are not less than all other losses, the certain consequences of the impoverishment of their soil.

If it were possible that, for all lower South Carolina, the system of improvement could be directed by one mind and will, as much as the operations of any one great individual estate, the most magnificent results could be obtained with great and certain profit, and in a few years. Without any additional labour or capital, more than now possessed, for beginning the improvement—and with only the subsequent increase of means which would be supplied by the clear profits of the improvements as they became productive—most of the lands accessible to marl or lime could be covered by these manures in ten years. In twenty years from this day, all such lands could be thus improved, and, by that time, might yield double or tripled general products, and would exhibit a proportionally greater increase of value as capital. The new clear profits of this one great improvement would be enough in amount to effect all the practicable drainage of inland and river swamps in twenty years more. Or, in that additional time, the increased revenue of the State treasury, from these new sources only, would suffice to construct all the great works of drainage, which would be beyond the means of individual proprietors.

In all opinions expressed as to the value and effects of the agricultural improvements proposed for South Carolina, my data are the experienced and unquestionable results of like labours in Virginia. The legitimate deductions, and the only one for untried operations is that like causes will produce like effects in both these different localities. I cannot conceive any reason, founded on existing differences of climate, soil or subjects of culture, that can make calcareous manures less efficient, or less profitable, with you than with us. Nevertheless, I have learned from mere rumour, that in the small extension of their use, by new operators, which occurred here, there was no general and important benefit obtained. And such, I must infer, was the conclusion reached by nearly all the makers and observers of these trials, from the irresistible, though negative evidence (which only is before me,) that nothing considerable of such improvements, or of public notoriety, has been effected in latter years. In the absence of all particular information of the actual trials, their results and the accompanying circumstances, of course I cannot pretend or be expected to explain the causes of disappointment, which must be the general result, as it seems that marling has languished, if not ceased, in general, after a few faint efforts.[5] But I infer that the main and usual cause of supposed failure, or of inconsiderable benefit, has been the same prevailing bad practice, before denounced, of incessant, or, at least, much too frequent tillage, which does not permit the fields to receive and retain organic matter from their own growths especially. This cause had operated on nearly all the trials of marl made previous to my service in South Carolina. Of all such cases of alleged failure, that I was enabled to see and investigate the circumstances, the causes were such as I now suppose of the still later failures. These cases of failure and of disappointment, and the known causes, were brought fully to view in my Report of the Agricultural Survey; and from the more extended remarks, I will quote a short passage, to show my then opinion of the facts and the causes of previous failures, and my earnest warning against the general course pursued. After reciting the general facts of failure of previous trials of marling, I proceeded in these words: "Can any opponent of marling desire more full admissions than these? And yet they all serve but to illustrate what I have continually striven to impress, *that without vegetable matter to combine with, calcareous manures will be of little value.* But, on the other hand, I have heard of no trial of marl on land in proper condition, that is, recently and sufficiently rested, and thereby provided with vegetable matter, in which the effect has not been very great *on the first crop*. And three or four of such results only, would be enough to explain the causes, (of failure in all other cases,) and to prevent all inferences unfavourable to marling, if from a hundred failures of early efforts under reverse circumstances." Then followed particular statements of two different experiments, carefully made that year, (and the circumstances noted at my request,) of marling on new land, and, therefore, not exhausted of its valuable matter, and in which the products (which were cotton) were nearly doubled in the first year of the application.

Here, then, even in the few lines quoted from the much more full precepts to the same purport, there is full evidence of my having stated, in advance of all later trials, the sure cause of failure; and in the warning against that cause, I may claim to have predicted all later failures of like occurrence. And if there had been thousands of failures, preceded and accompanied by very frequent and exhausting tillage, all of them would but the more strongly confirm my long entertained and often expressed opinions and instructions as to the action of calcareous manures; and all such cases would not detract a tittle from the alleged available values. When urging the use of lime, I have never omitted to state that it gave no fertility of itself, or by direct action; and that vegetable matter in sufficient quantity, and in conjunction, was essential to the beneficial operation of calcareous manures. The required organic matter may be supplied mainly in the growth of the land to be improved. But it *must* be supplied in some form, and in sufficient quantity—and, also, should be, in part, present in advance of the use of calcareous manures, to secure their best early effects.

Planters of South Carolina—I have offered to you in plain and unvarnished language, and, possibly, it may be in ungracious and distasteful terms, the last advice and admonition that I can expect to utter to you, or to any similar audience. My burden of years, and infirmities much greater than even suited to my age, admonish me that my labours may soon close. I would deem it a reward of more value to me than will be the short remainder of my life, if you and your fellow-labourers, even at this late time, (in reference to myself,) would heed my words, and fully profit by them. It is but little that a private individual can do, to warrant to a great commonwealth or community, the beneficial results predicted upon stated premises and conditions. But so perfect is my confidence in the general results I have predicted, that I would willingly hazard upon the issue all that I have, in property, reputation, and even life itself. For illustration, and in mercantile or business language—if I possessed hundreds of millions of dollars, to that full amount, for a premium of ten percent, I would insure as much clear profit to South Carolina, to be gained by conforming to my directions, for saving and increasing the fertility of her soil. As, however, it is impossible for me to offer any such guaranty, and for me either to incur risk or loss, or to derive pecuniary gain from the results, I can only offer my earnest verbal assurances of your available gain, as great and as sure to be obtained by your pursuing a proper course of improvement, as will be the growing loss and eventual ruin of your country, and humiliation of its people, if the long existing system of exhausting culture is not abandoned. It is not merely my feeble voice and my questionable personal testimony, but also thousands of unquestionable facts, and the sure experience and realized profits of thousands of farmers, which offer to your acceptance the highest agricultural prosperity in exchange for present decline and approaching exhaustion of the remaining fertility of your land. Choose, and choose quickly! And remember, as my last warning, that your decision will be between your purchasing, at equal rates of price, either wealth and general prosperity, of value exceeding all present power of computation, or ruin, destitution, and the lowest degradation to which the country of a free and noble minded people can possibly be subjected.

Notes

Introduction

1. The quoted description of Ruffin's library was that of an unidentified occupying Union soldier whose letter to relatives about Beechwood survives (William Kauffman Scarborough, ed., *The Diary of Edmund Ruffin*, vol. 2, *The Years of Hope: April 1861–June 1863* [Baton Rouge: Louisiana State University Press, 1976], 417 n. 32).

2. Ibid., 416–20. On Ruffin family farms in wartime, see also David F. Allmendinger Jr., *Ruffin: Family and Reform in the Old South* (New York and Oxford: Oxford University Press, 1990), 160–69.

3. On Julian's death and Edmund's suicide, see Scarborough, ed., *Diary*, vol. 3, *A Dream Shattered: June 1863–June 1865* (Baton Rouge: Louisiana State University Press, 1989), 435–38, 893 (on deafness and isolation), and 935–46 (suicide quotation 936). Allmendinger, *Ruffin*, 171–76, provides an insightful summary of Ruffin's health and infirmities as well as his long meditations on death.

4. See Avery O. Craven, *Edmund Ruffin, Southerner: A Study in Secession* (1932; Baton Rouge: Louisiana State University Press, 1966). On Ruffin as intellectual without institutional help, see especially Drew Gilpin Faust, *A Sacred Circle: The Dilemma of the Intellectual in the Old South, 1840–1860* (Baltimore: Johns Hopkins University Press, 1977); Eric H. Walther, *The Fire-Eaters* (Baton Rouge: Louisiana State University Press, 1992), esp. 228–69; and Allmendinger, *Ruffin*, esp. 109–14. On the Ruffin Flag Company, see Peter Applebome, *Dixie Rising: How the South Is Shaping American Values, Politics, and Culture* (New York: Times Books/Random House, 1996), 120.

5. See Allmendinger, *Ruffin*, 25, 122; David F. Allmendinger Jr., ed., *Incidents of My Life: Edmund Ruffin's Autobiographical Essays* (Charlottesville: University Press of Virginia for the Virginia Historical Society, 1990), 189–209; William M. Mathew, ed., *Agriculture, Geology, and Society in Antebellum South Carolina: The Private Diary of Edmund Ruffin, 1843* (Athens: University of Georgia Press, 1992); and [William Boulware], "Edwin [sic] Ruffin, of Virginia, Agriculturist," *De Bow's Review* 11 (1851): 431–36. (Boulware, of King and Queen County, Virginia, was an acquaintance of Ruffin.)

6. W. P. Cutter, "A Pioneer in Agricultural Science," in *Yearbook, Department of Agriculture, 1895* (Washington: U.S. Department of Agriculture, 1895), 493; "Agri-

cultural Teachers and Writers of the Present Day," *Southern Planter* 65 (January 1904): 1–18 (esp. 2–7, on Ruffin).

7. See Roy V. Scott, *The Reluctant Farmer: The Rise of Agricultural Extension to 1914* (Urbana: University of Illinois Press, 1970); Alfred Charles True, *A History of Agricultural Education in the United States, 1785–1925* (Washington, D.C.: U.S. Government Printing Office, 1929); and David B. Danbom, *Born in the Country: A History of Rural America* (Baltimore: Johns Hopkins University Press, 1995), 167–75. Ruffin was himself a precocious and eloquent advocate of government-sponsored agronomic education. See his "Sketch of the Progress of Agriculture in Virginia" (1836), below.

8. On Craven's background, see his entry in *Directory of American Scholars*, 5th ed. (New York: Jacques Cattell/R. R. Bowker, 1969), 108; his obituary in *Journal of Southern History* 46 (August 1980): 478; Avery O. Craven, "An Historical Adventure" (reprint of his Mississippi Valley Historical Association presidential address), in Craven, *An Historian and the Civil War* (Chicago: University of Chicago Press, 1964), 217–33; and John David Smith, "Avery Craven (12 August 1885 [sic]–21 January 1980)," in *Dictionary of Literary Biography*, vol. 17, *Twentieth-Century Historians*, edited by Clyde Wilson (Detroit: Gale Research/Book Tower, 1983), 126–31; Avery O. Craven, *Soil Exhaustion as a Factor in the Agricultural History of Virginia and Maryland, 1606–1860* (Urbana: University of Illinois Press, 1926).

9. A significant sample of Craven's influence includes Lewis Cecil Gray, *History of Agriculture in the Southern United States to 1860* (Washington, D.C.: Carnegie Institution of Washington, 1933); William B. Hesseltine, *The South in American History* (1936; reprint, New York: Prentice-Hall, 1943), esp. 276–78; Clement Eaton, *A History of the Old South* (1949; reprint, New York: Macmillan, 1966), esp. 216–17; Scarborough, ed., introduction to *Diary*, vol. 1, *Toward Independence, October 1856–April 1861* (Baton Rouge: Louisiana State University Press, 1972), xviii; William Edwin Hemphill, Marvin Wilson Schlegel, and Sadie Ethel Engelberg, *Cavalier Commonwealth: History and Government of Virginia* (New York: McGraw-Hill, 1957), a popular school text; Virginius Dabney, *Virginia, the New Dominion: A History from 1607 to the Present* (Garden City, N.Y.: Doubleday, 1971); and Betty L. Mitchell, *Edmund Ruffin: A Biography* (Bloomington: Indiana University Press, 1981).

10. See Donald Worster, ed., *The Ends of the Earth: Perspectives on Modern Environmental History* (Cambridge: Cambridge University Press, 1988), 318; and the 1997 conference program of the American Society for Environmental History.

11. A convenient comparison (in English) appears in *Rural Society in France: Selections from the* Annales: Economies, Societes, Civilisations, ed. Robert Forster and Orest Ranum, trans. Elborg Forster and Patricia M. Ranum (Baltimore: Johns Hopkins University Press, 1977).

12. On conservationism and the origins of environmentalism, see Donald Worster, *Nature's Economy: A History of Ecological Ideas* (Cambridge: Cambridge University Press, 1977), esp. 191–348; and Samuel P. Hays, *Beauty, Health, and Permanence:*

Environmental Politics in the United States, 1955–1985 (Cambridge: Cambridge University Press, 1987); Aldo Leopold, *A Sand County Almanac, and Sketches Here and There* (New York: Oxford University Press, 1949); Rachel Carson, *Silent Spring* (Greenwich, Conn.: Fawcett Publications, 1962).

13. William M. Mathew, *Edmund Ruffin and the Crisis of Slavery in the Old South: The Failure of Agricultural Reform* (Athens: University of Georgia Press, 1988); G. E. Mingay, "Introduction: Rural England in the Industrial Age," in *The Victorian Countryside*, ed. Mingay (London: Routledge and Kegan Paul, 1981): 1:3–16; and Georges Duby and Armand Wallon, eds., *Histoire de la France Rural*, vol. 3, *Apogée et Crise de la Civilisation Paysanne, 1789–1914* ([Paris]: Seuil, 1976), 9.

14. Mathew, *Edmund Ruffin*, 196.

15. Allmendinger, *Ruffin*, 120–22, 148–50 (on Malthus), 115 (on reclamation and the Charleston address).

16. See esp. William Gilmore Simms, *The Social Principle: The True Source of National Permanence* (Tuscaloosa, Ala.: n.p., 1843), esp. 36.

17. See Jack Temple Kirby, *Poquosin: A Study of Rural Landscape and Society* (Chapel Hill: University of North Carolina Press, 1995), esp. 61–91, 114–25.

18. I do concede one significant exception of disturbance, however, and a troubling assumption regarding human population: (1) ranging swine undoubtedly reduced stands of longleaf pine by consuming roots and especially terminal buds of infant trees; and (2) hinterland human populations were certainly kept relatively thin by migrations to the West. See Kirby, *Poquosin*, 205–6, 95–125.

19. In 1836, responding to an eccentric letter about Malthusian theory—see the second Ruffin text below—Ruffin wrote, "I have not read Malthus for nearly twenty years," so conceivably the actual reading was accomplished after his war service, when he also read Taylor's *Arator*.

20. On Ruffin's early education, formal and informal, see Allmendinger, ed., *Incidents*, 167–68, 190.

21. On Davy, see J. D. Sykes, "Agriculture and Science," in *Victorian Countryside*, ed. Mingay, 1:260–61. On Ruffin's reading and application, see Allmendinger, ed., *Incidents*, 194–97.

22. See Sykes, "Agriculture and Science," 1:262; and B. A. Holderness, "Agriculture and Industrialization in the Victorian Economy," in *Victorian Countryside*, ed. Mingay, 1:194; Justus von Liebig, *Die Organische Chemie In Ihrer Aawendung Auf Physiologie Vad Pathologie* (1835; Braunschweig: F. Viewig and Son, 1842); Justus von Liebig, *Organic Chemistry in Its Applications to Agriculture and Physiology*, trans. Lyon Playfair (London: Taylor and Walton, 1840).

23. See Sykes, "Agriculture and Science," 1:261–64. Liebig had contemporary French critics as well. See Robert Specklin, "Les Progres Techniques," in *Histoire de la France Rural*, ed. Duby and Wallon, 3:210–12.

24. See [Edmund Ruffin], "British Opinions on the 'Essay on Calcareous Manures' . . ." *Farmers' Register* 5 (10 June 1836): 380–82; [Edmund Ruffin], "Liebig's

Organic Chemistry . . ." *Farmers' Register* 9 (31 August 1841): 459–69 (Ruffin quotation 459). *Southern Planter* editor quoted in Albert Lowther Demaree, *The American Agricultural Press, 1819–1860* (1941; reprint, Philadelphia: Porcupine Press, 1974), 67. See also Margaret W. Rossiter, *The Emergence of Agricultural Science: Justus Liebig and the Americans, 1840–1880* (New Haven: Yale University Press, 1975). On marling in Europe, see Robert Specklin, "Victoire sur la Disette," in *Histoire de la France Rural*, ed. Duby and Wallon, 3:121–22. According to Specklin, "Le procédé etait connu dans l'Antiquité"—that is, it was a folkish practice, especially in Picardy, but during the eighteenth century the monarchy officially encouraged marling in central France.

25. See Kirby, *Poquosin*, esp. 61–91, on Ruffin's farming; and Sykes, "Agriculture and Science," 1:263–64, on Lawes and Gilbert.

26. Henry David Thoreau was a user of Michaux as well. See *The Maine Woods*, ed. Edward Hoagland (New York: Penguin, 1988), 57.

27. See Mathew, ed., *Agriculture*, 86–87, 194.

28. See Sykes, "Agriculture and Science," 1:261; and Nicholas Goddard, "Agricultural Societies," in *Victorian Countryside*, ed. Mingay, 1:245–59.

29. On the provenance of "Incidents," see Allmendinger, ed., *Incidents*, 1–2.

30. See Earl G. Swem, "An Analysis of Ruffin's Farmers' Register, with a Bibliography of Edmund Ruffin," *Bulletin of the Virginia State Library* 11 (July, October 1918), on microfilm at the head of the American Periodical Series, *Farmers' Register*. See also Judith Lee Hallock, "The Agricultural Apostle and His Bible: Edmund Ruffin and the *Farmers' Register*," *Southern Studies* 23 (1984): 205–15.

1. Production as Moral Imperative: The Morals of Agriculture

Baltimore American Farmer 4 (20 September 1822): 207–8. Ruffin recycled this essay verbatim in *Farmers' Register* 5 (1 September 1837): 304–6.

1. [John Taylor of Caroline (1753–1824) was the author of *Arator*, published first in book form in 1813. Ed.]

2. A Malthusian Crisis: A Letter from George Henry Walker and Ruffin's Response

Farmers' Register 3 (February 1836): 601–3. Walker lived at the time in Holmesburg, Pennsylvania (near Philadelphia). Ruffin identified him (in a note) as a British immigrant and excellent hay grower. Walker seems also to have been an eccentric. The meaning of his following letter may have been as obscure to Ruffin as it is to me, except that the hay grower from a beef-eating country reasoned that Malthus (Walker's countryman) somehow neglected modern meat production in his calculations of population's natural limits.

1. ["Address of Col. Thomas M. Bondurant, President of the Buckingham Agricultural Society, Delivered at Its Annual Meeting, on the 15th Oct. 1835," *Farmers' Register* 3 (December 1835): 503–7, which is contained within "Proceedings of the Buckingham Agricultural Society," 502–7. Ed.]

2. [Thomas Robert Malthus, *An Essay on the Principle of Population* (1798). Ed.]

3. [The U.S. slave population increase was indeed remarkable compared with the negative rates experienced throughout the rest of the Americas. Still, the black U.S. rate was less than that of the white population. See.U.S. Bureau of the Census, *Negro Population, 1790–1915* (Washington, D.C.: U.S. Government Printing Office, 1918); and Philip D. Curtin, *The Atlantic Slave Trade: A Census* (Madison: University of Wisconsin Press, 1969). Ed.]

3. The Old Dominion's Declension: Sketch of the Progress of Agriculture in Virginia, and the Causes of Its Decline

"Sketch of the Progress of Agriculture in Virginia, and the Causes of Its Decline: An Address to the Historical and Philosophical Society of Virginia," *Farmers' Register* 3 (April 1836): 748–60. It is prefaced by: "Prepared by request of the Executive Committee, for the Annual Meeting, in February 1836—and by resolution of the Society, ordered to be published in the Southern Literary Messenger, and the Farmers' Register."

1. [See Ruffin's long note following the appendix to this text. Ed.]

2. The biographer of Fletcher of Salton, though strongly disapproving many of his political opinions, and the devotion to liberty, which are treated as tending to the extreme of republicanism, still bears ample testimony to his talents and worth. The sketch of his life in the Edinburgh Encyclopaedia contains these passages. "Andrew Fletcher, of Salton, was a statesman and patriot of the highest order; and though Scotland, his native land, was the chief object of his exertions, yet, wherever the love of country and of liberty prevails, he deserves to be remembered with respect and gratitude."—"Fletcher was endowed with high talents, great courage, integrity, generosity, and temperance. On the purity of his intentions as a patriot, the exertions and sufferings of his life form the best commentary. He was a most elegant scholar, and an accomplished orator. His speeches are remarkable for their plainness and energy—and form, by their brevity, a striking contrast to the wordy eloquence of the present day."—"His *Two Discourses on the Affairs of Scotland* were written in 1693, when, in consequence of some years of barrenness, a scarcity, or rather a famine, existed through the land, and occasioned the most severe sufferings to the lower classes. The author declares, that besides those who were scantily provided for out of the church [or charity] boxes, there were at the time when he wrote, not less than 200,000 persons in Scotland begging from door to door; 'and though' he continues 'the number of these be perhaps double to what it was formerly, by reason of this present distress, yet in all time there have been about 100,000 of those

vagabonds, *who have lived without any regard either to the laws of the land, or to those of God and nature.*' He tells us also, that when he considers the many excellent laws enacted by former parliaments, for setting the poor to work, *contrasted with their utter inutility*—when he considered farther, that *all the other nations of Europe* (Holland alone excepted) *groaned under a similar pressure*, he was led to believe that neither the cause nor the remedy of the evil had been discovered. As no such evil had been complained of by the classical writers of antiquity, *and as much poverty was the consequence, in Europe, of the manumission of slaves by their Christian masters*, he gravely supposes that the *existence of slavery was the cause of the comfort and industry of the lower orders* in former times. It will hardly be credited by those who are acquainted with his high notions of political rights, and his constant jealousy of the power and ambition of princes, that he proposes *reducing all those persons, and their posterity, to slavery*, by a solemn act of the legislature, that on the one hand they might be *compelled to work*, and on the other, might be *insured the necessaries of life.*"

It is not in approbation of, or concurrence with, this remarkable opinion of Fletcher, that it is here quoted. That his view was erroneous, and that his proposed remedy would have been most impolitic, (without naming objections on other and higher grounds,) is readily admitted. But that such opinions should have been entertained, and zealously asserted, by one who so well deserved respect as a scholar, a statesman, a patriot, and a philanthropist, is enough to establish the fact that the abolition of personal slavery in Europe, and even in so inhospitable a region as Scotland, together with its undoubted advantages, was attended with many great and long continued evils to the community, and, more especially to the very class relieved from slavery. The evils of slavery, such as they are, bear most heavily on the master—and the evils of emancipation, on those who were the slaves.

[Ruffin has slightly altered this quotation, mostly by adding emphases where none exist in the original. It may be significant, however, that "the existence of domestic slavery was the cause of the comfort and industry of the lower orders in ancient times" has been changed by the deletion of "domestic." See "Fletcher, Andrew of Salton," in *The Edinburgh Encyclopedia; Conducted by Sir David Brewster, LL.D. with the Assistance of Gentlemen Eminent in Science and Literature*, 18 vols. (Edinburgh: William Blackwood, 1830), 9:368–71; Andrew Fletcher, *Two Discourses on the Affairs of Scotland* (1693). Allmendinger, *Ruffin*, 149, discusses Ruffin's anticipation by two decades of George Fitzhugh's Fletcherian defense of slavery in the abstract. Ed.]

3. Immense quantities of common tobacco were sold at less than $2 the 100 lbs.; and hay of very inferior quality, from natural wet meadows on the writer's farm, was bought and shipped to Norfolk during the war, and sold there at $2 50, the 100 lbs.

4. [James Watt (1736–1819), the Scottish inventor of the steam engine; Robert Fulton (1765–1815), the American inventor of the steamboat; and Sir Richard Arkwright (1732–92), the English textile industrialist who pioneered the use of power-driven machinery. Ed.]

5. [John Burk, *The History of Virginia, from Its First Settlement to the Present Day*, 4 vols. (Petersburg, Va.: Dickson and Pescud, 1804–16), 172. The fourth volume was not by Burk, but "continued by Skelton Jones and Louis Hue Girardin" and printed by M. W. Dunnavant, also of Petersburg. Ed.]

6. [Ibid., 1:172, quotes this in a note and cites "Purchas, V., 1766"—i.e., Samuel Purchas, *Hakluytus Posthumus, or Purchas His Pilgrimes: Contayning a History of the World in Sea Voyages and Lande Travells by Englishmen and Others*, 4 vols. (1625). Ed.]

7. [William Stith, *The History of the First Discovery and Settlement of Virginia: Being an Essay towards a General History of this Colony* (Williamsburg, Va.: William Parks, 1747), 131–32. In all his quotations from Stith, Ruffin has modernized the orthography. Here he has omitted a sentence after "must maintain them" that reads, "Even the most honest and industrious would scarcely take so much true Pains in a Week, as they would have done for themselves in a Day." The final word of the quotation should be "supplies," not "settlers." Ed.]

8. [Ibid., 60. There should be a period after "employment in England." Ed.]

9. [Ibid., 107. Ruffin has changed *muskets* to *musquets*. Ed.]

10. [Burk, *History of Virginia*, 1:157. Ed.]

11. [Stith, *History*, 116–17. Ruffin has altered STARVING TIME to *starving year*. Ed.]

12. [Burk, *History of Virginia*, 1:159–60. Ed.]

13. [Stith, *History*, 117. Ruffin has extensively altered this quotation, which reads, "At their Departure, many were importunate to burn the Houses and Fort at *James-Town*. But God, who did not intend, that this excellent Country should be so abandoned, put it into the Heart of Sir *Thomas Gates*, to save the Town and Fortifications." Ed.]

14. [Burk, *History of Virginia*, 1:163–65. Ruffin has omitted paragraphing after "should not be abandoned." Ed.]

15. [Ibid., 1:164, in a note citing "Purchas." Ed.]

16. [Ibid., 1:177. Ed.]

17. [Ibid., 1:178. Ruffin has removed ellipses between "the purse" and "Every person." Ed.]

18. [Stith, *History*, 140. Ruffin has inserted a period after "scorn and difficulty," where the original has a comma. Ed.]

19. [Burk, *History of Virginia*, 1:305–7. Ed.]

20. [Ibid., 1:307. Ruffin has removed several ellipses. Ed.]

21. [Stith, *History*, 147: "And this Year, one Mr. *Lambert* made a great Discovery, in the Trade of Planting. For the Method of curing Tobacco then was in Heaps. But this Gentleman found out, that it cured better upon Lines; and therefore the Governor wrote to the Company, to send over Line for that Purpose." Ed.]

22. [Burk, *History of Virginia*, 1:206. Ed.]

23. [Stith, *History*, 166. The original begins, "And that such of these Maids, as were married to the publick Farmers, should be transported." Ed.]

24. [Burk, *History of Virginia*, 1:207. Ed.]

352 Notes

25. See at page 11, vol. ii of the Farmers' Register some amusing passages extracted from King James' *Counterblaste to Tobacco*. ["Origin and Use of Tobacco," *Farmers' Register* 2 (June 1834): 11–12, taken "from the London Penny Magazine." Ed.]

26. [Burk, *History of Virginia*, 1:209–10. After the first word of the quotation, "tobacco," Ruffin has removed "a nauseous weed, which, according to some writers, was first discovered in the island of Tobago, had by accident or design." Ed.]

27. [Ibid., 1:210. Ed.]

28. [Stith, *History*, 197–98. Ruffin has divided into three quotations what forms part of a single paragraph in the original, has printed "multitude" followed by a comma instead of "multitudes," and has omitted a definite article between "granted" and "adventurers." Ed.]

29. [Burk, *History of Virginia*, 1:226. Ed.]

30. [Ibid., 1:280–81. Ed.]

31. [Ibid., 1:283. Ed.]

32. [Ibid., 1:285–86. Ruffin has added the emphases. Ed.]

33. [Ibid., 1:139–40. Ed.]

34. [Ibid., 1:228. Ed.]

35. [Ibid., 2:233. Ed.]

36. [Ibid., 2:240. Ed.]

37. [Robert Beverley, *The History and Present State of Virginia* (London: R. Parker, 1705), 90. The section number should be 143. Ed.]

38. [Burk, *History of Virginia*, 2:xxxii. Ed.]

39. ["Report on Agricultural Premiums: Submitted to and Approved by the Delegation of the United Agricultural Societies of Virginia, Dec. 6th, 1822," *Farmers' Register* 1 (August 1833): 147–49 (minutes by Ruffin); Ruffin, "The Advantages and Defects of Agricultural Societies," *Farmers' Register* 1 (September 1833): 201–2; "Supplement to No. 4, of Farmers' Register," *Farmers' Register* 2 (September 1834): 257–59 (editorial note by Ruffin [257] prefacing various extracts from British agricultural journals). Ed.]

4. On Fencing—Against the Open Range: Letters and Editorials

On Fencing and Other Enclosures

Farmers' Register 1 (November 1833): 338–40. Herbemont (d. 1838), tutor in French at South Carolina College (1807–17) and author of a treatise on viticulture, was the father of diplomat Alexander Herbemont (1790–1863).

1. ["Pisé," *American Farmer* 6 (26 March 1824): 6 (a letter from "W. W. A." of Statesburgh, S.C., dated 15 March 1824); *The Cyclopedia; or, Universal Dictionary of Arts, Sciences and Literature, by Abraham Rees; with the Assistance of Eminent Professional Gentlemen*, 47 vols. (Philadelphia: Samuel F. Bradford and Murray, Fairman and Company, 1810–24): vol. 28, n.p.; Cadet de Vaux, "Pisé," in M. M. Sonnini

et al., *Cours Complet ou Dictionnaire Universel d'Agriculture Pratique, d'Économie Rurale et Domestique, et de Médecine Vétérinaire, par l'Abbé Rozier*, 6 vols. (Paris: Desray, 1815), 5:391–96. Ed.]

On the Law of Enclosures

Farmers' Register 1 (December 1833): 396–98. The article is prefaced with "Respectfully addressed to the Members of the General Assembly of Virginia."

1. [Ruffin would shortly term this arrangement a "ring fence" association, and he persuaded his contiguous neighbors to form the first known example. The Virginia General Assembly finally formally authorized ring fences in 1857. See Allmendinger, *Ruffin*, 117; and Kirby, *Poquosin*, 76–78. Ed.]

2. [Trans. "to each his own." The phrase often occurs in classical writings: e.g., "iustitia, quae suum cuique distribuit" (justice, which apportions to each his own) in Cicero, *De Natura Deorum*, 3:38, trans. P. G. Walsh (New York: Oxford University Press, 1997), 120. Ed.]

On the Law of Enclosures in Virginia

Farmers' Register 1 (January 1834): 450–52. William Jones Dupuy (1792–1853), incorrectly rendered "W. I. Dupuy" in Swem, "Analysis," was a substantial Nottoway County farmer and Ruffin's acquaintance and correspondent.

1. [This editorial note from Ruffin follows the letter: "It is proper to state, that the foregoing communication was received, before the article on the same subject in No. 7 [Ruffin, "On the Law of Enclosures," reprinted above], had issued from the press, and that neither of the two writers knew that he had an auxiliary so near. We are gratified that this subject has already engaged the attention of several of our correspondents." Ed.]

The Right of Grazing in Common

Farmers' Register 1 (January 1834): 490. The author is probably William Henry Foote (1794–1869) of Fairfax County, who later wrote *Sketches of Virginia, Historical and Biographical* (1850). He is not listed among private correspondents in Ruffin's papers; however, Foote was a correspondent of Hill Carter, master of Shirley Plantation, upriver from Ruffin's James farms. Carter corresponded with Ruffin both privately and publicly. See Carter Family Papers, Virginia Historical Society, Richmond.

1. [John Selden, the master of Westover Plantation. The original text has "Sedden," which is a misprint. Ed.]

Enormous Losses Caused by the Fence Law of Virginia

Farmers' Register 1 (March 1834): 633–34. Assuming Ruffin correctly identified Theodore A. Field, he may have been a member of the family of Richard W. Field, who was the largest landholder in Brunswick County. (According to the manuscript agricultural census of 1850, Richard Field held one thousand acres of improved land and six hundred additional acres of unimproved land.) Neither Field was among Ruffin's private correspondents, and neither appears in the extensive files of the Virginia Historical Society.

Petition for a Change of the Law of Enclosures

Farmers' Register 2 (October 1834): 283–84. James Mercer Garnett (1770–1843) of Elmwood Plantation, Essex County, was author of "Lectures on Female Education" (3d ed., 1825) and a member of the Virginia Convention of 1829–30. See James Mercer Garnett Letters, Mercer Family Papers, Virginia Historical Society, Richmond.

Ruffin's endorsement of the petition prefaces this letter: "We recommend the subject of the following petition to all who have not already considered it maturely—and to those who are satisfied of the unjust and oppressive operation of the law of enclosures, that they will forthwith proceed to *act*, for the removal of the grievance. Let similar petitions (either in this or any other form that may be preferred,) be prepared and properly presented to the consideration of the agricultural community, and there can be but small ground for fear as to the result."

A Commentary on the Law of Enclosures of Virginia

Farmers' Register 2 (November 1834): 346–47.
1. [That is, Ruffin himself. The essay is reprinted above. Ed.]

On the Petition for a Change of the Law of Enclosures

Farmers' Register 2 (December 1834): 402–3.

Petition of Stock-Owners

Farmers' Register 2 (December 1834): 426–27.
1. [Trans. "The custom of a place should be respected." Ed.]

Petition for an Amendment of the Law Respecting Enclosures, on the Margins of the Navigable Tide Waters of James River

Farmers' Register 2 (December 1834): 450.

General Results of the Law of Enclosures in Virginia

Farmers' Register 2 (March 1835): 610–11.

The Policy of the Law of Enclosures Defended

Farmers' Register 3 (May 1835): 47–49. According to Swem, "Analysis," Ruffin could identify this correspondent, "Fencemore," only as "Scott," and lacking a hometown or county identification with the (rather common) surname, I am unable to discover further.

1. ["Systems and Products of Farming in Columbia County, New York," *Farmers' Register* 1 (February 1834): 542–45, which consists of two letters to Hill Carter of Shirley Plantation, Virginia, from J. P. Beekman, corresponding secretary of the State Agricultural Society of New York, and Teunis Harder. Ed.]

2. ["Report on Farms, of the Committee of the Agricultural Society of Albemarle: Made to the Society at Their Show and Fair, 21st October, 1828," *Farmers' Register* 2 (September 1834): 225–33. This document consists of John H. Craven, William H. Meriwether, and John Rogers's answers to a questionnaire about the condition of their farms. Ed.]

3. [Ruffin, "On the Law of Enclosures," reprinted above. Ed.]

4. [Presumably Ruffin, "General Results of the Law of Enclosures in Virginia," reprinted above. Ed.]

5. [Ruffin, "On the Law of Enclosures," reprinted above. Ed.]

6. [Ruffin, "A Commentary on the Law of Enclosures in Virginia," reprinted above. Ed.]

"Fence Less" and the Editor of the Register

Farmers' Register 3 (December 1835): 455–58.

1. [Fencemore, "The Policy of the Law of Enclosures Defended," reprinted above. Ed.]

2. ["Green Cropping," *Farmers' Register* 1 (April 1834): 674–76 (quotation on 674). This document is an extract from the *British Farmers' Register*. Ed.]

3. [Ruffin's response to Fencemore, "The Policy of the Law of Enclosures Defended," reprinted above. Ed.]

4. [John H. Craven of Penpark, Albemarle County, mentioned above. Ed.]

5. ["Report on the State of Agriculture in Prince George: Submitted to the Agricultural Society of That County and Approved March 9th, 1821," *Farmers' Register* 1 (September 1833): 232–35 (quotation on 234, emphasis added). Ed.]

6. [Ruffin, "On the Law of Enclosures," reprinted above, quotation on 397–98. Ed.]

5. Malaria—Against Mill Ponds: On the Sources of Malaria, or Autumnal Diseases, in Virginia, and the Means of Remedy and Prevention

Farmers' Register 6 (1 July 1838): 216–28.

1. [Ruffin, "General Remarks on the Causes, and Means of Preventing the Formation of Malaria, and the Autumnal Diseases Which Are the Effects of It, in Virginia," *Farmers' Register* 5 (1 May 1837): 41–43. Ed.]

2. [Historians of this century have not entirely affirmed Ruffin's argument in this instance, deducing instead that typhoid, salt-water poisoning, and melancholia were

principal causes of morbidity at Jamestown. See Carville V. Earle, "Environment, Disease, and Mortality in Early Virginia," in *The Chesapeake in the Seventeenth Century: Essays on Anglo-American Society*, ed. Thad W. Tate and David L. Ammerman (Chapel Hill: University of North Carolina Press, 1979), 96–125; and Albert E. Cowdrey, *This Land, This South: An Environmental History* (Lexington: University of Kentucky Press, 1983), esp. 25. Ed.]

3. [Ruffin was himself usually too preoccupied with work to join others of his class in highland retreats during malarial seasons, but in 1848, owing to the poor "health of one of [his] daughters," he fled to "the Springs" for August and September. See Allmendinger, ed. *Incidents*, 91. Ed.]

4. ["Remarks on the Soils and Agriculture of Gloucester County," *Farmers' Register* 6 (1 June 1838): 178–91. Ed.]

5. ["Malaria of Marshes: Substance of a Paper Read by M. Gaetano Giorgini, at the Academie Royal des Sciences, (Paris,) July 12, 1825, and recently published in the Annales de Chemie et de Physique," *Farmers' Register* 4 (1 December 1836): 502–4; "On the Insalubrity of the Air of Marshes in Communication with the Sea," *Farmers' Register* 5 (1 December 1839): 460–61. The latter is subscribed "by M. Georgini [sic], of Lucca" and is reprinted from the *Medico-Chirurgical Review*. Gaetano Giorgini (1795–1874) was an archivist in Lucca, where he kept the records of the *catasto* (property census) for the Grand Duchy of Tuscany. He also held an honorary university appointment at Pisa and was a professor of applied mathematics at the Imperiale e Reale Accademia delle Belle Arti in Florence. See Giorgini, *Elementi di Statica* (Florence: Luigi Pezzati, 1835); Giorgini, *Ragionamento sopra il Regolamento Idraulico della Pianura Lucchese e Toscana Interposta fra l'Arno e il Serchio* (Pisa: Stamperia Peraccini, 1839). Ed.]

6. See Farmers' Register, p. 107, and 129, Vol. I. and p. 41, 42, Vol. V. ["Leaves from a Traveller's Note Book: A Walk through Shirley Farm, Nov. 28th, 1832," *Farmers' Register* 1 (July 1833): 105–7; Hill Carter, "Account of the Embankment and Cultivation of the Shirley Swamp," *Farmers' Register* 1 (August 1833): 129–31; Ruffin, "General Remarks." Ed.]

7. See "Essay on Calcareous Manures," 2d ed., chap. xix., and "Essay on the Police of Health, commencing p. 154, Vol. V. of Farmers' Register. [Ruffin, "Desultory Observations on the Police of Health, in Virginia," reprinted below. Ed.]

8. Essay on Calc. Man., ch. iii., viii., and xii.

9. See the facts and reasons stated more fully at p. 179 and 190, in the last number of Farmers' Register. ["Remarks on the Soils and Agriculture of Gloucester County." Ed.]

10. The same, p. 190. [Ibid., 190. Ed.]

11. See Far. Reg. vol. I., pp. 232, 386, 518, 733, 734. [Ruffin, "Remarks Suggested by a Visit to Warren County," *Farmers' Register* 5 (1 August 1837): 228–31; Ruffin, "On Draining: Addressed to Young Farmers," *Farmers' Register* 1 (December 1833): 385–90, reprinted below; "A Citizen of Nottoway," "Necessity for a Law to Permit

Draining in Virginia," *Farmers' Register* 1 (February 1834): 518–20, containing a second letter signed by P. W. Harper of Green Field, Nottoway County (519–20); "On Straightening the Bed of Streams, and the Need of Legislative Aid for Draining," *Farmers' Register* 1 (May 1834): 733–34. Ed.]

12. See facts and statements on this subject at p. 231, vol. V., pp. 1 to 3; p. 579, vol. II.; p. 374, vol. IV. [Ruffin, "Remarks Suggested by a Visit to Warren County"; H. B. Croom (Lake Lafayette, Tallahassee), "Some Account of the Agricultural Soil and Products of Middle Florida, in a Letter to the Editor," *Farmers' Register* 2 (June 1834): 1–3; "T. J. P." (of Amelia), "A Profitable Change of a Mill Pond for a Canal," *Farmers' Register* 2 (February 1835): 579, letter dated 30 December 1834; Ruffin, "Rough Notes upon Some of the Agricultural Improvements of Charlotte, and the Adjacent Counties," *Farmers' Register* 4 (October 1836): 374–77, signed "A Gleaner." Ed.]

6. Public Health and Recycling: Desultory Observations on the Police of Health, in Virginia—As It Is, and as It Ought to Be

Farmers' Register 5 (1 July 1837): 154–71.

1. [Ruffin, "Supplementary Chapter to 'An Essay on Calcareous Manures,'" *Farmers' Register* 1 (July 1833): 76–79. Ed.]

2. Darwin's Phytologia, pp. 210 and 224. Dublin edition. [Erasmus Darwin, *Phytologia, or, The Philosophy of Agriculture and Gardening: With the Theory of Draining Morasses; and with an Improved Construction of the Drill Plough* (Dublin: P. Byrne, 1800). Darwin (1731–1802), a physician in Lichfield, was a friend of Joseph Priestley and Jean-Jacques Rousseau. Darwin was the grandfather of Charles Darwin and was perhaps best known for *Zoonomia; or, The Laws of Organic Life* (1794–96), which argued for evolution. Ed.]

3. [I find no direct confirmation of Ruffin's illustration of unsentimental practicality. However, Jac Weller, a mid-twentieth-century expert on Waterloo, deduced from old maps and eyewitness accounts that much of the principal battlefield, bulging with mounds of shallowly buried men and horses on 19–20 June 1815, soon returned to farmlands, new or broadened roadways, and finally housing developments, suggesting that corpses were removed. See Weller, *Wellington at Waterloo* (New York: Crowell, 1967), esp. 213–21, the battlefield topography. In addition, British and continental companies offering crushed-bone fertilizers to farmers were flourishing by the 1830s, when Ruffin wrote. See Holderness, "Agriculture," 1:179–99, esp. 193–95. Ed.]

4. [It is unclear why "(1833)" appears in the text. Ed.]

5. [The purpose of "(1835)" is opaque. Ed.]

6. [Ruffin moved with his family to Petersburg. Ed.]

7. [Marc-Antoine Puvis (1776–1851), a notable agronomist, served as a public-

health officer in an occupied zone toward the end of the Napoleonic Wars and in 1842 became a member of France's General Agricultural Council. In addition to the articles Ruffin translated and published, Puvis was a prolific author of books including *Voyage Agronomique en Beaujolais, Forez et Limagne* (1821) and *De l'Agriculture du Gatinais, de la Salogne et du Berri* (1833). Ed.]

8. See Puvis, "*On Lime as manure*," translated, in Farmers Register, vol. iii. pp. 359 to 366, and 385 to 392—and "*On Marl*," pp. 690 to 696, and 705 to 706, vol. iii— and pp. to , vol. iv. [Ruffin evidently intended to supply page numbers but did not do so, perhaps because there does not seem to be anything translated from Puvis in vol. 4 of the *Farmers' Register*. "On the Use of Lime as a Manure: By M. Puvis," *Farmers' Register* 3 (October 1835): 359–66, noted as being "translated for the Farmers' Register from the *Annales d'Agriculture Francaise* of 1835"; "On the Use of Marl as a Manure: By M. Puvis," *Farmers' Register* 3 (March 1836): 690–96, translated from the same source, *Farmers' Register* 3 (April 1836): 705–10. Ed.]

9. [In the following text quoted by Ruffin, he followed the contemporary custom of preceding each paragraph with quotation marks, but since there are quotations within the text, he sometimes grew confused about the usage of single and double quotation marks. To simplify matters, this text has been indented in the modern manner. Ed.]

10. [The next eleven paragraphs are a translation of Donat. Ed.]

11. [Trans. "comfort pit." Ed.]

12. [Probably François Rozier (1734–93), the great French agronomic scholar of the generation before Puvis. Rozier's *Cours Complet d'Agriculture, Theorique, Practique . . .* , a magnificent Enlightenment "encyclopédie rurale," was first published in nine volumes, Paris, 1781–93, but Ruffin seems to be using the 1815 edition by Sonnini and others cited above. One wonders if Ruffin were also acquainted with Rozier's *Traité sur le Meuilleure Maniere de Cultiver la Navette et le Colsat, et d'en Extraire une Huile Dépouillée de Son Mauvais Goût* (1774). Ed.]

13. It is the writer of this article, Cadet de Vaux, who speaks, and probably in reference to the previous edition of Rozier's *Cours Complet*.—Ed. Far. Reg. [See Cadet de Vaux, "Fosse d'Aisance," in Sonnini et al., *Cours Complet*, 3:428–35. Ruffin's conjecture was wrong. Antoine Cadet de Vaux (1743–1828) was an important French chemist and student of agriculture who wrote extensively on subjects such as wheat germ, flooding, vines, coffee, and fruit trees (see n. 14). Ed.]

14. [Louis Guillaume Laborie (d. 1800), a Parisian apothecary and chemist, author of *Analyse de la Mine de Plombe Blanche* (Paris: Imprimerie Royale, 1780), and Antoine-Augustin Parmentier (1737–1813), a pharmacist and subsequently an agronomist beloved among the French for his experiments and disquisitions on potatoes. (A Paris metro station containing a potato exhibit is named in Parmentier's honor.) These two men and Cadet de Vaux had published together *Observations sur les Fosses d'Aisance: Et Moyens de Prévenir les Inconvéniens de Leur Vuidange* (Paris: Imprimerie de Ph.-D. Pierres, 1778). However, it may be relevant to note that biblio-

graphical sources attribute the 1778 publication to Antoine's brother, Louis-Claude Cadet de Gassicourt (1731–99); the title page reads "Cadet le jeune." Ed.]

15. [Baron Guillaume Dupuytren (1777–1835), a French surgeon and pathologist who was on the staff of the Hôtel Dieu in Paris from 1802 and who developed several pioneering surgical procedures. Ed.]

16. [Antoine-Laurent Lavoisier (1743–94), a principal founder of modern chemistry, had been a contributor to the 1778 work on cesspools. Ed.]

17. [Possibly Louis, marquis de Fontanes (1757–1821), a notable French Catholic intellectual who was appointed Grand Master of the University of Paris in 1808 but who survived into the Restoration. Louis XVIII made him a marquis in 1817. Ed.]

18. These are provincial names given to the different exhalations, (or *mofettes,*) as distinguished by their different poisonous effects on those exposed to their greatest power.—ED. [Ruffin has inserted asterisks at both *mitte* and *plomb* in the original text and furnished the preceding note for both. Ed.]

19. [The next paragraph is Ruffin. Ed.]

20. [The article is followed by "To be continued." Ed.]

7. Reclamation—Against Erosion: On Draining: Addressed to Young Farmers

Farmers' Register 1 (December 1833): 385–90.

1. [Ruffin continued instruction on draining cropland, offering more than five more large pages of text and diagrams in *Farmers' Register* 1 (May 1834): 705–10. The installment presented above, meanwhile, elicited only two responses (February 1834: 518–20; May 1834: 729–30), although both were lengthy and approving. Ed.]

8. Farming the Great Dismal Swamp: Hasty Observations on the Agriculture of the County of Nansemond

Farmers' Register 4 (1 January 1836): 524–27.

1. [That is, Norfolk. Ed.]
2. [The Roanoke River in North Carolina. Ed.]
3. [Probably of Athens, Georgia. Ed.]
4. This island, and the western shore, on which stood the Indian town, and the abundant product of the large corn fields thereon, are mentioned in the following passage in Smith's History of Virginia. The incidents occurred in Sept. 1608, during Captain Smith's second expedition of discovery of the waters of Chesapeake Bay—in which adventure he was accompanied by 12 men only. See *True Travels, Adventures,* &c. ch. vi. [A long quotation from Smith follows, beginning "In a fayre,

calme, rowing towards Poynt Comfort" and ending "we returned to Iames Towne, where we safely arrived the 7. of September, 1608." This quotation does not, in fact, come from John Smith's *The True Travels, Adventures, and Observations of Captaine John Smith* (London: Slater, 1630), but from Smith's *The General Historie of Virginia, New-England, and the Summer Isles; with the Names of the Adventurers, Planters, and Governeurs from their First Beginning, Ano. 1584 to This Present 1624* (London: Michael Sparkes, 1624), reprinted in Edward Arber, ed., *Works, 1608–1631/John Smith*, 2 vols. (Westminster: Constable, 1895), 1:430–32. Ed.]

5. [Mills Riddick Sr. was dean of northern Dismal Swamp shingle and timber cutters and merchants, an occupation perhaps contradicting Ruffin's generalizations regarding Nansemond's shabby agriculture, below. See Riddick's accounts in the Dismal Swamp Land Company Papers, Perkins Library, Duke University, Durham, N.C.; and Kirby, *Poquosin*, 148, 150. Ed.]

6. [Ruffin described the beginnings of a soon-to-be far more extensive market-gardening enterprise in Nansemond and Norfolk Counties. In addition to sweet potatoes and watermelons, commodities would include peas and beans, kale and other greens, onions, and Irish potatoes. See the U.S. censuses of agriculture (tables for these two counties), 1830 and 1840; and Col. William H. Stewart, ed. and comp., *History of Norfolk County, Virginia and Representative Citizens* (Chicago: Biographical Publishing, 1902), esp. 156. Ed.]

9. The Great Dismal: "Jottings Down" in the Swamp

This essay was published serially in *Farmers' Register* 7 (30 November 1839): 698–703 (sections 1–3); 7 (31 December 1839): 724–33 (sections 4–9). The essay is signed below the title, "By the Editor." The first three sections are headed with Roman numerals, the last six with Arabic: they have been standardized to the latter.

1. Farmers' Register, p. 532, vol. iv. [Ruffin, "The Portsmouth and Roanoke Railway, the Navigation of the Meherrin, Nottoway, and Blackwater Rivers," *Farmers' Register* 4 (1 January 1837): 532–33. Ed.]

2. [Anthony Armistead is the only person with this surname in the 1830 manuscript census for Washington County. He lived in the town of Plymouth, heading a household with one other white person and fourteen slaves. Ed.]

3. [This is surely an exaggeration, one Ruffin was fond of repeating. Such ordinary farmers burned over corn and pea stubble in winter or plowed under (probably) in March, returning valuable humus (pea vines are rich in nitrogen) to the soil. See Carville Earle, "The Myth of the Southern Soil Miner: Macrohistory, Agricultural Innovation, and Environmental Change," in *Ends of the Earth*, ed. Worster, 175–210; and Kirby, *Poquosin*, esp. 95–125. Ed.]

4. [Ruffin was correct in the first instance—the Great Dismal is indeed higher than surrounding land on all but its northwestern extremity. See Robert Q. Oaks Jr.

and Donald R. Whitehead, "Geologic Setting and Origin of the Dismal Swamp, Southeastern Virginia and Northeastern North Carolina," in *The Great Dismal Swamp*, ed. Paul W. Kirk Jr. (Charlottesville: University Press of Virginia, 1979), 1–24. Ed.]

5. [Washington County's 1830 manuscript census contains two Swifts, Joshua and Swain, who are listed consecutively, probably indicating that they were neighbors. Joshua Swift owned seventeen slaves, Swain Swift nineteen. Ed.]

6. [Josiah Collins III (1808–63) was born in Edenton, was educated at Harvard, and studied law in Lichfield, Connecticut. He was later a planter, a member of the state senate as a Whig, and one of the largest slaveholders in North Carolina, with 328 slaves enumerated in the 1860 census. Ed.]

7. [Pungo Lake is actually more than fifteen miles from the nearest part of Lake Mattamuskeet. Ed.]

8. [Ruffin, "Observations Made during an Excursion to the Dismal Swamp," *Farmers' Register* 4 (1 January 1837): 513–21. Ed.]

9. [Ruffin's "savanna lands" are probably pocosins—a name of Algonquian origin for "upland swamp" commonly used in Ruffin's time (and ours) by mapmakers and natives of both southeastern Virginia and eastern North Carolina. Ruffin attempted to sort out vernaculars for various wetlands—a valuable exercise both in his day and today, when scientific specialists remain flummoxed at definitions. See, e.g., Curtis R. Richard, ed., *Pocosin Wetlands: An Integrated Analysis of Coastal Plain Freshwater Bogs in North Carolina* (Stroudsburg, Pa.: Hutchinson Ross Publishing, 1981), esp. 3–19. Ed.]

10. [Josiah Collins Sr. (1735–1819), was born in Taunton, Somersetshire, England, moved to the American colonies in 1773, and moved to Edenton in 1777. He was one of North Carolina's most important merchants, trading internationally in a wide variety of goods, such as tobacco, lye, staves, sugar, molasses, and rum. His son, Josiah Collins Jr. (1763–1839), built up a notable rope-making business. The elder Collins was a great land speculator in the region of the Dismal Swamp and founded the Lake Company in 1787 to drain Lake Phelps and build canals. Ed.]

11. [See, e.g., Hill Carter, "Account of the Embankment and Cultivation of the Shirley Swamp," *Farmers' Register* 1 (August 1833): 129–31; and Carter, "The Progress of Sinking and Loss in the Embanked Marsh of Shirley," *Farmers' Register* 5 (May 1837): 40–41, on an expensive failure at reclamation of tidal lands. Ed.]

12. [The quotation is from Jonathan Swift, *The Portable Swift*, ed. Carl van Doren (New York: Penguin Books, 1984), 345. Ruffin employed the expression as an epigraph for the first issues of the *Farmers' Register*. Ed.]

13. [Charles Lockhart Pettigrew (1826–87), a planter, was the brother of James Johnston Pettigrew, later a Charleston lawyer and Confederate general. Charles married Jane Caroline North of South Carolina in 1853. On the Pettigrews of this generation, see Sarah McCulloch, ed., *The Pettigrew Papers*, vol. 2, 1819–43 (Raleigh: North Carolina Department of Cultural Resources, 1988); Clyde N. Wilson,

Carolina Cavalier: The Life and Mind of James Johnston Pettigrew (Athens: University of Georgia Press, 1990); Michael O'Brien, ed., *An Evening When Alone: Four Journals of Single Women in the South, 1827–1867* (Charlottesville: University Press of Virginia, 1993); and Wayne K. Durrill, *War of Another Kind: A Southern Community in the Great Rebellion* (New York: Oxford University Press, 1990). Ed.]

14. [This chapel was built in 1836 for the purpose of converting Collins's slaves from Methodism to Episcopalianism.]

15. [One barrel of corn consisted of five bushels. Ed.]

16. [Thomas Fuller (1608–61), M.A. Cambridge, a clergyman, historian, and author of *The Worthies of England* (1662). Ed.]

10. Carolina Swamps and Heroic Reclamation: Notes of a Steam Journey

Farmers' Register 8 (30 April 1840): 243–54. The essay is subscribed under the title "By the Editor."

1. [Aaron Lazarus (1777–1841), a native of Charleston, S.C., ran a Wilmington shingle and naval stores enterprise that included a steam mill. In 1821 he married Rachel Mordecai (1788–1838), well known for her long correspondence with Irish author Maria Edgeworth. Ed.]

2. [Dr. Frederick Jones Hill (1790–1861) of Orton Plantation, Brunswick County, N.C. He enrolled at the University of North Carolina before studying medicine at the College of Physicians and Surgeons of the State of New York (1811–12). He was a planter, state legislator, and member of the state convention of 1833. Ed.]

3. [Probably turkey oaks, common understory trees in South Atlantic pine forests. Many would probably remain, even as the natural succession of taller pines transpired. Ed.]

4. [Stephen Elliott (1771–1830), *A Sketch of the Botany of South Carolina and Georgia*, 2 vols. (Charleston: J. R. Schenck, 1821–24). Elliott was a banker and botanist as well as the first editor of the *Southern Review* (1828–32). Ed.]

5. [James Fergus McRee (1794–1869) was a Wilmington physician born in Brunswick County, N.C., and educated in New York. He was a botanist and is said to have introduced to North Carolina crops such as tapioca, rhubarb, cantaloupe, and artichokes. In the 1830 manuscript census for New Hanover County, McRee headed a household of four whites and forty-one slaves. Ed.]

6. [Walter Gwynn (1802–82) was a civil engineer born in Virginia and educated at West Point. He worked extensively as chief engineer for railroads in Virginia and North Carolina in the 1830s and 1840s. In 1839 a Russian delegation representing Czar Nicholas I offered him a similar position building the railway from Saint Petersburg to Moscow, but he declined, thinking the Russian winters too cold. He later served as a general in the Confederate army. Ed.]

7. [Ruffin, "Inquiry in the Causes of the Formation of Prairies, and of the Peculiar Constitution of Soil, Which Favors or Prevents the Destruction of the Growth of Forests," *Farmers' Register* 3 (October 1835): 321–36. Ed.]

8. ["R. N." [Edmund Ruffin], "Economical Method of Excavating Ditches, Canals, Ponds, &c," *Farmers' Register* 7 (December 1833): 435, noted as being "translated for the Farmers' Register from the *Journal d'Agriculture* &c, des Pays-Bas"; Ruffin, "On Draining: Addressed to Young Farmers: No. II," *Farmers' Register* 1 (May 1834): 705–10; Ruffin, "Remarks on the Soils and Agriculture of Gloucester County." Ed.]

9. [Thrown up with a turning plow. Ed.]

10. [Almost certainly the loblolly pine. Ed.]

11. [Perhaps Virginia pine or pond pine, but see "Notes on the Pine Trees of Lower Virginia and North Carolina," below. Ed.]

12. [Hogs also consumed the terminal buds (which are virtually pure protein). See Cecil C. Frost and Lytton J. Musselman, "History and Vegetation of the Blackwater Ecologic Preserve [Isle of Wight County, Virginia]," *Castenea* 52 (1987): 16–46. Ruffin considers the turpentine industry's role in decimating longleafs below. Ed.]

13. [Ruffin enjoyed the company of the distinguished William Gaston (1778–1844) of New Bern and Craven County. A Roman Catholic and advocate of toleration of Jews as well as his coreligionists, Gaston served as a member of the North Carolina legislature early in the nineteenth century. During the 1830s he helped found the state's Whig Party, and the General Assembly elected him an associate justice of the North Carolina Supreme Court. See William Stevens Powell, *North Carolina through Four Centuries* (Chapel Hill: University of North Carolina Press, 1989), 253, 270–76; or, more elaborately, J. Herman Schauinger's biography, *William Gaston, Carolinian* (Milwaukee: Bruce Publishing, 1949). Ed.]

14. [James West Bryan (1805–64) was educated at the University of North Carolina and became a lawyer in New Bern and a prominent Whig politician. For Bryan's letter, see n. 19. Ed.]

15. [Four of these five Carolina marlers (or limers) appear in Mathew, *Edmund Ruffin*, 105, 114, 121, 155, 177, 184: Lucas Benner (the pioneer marler) lived on the Neuse in Craven County, as did James Bryan and Isaac Taylor; Burgwyn lived in Jones County. Mathew agrees with Ruffin that these planters were exceptional, proving the rule that Carolina possessed little good marl and generally lagged behind Virginia in the use of calcareous manures. Ed.]

16. [Nicholas Herbemont, "A Treatise on the Culture of the Vine and on Wine Making in the United States (Baltimore: I. I. Hitchcock, 1833). Ed.]

17. [The 1840 manuscript census for Norfolk County includes one "Wm Dusberry," head of a household of four, including one slave. Nothing more has been discovered of this man Ruffin thought so ingenious. Ed.]

18. [The Chinese mulberry received frequent attention in the *Farmers' Register*; see, e.g., "Account of the Chinese Mulberry Tree (*Morus Multicaulis*)," *Farmers' Register* 3 (December 1835): 481–83. Ed.]

19. The letter of J. W. Bryan, esq., spoken of in the foregoing article, as to follow, though in type, is compelled to be postponed, for want of space for its insertion. It will appear in the next number.—Ed. F.R. [James W. Bryan, "The Marl and Limestone of the Borders of the Neuse and Trent," *Farmers' Register* 8 (31 May 1840): 257–59, a letter written 24 March 1840, from Egypt, Craven County, N.C. Ed.]

11. Pines as Monuments and Commodities: Notes on the Pine Trees of Lower Virginia and North Carolina

Russell's Magazine 2 (October 1858): 34–42; 2 (November 1858): 139–51. It was reprinted in the *Southern Planter* 19 (February 1859): 81–96. The first version is anonymous, but the *Southern Planter* reprinting is subscribed under the title "By Edmund Ruffin."

1. [Ruffin referred to the authority of François-André Michaux (1770–1855), *The North American Sylva; or, A Description of the Forest Trees of the United States, Canada and Nova Scotia* (1810, with several subsequent editions). François-André was the son, companion, student, and translator of the remarkable globe-trotting botanist André Michaux, who in his native France studied with Bernard de Jussieu. Jussieu developed and the elder Michaux adopted a "natural" classification system that rivaled Carl Linnaeus's method. The younger Michaux persisted, and, as seems obvious below, Ruffin found more virtue in the French system while nonetheless occasionally taking issue with some of Michaux's descriptions. Ed.]

2. [Plus Isle of Wight County, the southern end of which separates Southampton and Nansemond. See Frost and Musselman, "History." Ed.]

3. I have since found and measured leaves 19½ inches long, in Barnwell, S.C.

4. [Ruffin again employed Michaux's Latin names here. The shortleaf's Linnean designation is *P. echinata*. Ed.]

5. [Neither the common nor the Latin name survives, to my knowledge. Ruffin's following description, however, most closely resembles what is now called the Virginia pine (*P. virginiana*), whose range extends northward through New Jersey. Ed.]

6. [Willoughby Newton (1802–74) was one of Ruffin's oldest friends. Newton was a noted agriculturalist, lawyer, and political figure in northern Virginia. He pioneered the introduction of Peruvian guano in the 1840s and served as president of the state's agricultural society in 1852. Ed.]

7. [William Darlington, *American Weeds and Useful Plants: Being a Second and Illustrated Edition of Agricultural Botany*, rev. with additions by George Thurber (New York: A. O. Moore and Company, 1859). Ed.]

8. [If not meaning, the name does possess historical associations: *loblolly* approximates the word used by aboriginals along the Gulf Coast for low, wet landscapes and apparently the pines that preferred such sites. See Joshua A. Cope, *Loblolly Pine in Maryland*... (Baltimore: Maryland Board of Forestry, 1925), 1–3. Ed.]

9. [Ruffin was correct that loblolly and slash pines are different species. Ed.]

10. [According to the manuscript agricultural and slave population censuses, in 1850 Edward H. Herbert owned a farm of two hundred acres (all improved land valued at nine thousand dollars) and fifty-six slaves. Ed.]

11. [Slash pine's proper Linnean designation is *P. elliottii*. The species' range today hardly extends northward to North Carolina, and its ultimate height seldom exceeds one hundred feet. So as seen below, Ruffin was probably correct that what Carolinians called "slash" were actually loblollies. Ed.]

12. [François-André Michaux, *Histoire des Arbres Forestiers de l'Amerique Septentrionale*, 3 vols. (Paris: Haussmann and d'Hautel, 1810–13), translated by Augustus L. Hillhouse and first published in English in 1817 as *The North American Sylva; or, A Description of the Forest Trees of the United States, Canada and Nova Scotia; Considered Particularly with Respect to Their Use in the Arts and Their Introduction into Commerce: To Which Is Added a Description of the Most Useful of the European Forest Trees*. Editions later multiplied, with added notes by J. Jay Smith, any one of which might have been used by Ruffin. The edition on hand for locating these quotations is *North American Sylva*, 5 vols. (Philadelphia: D. Rice and A. N. Hart, 1857), 3:123–25, on the loblolly pine. Ed.]

13. [The parentheses in this paragraph are Ruffin's interpolations. However, this quotation does not appear in the 1865 edition. Ed.]

14. [Michaux, *North American Sylva*, 3:125. Ed.]

15. Whilst engaged in the investigation of this subject, and particularly as to the question of the species of the valuable "swamp pine," and its being identical in species, or not, with the worthless "old field" or loblolly pine, I sought scientific information from Dr. James F. McRee, of Wilmington. No person was better qualified to instruct, and to decide doubts, on this question, than Dr. McRee—not only because of his extensive botanical knowledge, but, also, as being a native and long resident of the region in which these pines (generally supposed of two different kinds) grow in great number and in their greatest perfection of size and luxuriance. Failing to find him at home, I made my inquiries by letter, and subsequently received from him, though after this writing was completed, full confirmation of the correctness of my position—that the above trees, deemed so different by all lumber-cutters, are the same. The question of identity had previously attracted Dr. McRee's attention, not only as a botanist, but as a proprietor of pine forest, in which these trees were abundant, and of which it was important to designate those best for timber and for sale. He says, in his letter, that "both kinds [deemed the most distinct and altogether different by all lumber-cutters and carpenters,] when subjected to the closest botanical scrutiny, show no signs of specific difference. Of this you will

be better assured, when I inform you that I have recently had the pleasure of a visit from the Rev. M. A. Curtis, (than whom there is no better botanist South of the Potomac) when we examined together two varieties of the *p. taeda* spoken of, and he unhesitatingly agrees in opinion with me as to their identity." "You will find the two varieties of the *p. taeda* recognized by Elliot, who calls the 'swamp pine' *p. taeda*, and the 'loblolly' *var. Heterophylla*"—[which latter is recognized by all other botanists as simply *p. taeda*.]

Dr. McRee says that the experienced timber-cutters profess to be able to distinguish, at the first glance, the difference between the two (so-called) kinds of pine. And this they can generally do, from external signs—that is, they can judge whether a standing tree has much heart, [which they would call "swamp pine" generally, but to which, near Wilmington, they give the name of "rosemary pine," which elsewhere is given exclusively to the *p. variabilis*,] or but little heart, in which case they call it loblolly. But, by external examination, with the aid and direction of one of the most experienced and intelligent lumberers, who was fully satisfied of the difference of these trees, and of his ability always to designate them, Dr. McRee found that even the actual and *only* difference, as to the size of heart-wood and the comparative value for timber, in numerous cases, could only be determined by applying the axe, and so reaching the heart.

[Moses Ashley Curtis (1808–72) was born in Massachusetts and educated at Williams College. He migrated to North Carolina in 1830 and became an Episcopal priest. He studied botany with James F. McRee in Wilmington, wrote much on the subject, and in the 1840s was involved in the controversy over polygenesis. Ed.]

16. [Michaux, *North American Sylva*, 3:118. Ed.]

17. [Ruffin's naming and description of pond pine comports with contemporary usage. I have, however, seen pond pines in southern Isle of Wight County, Virginia, west of the Dismal Swamp. Ed.]

18. I have since seen a few young trees of this species in Albemarle, on the road from Charlottesville to Ridgeway on the Rivanna. These compared to the surrounding and ordinary growth of *pinus variabilis*, were very different—and especially in the much thicker and more rigid leaves of the *p. regida*—and also in the general appearance, in tint and outlines, of the two kinds of young trees.

19. [Michaux, *North American Sylva*, 3:118. Ed.]

20. [Ibid., 3:119, though this passage is much rephrased by Ruffin. Ed.]

21. [Ibid., 3:118, rephrased. Ed.]

22. [The interpolations in square brackets in this paragraph are Ruffin's. Ed.]

23. [These three quotations all appear in Michaux, *North American Sylva*, 3:119. Ed.]

24. This statement of sizes, induces a suspicion that the writer, (Smith,) had mistaken the great swamp pine (*p. taeda*,) for the *p. rigida*.

25. [Michaux, *North American Sylva*, 3:122. Ed.]

26. [In fact, the pitch pine's range does not usually include tidewater Virginia or North Carolina. Ed.]

27. [Michaux, *North American Sylva*, 3:118. Ed.]

12. The Geology of Low Country Progress: Agricultural Features of Virginia and North Carolina

The essay was published in two parts: the first (sections 1–7) as "Agricultural Features of Virginia and North Carolina: Low-Lands between Albemarle Sound and Chesapeake Bay," *De Bow's Review* 22 (May 1857): 462–79; the second (sections 8–14) as "Agricultural Features of Virginia and North Carolina," *De Bow's Review* 23 (July 1857): 1–20. The second part is prefaced by a note from J. D. B. De Bow: "In the May number of the Review Mr. Ruffin discusses the subject of the Low-Lands between Albemarle Sound and the Chesapeake Bay under several leading heads, to wit: their surface, soil, rivers, drainage, etc. He now completes the discussion, with other practical and striking arguments and illustrations, having prepared his very full notes at the instance of the State Agricultural Society of Virginia, of which he is the able and influential head."

1. [Of course, Ruffin had traveled down the Chowan and visited Plymouth and the countryside around the Roanoke's delta during his 1840 "steam journey." He apparently refers here only to the landscape between the Perquimans and the Chowan—the three counties of Perquimans, Chowan, and (to the north, bordering Virginia) Gates. Ed.]

2. There may be, and probably are exceptions, as higher in the tide-water region, in some coarse and imperfect sand-stone, recently formed, by ferruginous spring-water filtrating through coarse sand, and, in the course of time, cementing with a deposit of iron the before separate and loose grains of sand. There are many such recent formations of stone.

3. [Josiah T. Granberry was a prominent Perquimans Democrat and a planter on Durant's Neck. See Alan D. Watson, *Perquimans County, a Brief History* (Raleigh: North Carolina Division of Archives and History, 1987), 39, 46–47. Ed.]

4. [Cross-plowing was a common labor-saving method of cultivation (aeration and weed removal) practicable in an age when corn and other row crops were sparsely seeded, leaving considerable spaces between individual plants. Ed.]

5. [William Sayre (b. 1814), Ruffin's son-in-law and sometime manager of Marlbourne. See Allmendinger, *Ruffin*, 96–99. Ed.]

6. [Francis Nixon was a planter neighbor of Granberry on Durant's Neck. See Watson, *Perquimans County*, 47. Ed.]

7. [Ruffin refers to "ring-fence" associations of his design and initiative (during the 1830s), in which contiguous planters agreed to enclose their collective estates

with one fence, enclose their livestock, and dispense with the fencing of crop fields. Ed.]

8. [Ruffin, *Essays and Notes on Agriculture* (Richmond: J. W. Randolph, 1855). Ed.]

9. When I first began this manner of flush ploughing of land and bedded land, and with considerable apprehension as to its complete success, it was not known to me that any other farmer had either used or thought of the same method. But, subsequently, when recommending it to the trial of E. H. Herbert, esq. of Princess Anne, as an important aid to his usually efficacious practice of draining, he informed me that he had already introduced and used this plan of flush ploughing on his land earlier than my first trial of it, and had found the results entirely satisfactory.

13. Toward the Managed Landscape: An Address on the Opposite Results of Exhausting and Fertilizing Systems of Agriculture

Published in two parts as "An Address on the Opposite Results of Exhausting and Fertilizing Systems of Agriculture, Read before the South Carolina Institute, at its Fourth Annual Fair, November 18th, 1852," *Southern Planter* 20 (July 1860): 401–8; and 20 (August 1860): 481–86, and subscribed "By Edmund Ruffin, Esq." The first paragraph of the second part begins, "From this digression to a particular branch, I will now return to the general subject, of the neglect of rest and manuring crops, for land." A note from the editor of the *Southern Planter* prefaces the essay: "We omit the introductory part of this admirable address as mainly applicable to the time, place and circumstances of its delivery, but the subject discussed is one of general and permanent interest to all intelligent improvers of the soil. We, therefore, desire to rescue it from the oblivion to which it might be destined, if not committed to some more enduring form of publication than the fugitive pages of the merely occasional pamphlet in which we find it."

1. [He refers to his geological survey of South Carolina (at the invitation of his friend, Governor James H. Hammond), in 1843. Ed.]

2. In a communication recently made to the State Agricultural Society of Virginia, on "Some of the Results of the Improvement of lands, by Calcareous Manures, on Public Interests in Virginia, in the increase of Production, Population, General Wealth and Revenue to the Treasury." ["Communication to the Virginia State Agricultural Society: Some of the Results of the Improvements of Land by Calcareous Manures," *Southern Planter* 12 (1852): 258–70. Ed.]

3. In a recent communication in the Virginia State Agricultural Society entitled "New Views of the Theory and Laws of Rotation of Crops and their practical application." These views I deem especially applicable to the agricultural condition of South Carolina and of importance next to the main subject of the present address. ["New Views of the Theory and Laws of Rotation of Crops and Their Practical Ap-

plication," *Virginia State Agricultural Society Journal of Transactions* 1 (1853): 23–39. Ed.]

4. A condition made by the Government of Virginia, in the act of cession, to the United States of all her north-western territory, was that this territory should afterwards be divided into not more than five new States. Five have already been carved out of this great doma[i]n, Ohio, Indiana, Illinois, Michigan, and Wisconsin, and a space of 22,336 square miles remains, in the new territory of Minnesota, which will hereafter constitute so much of another State, in violation of the act of cession by Virginia, and of the faith of the present Federal Government, and in which space, with all the north-western territory, slavery was interdicted by the ordinance of 1787, of the Confederation. This space of 22,336 square miles, which ought to have been included in the five anti-slavery States already formed, but which will go to constitute a sixth, is nearly as large as South Carolina, and larger, by nearly 1000 square miles, than the united surfaces of New Hampshire, Massachusetts and Connecticut.

5. There is, however, one important case known to me, of at least partial exception to the general rule of failure in marling in South Carolina, in the very extensive and also profitable labours and improvements of Gov. Hammond, on his estate bordering on the Savannah.

Index

Agricultural Society of Prince George County, Va., xiv
Agriculture: in Cape Fear Valley, 245–46; clover, 189, 316; corn, 18, 38–39, 186–87, 197, 202, 204, 217, 311; cotton, 16, 18–19, 42, 313; education of farmers, 21–25; farm boundaries, 25–26; manures, 19–21; and morality, 3–8; peas, 186, 197, 246, 315–16, 334–35; potatoes (sweet), 191, 246; rice, 216, 230–31, 300, 324–25; in Scotland, 79–80; sheep husbandry, 20; silk, 20, 255–57; in Spain, 80; tobacco, 15–17, 35–37, 40–42; truck farming, 313–14; watermelons, 191; wheat, 17–18, 189, 202, 217, 311–13, 315. *See also* Drainage; Marl; Reclamation; South Carolina
Albemarle Sound, 195–96, 202, 288, 304
Alligator Lake, N.C., 205
Alligator River, 206, 277
Alligators, 252
Allmendinger, David F., Jr., xvii–xviii, xxix
Appomattox River, 78–79, 194, 263
Armistead, Anthony, 196, 198, 200–201, 277, 360 (n. 2)

Bears, 201, 252
Beaufort County, N.C., 196, 266
Bedford, Pa., 278
Belgrade Farm, 224–26
Benner ["Benners"], Lucas, 253, 363 (n. 15)
Bertie County, N.C., 269–70
Biddle, S., 253
Blackwater River, 193–94, 255
Bladensburg, Md., 266
Boulware, William, xxix
Brandon (plantation), 297
Brunswick County, N.C., 230
Brunswick County, Va., 102
Bryan, James West, 253, 363 (n. 14)
Bunch, J., 190–91
Burgwyn, Henry, 269
Burgwyn, J. C., 253, 363 (n. 15)
Burton, James W., 253
Byrd, William II, xix

Cabell, N. F., xxx
Camden County, N.C., 313
Cape Fear River, 230, 262
Carson, Rachel, xvii
Carter, Hill, 331
Charleston, S.C., xii, xviii, xxx–xxxi, 231–33
Chowan River, 131, 193, 238, 311
Cocke, Thomas, xxix

371

Collins, Josiah, Sr., 208, 361 (n. 10)
Collins, Josiah, III, 201–24
Conservationism, xvi–xviii, xx
Cowper, Wills, 187–88
Craven, Avery Odell, xiii, xv–xvii, xix
Craven, John H., 82, 93, 355
Culpeper, Va., 278
Currituck County, N.C., 311, 313
Currituck Court House, 289
Currituck Inlet, 289
Currituck Sound, 288–89
Curtis, Moses Ashley, 365–66 (n. 15)
Cutter, W. P., xiv, xxii

Darwin, Erasmus, xxv, 130, 357 (n. 2)
Davy, Sir Humphry, xxii–xxv
De Bow, J. B. D., xiv, xxix
Dinwiddie County, Va., 116
Dismal Swamp, 190–91, 193–227, 254, 289
Dismal Swamp Canal, 195–96
Donat, Joseph Etienne-Victor Gabriel, 152–55
Drainage, 170–82, 290–310. *See also* Mill ponds, Reclamation
Drummond, Lake (Va.), 204
Duplin ["Dumplin"] Island, Va., 187–88, 359–60 (n. 4)
Dupuy, W. J., 54–58, 353
Dupuytren, Baron Guillaume, 160, 359 (n. 15)
Durant's Neck, N.C., 304, 316
Dusberry ["Duesberry"], William, 255–56, 363 (n. 17)

Elizabeth City, N.C., 289
Elizabeth River, 289
Elliott, Stephen, 233, 362 (n. 4)

Environmentalism, xvi–xvii
Epes, John S., 143–44

Fairfax County, Va., 55, 353
Fencemore, 81–99, 354–55
Fences, xix–xx, xxx, 47–99, 316–17; earthen, 49–51; laws of, 51–99; live (hedges), 48–49
Field, Richard W., 353–54
Field, Theodore A., 59–62, 353–54
Fletcher, Andrew (of Salton), 17, 349–50 (n. 2)
Fontanes, Louis Marquis de, 160–61, 359 (n. 17)
Foote, William Henry, 58–59, 353
Fortress Monroe, Va., 191
Franklin, Va., 193
Fuller, Thomas, 226, 362 (n. 16)

Garnett, James M., 63–67, 74–76, 354
Gaston, Judge William, 250–52, 363 (n. 13)
Gilbert, Joseph Henry, xxiii–xxv
Giorgini, M. Gaetano, 112, 356 (n. 5)
Gloucester County, Va., 108–9, 123, 237
Goldsboro ["Goldsborough"], N.C., 249
Granberry, Josiah T., 290, 300, 303, 311, 367 (n. 3)
Gwynn ["Gwynne"], Major Walter, 239, 362 (n. 6)

Hanover County, Va., xi, 265
Harrison, William B., 331
Herbemont, Nicholas, 47–51, 255, 352, 363 (n. 16)

Herbert, Edward H., 269–73, 287, 303, 313, 365 (n. 10)
Hill, Frederick Jones, 231, 362 (n.2)
Hindus, 3–4
Historical and Philosophical Society of Virginia, xix, 14, 43
Hogs, 51–99, 315; and long leaf pines, 247–48
Hyde County, N.C., 196

James River, xi, 18, 62–63, 134, 297, 309; fences by, 77–79
Jamestown, Va.: illness in, 103–4, 355–56 (n. 2)
Jones County, N.C., 353

Kilby, J. T., 190–91

Laborie, Louis Guillaume, 160, 162–63, 358 (n. 14)
Lavoisier, Antoine-Laurent, xxvi, 160–61, 359 (n. 16)
Lawes, John Bennett, xxiii–xxv
Lazarus, Aaron, 230, 362 (n. 1)
Leopold, Aldo, xvii
Liebig, Baron Justus von, xxiii–xxv
Little River, 304
Lyell, Sir Charles, xv

Malaria, xx, xxx, 100–127, 189–90
Malthus, Thomas Robert, xviii–xiv, xxiv, xxvii, xxix; and agriculture, 10–13
Marl, 188–89, 237–38, 240, 253–54, 311, 326, 336, 338, 341–42. *See also* Agriculture; Drainage; Public Health; Recycling
Marlbourne (farm), xii, xvii, 265, 301, 318

Mathew, William M., xvii–xviii, xxix
Mathews ["Matthews"] County, Va., 123
Mattamuskeet, Lake (N.C.), 205–7, 277
McClellan, General George, xi
McRee, James Fergus, 236–41, 245, 256, 277, 362 (n. 5)
Meherrin River, 193, 195
Meriwether, William H., 82, 355
Michaux, François-André, xxv, 260–61, 272, 276, 364 (n. 1)
Mill ponds, 100–127, 189–90, 194
Mobile, Ala., 137–38
Mustaches ["moustaches"], 232

Nansemond County, Va.: agriculture in, 185–91; timber-getting in, 190–91, 260
Nansemond Indians, 187
Nansemond River, 118, 185–91
Neuse River, 238–40, 250, 253–54
New Bern ["Newbern"], N.C., 249–50, 253–54, 257
New Hanover County, N.C.: lands and soils of, 236–45
New Jersey, 62–63
Newton, Willoughby, 264, 331, 364 (n. 6)
Nixon, Francis, 311, 367 (n. 6)
Norfolk, Va., 186, 255–56, 286
Norfolk County, Va., 303, 305
Nottoway River, 193, 195

Okracoke Inlet, N.C., 195
Open range. *See* Fences
Orange County, N.C., 265
Oysters, 187

374 Index

Pamlico Sound, 196–97
Pamunkey River, xi, 287, 292
Parmentier, Antoine-Augustin, xxvi, 160, 162–63, 358–59 (n. 14)
Pasquotank County, N.C., 313
Pasquotank River, 289
Perquimans ["Perquimons"] County, N.C., 288, 290, 299–301, 303, 305, 311–12, 319
Perquimans River, 287, 304
Petersburg, Va., xi, xiv, 149, 164, 255
Pettigrew, Charles Lockhart, 207, 218, 277, 361 (n. 13)
Pettigrew, Ebenezer, 209, 217–21, 224–26
Phelps, Lake (N.C.), xxxi, 201–24, 277
Pines: characteristics and succession, 258–60, 279–83; loblolly, 266–68, 271–76; long leaf, 260–62; pitch, 278–79; pond, 276–78; short leaf, 264–66; slash, 268–76; Virginia, 262–64; white, 264. *See also* Trees; Turpentining
Pitt County, N.C., 266
Playfair, Lyon, xxiii–xxiv
Plymouth, N.C., 193, 196, 202, 254, 277
Pocosins, 206–7, 241–45, 250–53
Poplar Swamp, N.C., 202
Portsmouth, Va., 187, 255–56, 286; navy yard of, 271
Portsmouth and Roanoke Railway, 193, 228, 255
Prince George County, Va., xi, xx, xxiii, 93, 116, 263, 316
Prince Georges ["George's"] County, Md., 278
Princess Anne County, Va., 269, 303, 305, 311, 313, 317

Public health, xx, xxviii, 100–169
Pungo, Lake (N.C.), 205–6
Pungo River, 206
Puvis, Marc-Antoine, xxvi, 144, 357–58 (n. 7)

Quicksand, 305–7

Rattlesnakes, 252
Reclamation, 170–82, 193–227, 250–53, 288–310, 336–37
Recycling, xx, xxvi, 128–69
Richmond, Va., 136–37, 149, 165–66
Riddick, Mills, Sr., 190–91, 360 (n. 5)
Roanoke Island, N.C., 284
Roanoke River, 187, 196, 231
Rockbridge County, Va., 104–5
Rogers, John, 82, 355
Rozier, François, xxvi, 50, 158–63, 358 (n. 12)
Ruffin, Edmund: as celebrity, xii; as editor of *Farmers' Register*, xiv, xix, xxiv, xxix, 47–99; education of, xxi–xxii; and European science, xxi–xxvii; as farmer in wartime, xi; library of, xi–xii; on Malthus, 10–13; postmortem reputation of, xiii–xx; soil research of, xiii–xiv; suicide of, xii–xiii; as writer, xxx–xxxi; writings of, xxvii–xxix
Ruffin, Edmund, Jr., xi, xiii, xxvi
Ruffin, Julian Calx, xi; death of, xii; son of, xiv
Ruffin, Susan, xxi
Ruffin Flag Company, xiii

Sampson, Richard, 331
Saracenia (plant), 235
Savanna lands. *See* Pocosins

Sayre, William, 303, 367 (n. 5)
Scarborough, William Kauffman, xxviii
Scuppernong, Lake (N.C.) *See* Lake Phelps
Scuppernong grapes, 202–3, 254–55
Scuppernong River, 203
Selden, John A., 58–59, 331, 353
Simms, William Gilmore, xix
Sitterson, J. Carlyle, xxviii
Skinner, John, xiv
Slaves: as benefit, 330–31; and fences, 68–69; health of Ruffin's, 141–42; as laborers, 16–17, 257; population growth, 12; religious instruction of, 221–24
Smith, J. J., 279
Smith-Lever Act, xv
Somerset Place (plantation), 201–24
Southampton County, Va., 195, 260, 303
South Carolina, 323–44; Ruffin's geologic survey of, xiv, xxv
Southern Planter, xiv, xxiv
Street, W. R., 257
Surry County, Va., 303
Swem, Earl G., xxx
Swift, Mr., 198, 361 (n. 5)

Taylor, John (of Caroline), xxii, 7, 19, 59, 83, 348 (n. 1)
Trees, 251, 254; cypress, 208, 213–14; of southeastern N.C., 247–48. *See also* Pines; Turpentining

Trent River, 239–40, 253–54
Trumpet flower, 234–35
Turner, Frederick Jackson, xv
Turpentining, 248–49
Tyrell County, N.C., 196, 277

Vaux, Antoine Cadet de, 358 (n. 13)
Venus fly-trap, 234
Virginia: increased land values in, 326–27; population of, 7; Ruffin's agricultural history of, 14–44
Virginia Department of Highways, xiii
Virginia General Assembly, xx, 28–29, 51–99; law on mill ponds, 114–16, 126–27

Walker, George Henry, 9–11, 348
Washington County, N.C., 196, 201–24, 266
Waterloo: battle dead of, xxvi, 134–35, 357 (n. 3)
Westmoreland County, Va., 263
Wilmington, N.C., 228–31; botany of environs, 233–35; rattlesnakes near, 252
Wilmington Railway, 228–29

Yorktown, Va.: illness in, 104–5, 108

www.ingramcontent.com/pod-product-compliance
Lightning Source LLC
Chambersburg PA
CBHW030125240426
43672CB00005B/26